An introduction to
dynamics and control

An introduction to dynamics and control

R. J. Richards

University Lecturer in Engineering
University of Cambridge

Longman
London and New York

Longman Group Limited London

Associated companies, branches and representatives throughout the world

Published in the United States of America by Longman Inc., New York

© *Longman Group Limited* 1979

First published 1979

Library of Congress Cataloging in Publication Data

Richards, R. J.
    An introduction to dynamics and control.

    1. Automatic control. 2. Dynamics.
3. Engineering models. I. Title.
TJ213.R52   629.8   77-21903
ISBN 0-582-44182-X
ISBN 0-582-44183-8 pbk.

Printed in Great Britain by
Richard Clay (The Chaucer Press) Ltd,
Bungay, Suffolk

Preface

Control engineering has developed within universities and colleges largely as a postgraduate subject and it is a subject which forms the basis of many masters and diploma courses at the advanced level. Although there has been some element of control in many undergraduate courses it has often, until comparatively recently, played a minor role with little more than the briefest introduction being given. This position has changed and alongside the general dynamics of processes and plant the study of control now features much more widely in first degree and similar courses. Based on a survey of university and polytechnic syllabi it has been the intention in this book to cater for the student, in the widest sense, who wishes to cover the basic material of modelling, dynamics and control and to be introduced briefly to extensions of it.

The motivation behind this book has been to provide a text giving adequate coverage of the subject within a reasonable length which enables the student, who may be carrying a load of several courses, to obtain full use of the contents. It is hoped that the sought-after conciseness and the selected examples illustrating the methods of modelling, dynamics and control will achieve this aim. It is realized that courses will continue to develop rapidly but it is expected that this material will maintain its essential place and aid the student in progressing, if and as need arises, to more complex system studies.

The theme of the text is developed in the following way. An 'understanding' of the dynamic behaviour of plant, within any specific discipline, requires some form of representation or model. By understanding one means here a knowledge of the plant which enables us to predict in some manner to a greater or less extent the behaviour of the plant. Such a knowledge of plant dynamics, or indeed lack of knowledge, guides us in our ways of controlling the plant such that it performs in a desired way. We add to, or design into, a plant controllers to improve both dynamic and steady state operation. The modelling, dynamics and control thus form an integrated area of study.

Modelling generally requires some specific knowledge of the system under study if the full advantage is to be obtained. Here it is intended that the basis of deterministic modelling is made clear so that when used in conjunction with this specialist knowledge a useful model may be achieved. Such a dynamic model, if it is an adequate representation of the plant or system, enables us to study the expected system behaviour. Control may be seen as a major subset of dynamic system study and as such it develops within the framework of the results of general dynamic systems theory. The pattern of development in this text is thus modelling to general dynamic theory to control.

Contents

Preface

Part 1 General system dynamics 1

Part 3 Control – B – Multivariable, nonlinear and sampled control systems 309

Chapter 7 Multivariable control systems 311

Chapter 8 Nonlinear control systems 360

Chapter 9 Sampled-data systems 404

Part 4   Control – C – Introduction to optimal and stochastic control systems   441

Chapter 10   Optimal systems   443

Chapter 11   Introduction to adaptive and stochastic systems   484

Acknowledgements

We are grateful to the McGraw-Hill Book Company for permission to reproduce Fig. 11.2 from Fig. 10.11 in *Control Systems Theory* by Elgerd, 1967 and to Prentice-Hall Inc. for Fig. 9.11 from the table 'The Effects of Samplers on z-Transform of System Output $C(x)$' in *Modern Control Engineering* by Katsuhiko Ogata. Copyright © 1970. Adapted by permission of Prentice-Hall Inc.

Part 1
General system dynamics

Chapter 1
Introduction

1.1 Some initial considerations

Within the following text we shall normally think of control as being related to the dynamics of the process. Thus, although the prime purpose of control usually will be to keep a device or process operating at some desired condition we shall not be concerned with the selection of that steady state which it is sought to maintain. The 'optimization' of the steady state during the design procedure may, however, require a knowledge of the control which is available, since, using the control it may be possible to operate at a more profitable condition, which would otherwise be unstable. In some instances it may be possible to cycle the plant in a manner which gives overall better average performance than we can obtain at the steady state condition.

When we refer to a 'plant' or 'process' we shall generally include mechanical, electrical and general physical and chemical processes in whose dynamics we are interested, but the study of dynamics and control now reaches beyond the traditional engineering fields. The performance of such a process is assessed by the quality of its outputs, e.g. purity, strength, speed, position. In some cases these may be measured directly or it may be more convenient either to measure some other property of interest, or measure the property of some intermediate variable in the process. The deviation of quality which can be tolerated will depend on the cost incurred by moving off specification and we shall have to take into account the cost of the control required to keep the process outputs within a tolerable band of quality and quantity combinations. One of the first requirements then in any control scheme will be the ability to choose and make measurements relevant to the purpose of the plant. The word plant may be used in its wider sense to mean some body or system of components which has one or more inputs and outputs and which may be subject to outside influences or disturbances. Obviously the idea of a chemical process plant or servomechanism falls within this general classification.

Having obtained a measurement it may be used in comparison with the desired performance to give an error, the difference between the desired and the actual value, the magnitude of which is a possible measure of the overall plant performance in meeting a specification. To do this it may be necessary to change the information contained in the measurement from one physical form to another by the use of a suitable transducer.

Before the information gained from a plant measurement can be used to improve the performance of the plant, i.e. to control the plant, it is necessary to modify it according to certain 'control laws'. The output from the controller which does this is itself put into a form which ensures that certain plant inputs, e.g. flowrate or power, can be suitably changed. This requires a source of power both for the controller and for the implementation of the control actions. Some of this power may be available directly from the process itself. The response of the plant to the new inputs will then show itself in further measurement. Control is thus essentially a continuous procedure, although as will be seen later, discretization may be introduced.

What is the basis for the choice of measurement and the nature of the control laws involved? The measurements are those which most adequately and practically impart the information that is required. These may be chosen on the basis of experience in a practically developed plant or from conclusions drawn from a model, i.e. a mathematical description in some form, of the plant. Similarly suitable control laws may be developed from experience or by a more rigorous theoretical approach. Thus the control theory and implementation are very strongly tied together and this may be illustrated by Fig. 1.1.

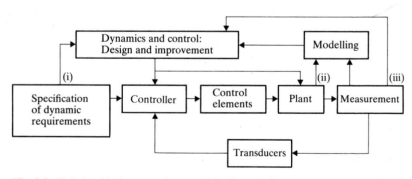

Fig. 1.1   Relationship between theory and implementation

Thus there are three routes: (i) the design of process and control together from a specification and the practical implementation of this; (ii) the design of control for a given plant via modelling of the plant; and (iii) an entirely empirical determination of control strategy using controller improvement based on plant performance (measurements).

1.2 Servo control and regulator control

Most conventional control loops are required to maintain certain plant outputs at, or close to, desired steady values. This is regulator control. If the output is intended to follow a time-varying reference input then this is referred to as a servo. The general servo problem is reduced to the regulator problem by making the reference input, or inputs, constant and it is further simplified by working in terms of new variables defined as deviations from the desired steady conditions. With the 'linear' systems which will occupy most of our effort this leads to easier analysis and less difficulty with numerical accuracy. With 'nonlinear' systems it will be seen that the problem can be more acute.

1.3 Open and closed loops

In the initial paragraphs of Section 1.1 the correction of a process by utilizing the measurement of the outputs was discussed. This is the basis of a feedback control system in which the output of the process is fed back to the process via the controller. It is conventional to include the measurement (and transducers) in the feedback loop and the basic feedback system in the absence of external disturbances to the plant is then shown as in Fig. 1.2.

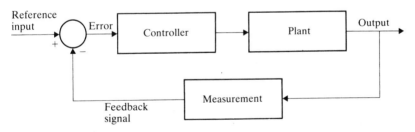

Fig. 1.2 Basic feedback control loop

The reference input and negative feedback signal are summed to give the error input to the controller. The loop is 'closed' by the feedback path. The term 'negative' feedback is used because of the sign change associated with the generation of the error signal. It should be noted that the error is given by 'reference minus output' in contrast to the conventional process error which has been described by the difference between the output and the reference (or desired) condition. The simplest form of representation by a block diagram is when unity (or direct) feedback is considered between process and controller, and the controller and process are combined in an overall plant description to give the representation of Fig. 1.3.

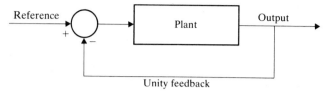

Fig. 1.3   Simple representation of unity feedback loop

Although a single input-single output case has been discussed the general case of multiple inputs and outputs may be similarly represented. We shall see also that plants may be represented by the alternative use of state variables and these will form the basis of much of the later analysis. In this case only a limited number of the plant outputs may in fact be used for feedback control purposes and the unity feedback path will not ordinarily be adequate.

If there is no closing of the loop as indicated in Fig. 1.4 then the system is referred to as open loop. We are normally concerned with open loop behaviour when we are principally interested in the behaviour or dynamics of the process itself or when we have an adequate knowledge of the process itself to establish a controller which does not rely on feedback knowledge. Such an ideal cannot be realized in practice although it does form the basis for 'feedforward' control concepts.

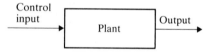

Fig. 1.4   Open loop control

An open loop situation may arise when the feedback path is deliberately broken for plant tests, e.g. to establish empirically controller settings, or under certain failure conditions. The open loop characteristics are then the only determining factor and they may be used to determine the fail–safe nature of the control systems.

1.4 Mathematical requirements

It is necessary that the user is familiar with the two most common forms of mathematical representation in control; the Laplace transformation and the elements of matrix algebra. In order to retain a certain compactness and in view of the now more general usage of these techniques by engineers the introduction to them here will be one of a statement of the rules and basic manipulation in each of these areas. It is intended that these will enable the reader to develop sufficient familiarity for their adequate use in the way in which they feature in the basic control field. For a full treatment of these topics, e.g. the rigorous limitations on Laplace transforms, the literature should be consulted, (e.g. Fuchs and Levin 1961; Kreyszig 1968).

The Laplace transformation

Within control work at all levels the Laplace transformation has replaced, or is tending to replace, other forms of differential equation representation, although these are still used to some extent, in mechanical systems in particular. The use of different notations does not in general cause confusion in changing from one form to the other. The Laplace transform and the use, for example, of the D-operator to represent d/dt have a common result in that differential equations which represent the dynamic behaviour of systems are reduced to an algebraic form which is susceptible of normal algebraic manipulation. The 'transformed variables' then yield (upon inversion) an explicit relationship between the original variables appearing in the differential equation. This inversion is aided by the use of standard tables. In the study of the dynamics of systems variables will be functions of time and the transformation will thus be based on such functions. However, although this is the common basis, the transformation can obviously be made when dealing with formations in which the independent variable is one other than time, e.g. a spatial dimension.

Let $f(t)$ be a function of time t, with $f(t) = 0$, $t < 0$. The Laplace transform of $f(t)$, $F(s)$, is defined by the relationship

$$\mathscr{L}[f(t)] = F(s) = \int_0^\infty f(t)e^{-st}\, dt \qquad (1.1)$$

where s is the complex variable, $\sigma + j\omega$. Note that the function $f(t)$ exists only for $0 \leqslant t < \infty$. Thus to form the Laplace transform the function $f(t)$ is multiplied by e^{-st} and then integrated over the time interval 0 to ∞. The usefulness of the transform lies in its providing the solution to the linear constant (time-invariant) coefficient differential equations which form the basis of simplified modelling or system behaviour descriptions. Its power for our purpose lies in the establishment of tables of transform pairs and the manipulation theorems which can be established on account of the form of the definition and in particular its linear characteristics. It is of maximum flexibility when all initial conditions at time $t = 0$ are zero as the concept of the transfer function between system input and output may then be established.

Consider the simple first order constant coefficient differential equation relating the input $u(t)$ of a system and its output $x(t)$, say

$$\frac{dx}{dt} = ax + bu.$$

Taking Laplace transforms of both sides of this equation gives

$$\int_0^\infty \frac{dx}{dt} e^{-st} dt = \int_0^\infty [ax + bu] e^{-st} dt$$

$$= \int_0^\infty axe^{-st} dt + \int_0^\infty bue^{-st} dt.$$

This is just the direct use of the definition, equation (1.1). The left-hand side may be integrated by parts to give

$$\mathscr{L}\left[\frac{dx}{dt}\right] = \int_0^\infty \frac{dx}{dt} e^{-st} dt$$

$$= [xe^{-st}]_0^\infty + \int_0^\infty x \cdot se^{-st} dt.$$

If the initial value of x is x_0 at $t = 0$, then we have

$$\mathscr{L}\left[\frac{dx}{dt}\right] = -x_o + s \int_0^\infty xe^{-st} dt$$

$$= -x_o + sX(s)$$

where $X(s)$ is the Laplace transform, by definition (1.1), of $x(t)$. (Although the upper case $X(s)$ will in general be used to signify the Laplace transform of the lower case variable $x(t)$ this will not always be the most convenient. Where the argument of the function t is replaced by s, e.g. $x(t)$ by $x(s)$, this will also infer that transformation has occurred and $x(s) = X(s)$.)

Thus we obtain from the transform of the initial equation, using this notation,

$$sX(s) = aX(s) + bU(s) + x_o$$

or

$$(s - a)X(s) = bU(s) + x_o$$

where $U(s)$ is the Laplace transform of $u(t)$. If the initial condition is $x_o = 0$ then the relation between the output and the input may be expressed as

$$\frac{X(s)}{U(s)} = \frac{b}{(s-a)} = G(s), \text{ say.}$$

The ratio of the Laplace transform of the output to the Laplace transform of the input, with zero initial conditions is the 'transfer function' of the system.

To obtain the function of time $f(t)$ from the function of the complex variable $F(s)$ involves an inverse transformation

$$\mathscr{L}^{-1}[F(s)] = f(t).$$

This requires a complex contour integration of the form

$$f(t) = \frac{1}{2\pi j} \int_{\sigma - j\omega}^{\sigma + j\omega} F(s)e^{+st} \, ds, \qquad t > 0.$$

Fortunately because of the properties of the transform it is not usually necessary to carry out such an integration. As it is comparatively easy to carry out the initial transformation, $\mathcal{L}[f(t)]$, a table of such functions may be established and the inverses then obtained in a look-up fashion from this table. This may require some algebraic manipulation, e.g. partial fraction expansion, but this is still the most rapid means of inversion.

Considering the example above

$$X(s) = \frac{b}{(s-a)} U(s).$$

If $u(t)$ is a unit step function, i.e. $u = 1$ for $t \geq 0$ and $u = 0$ for $t < 0$, then

$$U(s) = \int_0^\infty 1 \, . \, e^{-st} \, dt = \frac{1}{s}$$

and

$$X(s) = \frac{b}{(s-a)} \cdot \frac{1}{s} = \frac{b}{a} \left(\frac{1}{(s-a)} - \frac{1}{s} \right).$$

From the table below the inversion is obtained as

$$x(t) = \frac{b}{a} (e^{at} - 1).$$

This function is shown in Fig. 1.5 for both $a > 0$ and $a < 0$, $(b > 0)$.

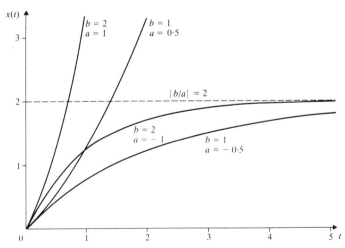

Fig. 1.5 Response of the system $\dot{x} = ax + bu$ for $u(t) = 1$, $t \geq 0$

Some of the more common Laplace transforms are given in Table 1.1. An extensive list is given in Bateman (1954). The form of $f(t)$ in some instances can be seen by inspection to be directly obtainable from the combination of the simple examples listed.

Table 1.1   Laplace transform pairs

$f(t)$	$F(s)$
Unit impulse, $\delta(t)$	1
Unit step, $1(t)$	$1/s$
t	$1/s^2$
t^2	$2/s^3$
$t^n (n = 1, 2, 3, \ldots)$	$n!/s^{n+1}$
$\dfrac{df(t)}{dt}$	$sF(s) - f(0)$
$\dfrac{d^nf(t)}{dt^n}$	$s^nF(s) - s^{n-1}f(0) - s^{n-2}\dot{f}(0) \ldots \overset{n-1}{f}(0)$
e^{-at}	$1/(s+a)$
te^{-at}	$1/(s+a)^2$
$t^n e^{-at}$	$n!/(s+a)^{n+1}$
$\dfrac{1}{(b-a)} \cdot (e^{-at} - e^{-bt})$	$\dfrac{1}{(s+a)(s+b)}$
$\dfrac{1}{(b-a)} \cdot (be^{-bt} - ae^{-at})$	$\dfrac{s}{(s+a)(s+b)}$
$\dfrac{1}{ab} \left\{ 1 + \dfrac{1}{(a-b)} \cdot (be^{-at} - ae^{-bt}) \right\}$	$\dfrac{1}{s(s+a)(s+b)}$
$\sin \omega t$	$\dfrac{\omega}{(s^2 + \omega^2)}$
$\cos \omega t$	$\dfrac{s}{(s^2 + \omega^2)}$
$\sinh \omega t$	$\dfrac{\omega}{s^2 - \omega^2}$
$\cosh \omega t$	$\dfrac{s}{s^2 - \omega^2}$
$e^{-at} \sin \omega t$	$\dfrac{\omega}{(s+a)^2 + \omega^2}$
$e^{-at} \cos \omega t$	$\dfrac{s+a}{(s+a)^2 + \omega^2}$
$\dfrac{\omega_n}{\sqrt{1-\zeta^2}} \cdot e^{-\zeta\omega_n t} \sin (\omega_n \sqrt{1-\zeta^2}\, t)$	$\dfrac{\omega_n^2}{s^2 + 2\zeta\omega_n s + \omega_n^2}$

$f(t)$	$F(s)$
$\dfrac{-1}{\sqrt{1-\zeta^2}} \cdot e^{-\zeta\omega_n t} \sin(\omega_n\sqrt{1-\zeta^2}\,t-\phi)$	$\dfrac{s}{s^2+2\zeta\omega_n s+\omega_n^2}$
where $\phi = \tan^{-1}\dfrac{\sqrt{1-\zeta^2}}{\zeta}$	
$1-\dfrac{1}{\sqrt{1-\zeta^2}} \cdot e^{-\zeta\omega_n t} \sin(\omega_n\sqrt{1-\zeta^2}\,t+\phi)$	$\dfrac{\omega_n^2}{s(s^2+2\zeta\omega_n s+\omega_n^2)}$
where $\phi = \tan^{-1}\dfrac{\sqrt{1-\zeta^2}}{\zeta}$	

The usefulness of the Laplace transform lies in the way the transformed variables can be manipulated and in the rules governing their formation and inversion. The following theorems form the backbone of this usefulness. In most cases derivation can be shown from the definition and simple example already considered.

1. *Linearity*

$$\mathscr{L}[f_1(t)+f_2(t)]=\mathscr{L}[f_1(t)]+\mathscr{L}[f_2(t)].$$

2. *Constant multiplication*

$$\mathscr{L}[af(t)]=a\mathscr{L}[f(t)]$$

3. *Differentiation*

$$\mathscr{L}\left[\frac{df(t)}{dt}\right]=s\mathscr{L}[f(t)]-f(0)$$

$$\mathscr{L}\left[\frac{d^2f(t)}{dt^2}\right]=s^2\mathscr{L}[f(t)]-sf(0)-\dot{f}(0).$$

4. *Integration*

$$\mathscr{L}\left[\int f(t)\,dt\right]=\frac{F(s)}{s}+\frac{\left[\int f(t)\,dt\right]_{t=0}}{s}.$$

5. *Translated function*

$$\mathscr{L}[f(t-\alpha)]=e^{-\alpha s}\mathscr{L}[f(t)] \quad \text{with} \quad f(t-\alpha)=0 \quad \text{for} \quad t<\alpha.$$

6. *Multiplication of $f(t)$ by $e^{-\alpha t}$*

$$\mathscr{L}[e^{-\alpha t}f(t)]=F(s+\alpha).$$

7. *Time scale change*

$$\mathscr{L}\left[f\left(\frac{t}{a}\right)\right]=aF(as).$$

8. Final value theorem

$$\lim_{t \to \infty} f(t) = \lim_{s \to 0} sF(s).$$

9. Initial value theorem

$$\lim_{t \to 0} f(t) = \lim_{s \to \infty} sF(s).$$

10. *Heaviside expansion formula*: For the rational algebraic function $F(s) = \dfrac{A(s)}{B(s)}$

$$f(t) = \mathscr{L}^{-1}[F(s)] = \sum_{i=1}^{n} \frac{A(s_i)}{B'(s_i)} e^{s_i t}.$$

where s_i are the n distinct roots of the characteristic equation $B(s) = 0$ and B' signifies differentiation with respect to s.

11. *Convolution integral.* The Laplace transform of the convolution integral (see section 2.7) is given by

$$\mathscr{L}\left[\int_0^t f_1(t-\tau)f_2(\tau)\, d\tau \right] = F_1(s)F_2(s).$$

The Laplace transform will feature largely both in the study of single input–single output plant and in the multivariable case.

Example A component in a mechanical system is constrained to move in a straight line under the action of an external force u and viscous damping. The velocity v is given by the equation

$$\dot{v} = -2v + u$$

If the component starts from rest, when $t = 0$, what is the transfer function relating the velocity and the force? If the input u for $t > 0$ has the form $u = at + b$ express the velocity as a function of time.

The relationship between input u and output v may be represented by a simple diagram (Fig. 1.6(a)).

Taking Laplace transforms for the given relationship

$$\dot{v} = -2v + u$$

we have

$$sV(s) = -2V(s) + U(s)$$

Therefore

$$\frac{V(s)}{U(s)} = \frac{1}{s+2}$$
$$= G(s).$$

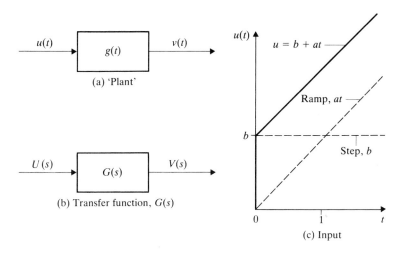

(a) 'Plant'

(b) Transfer function, $G(s)$

(c) Input

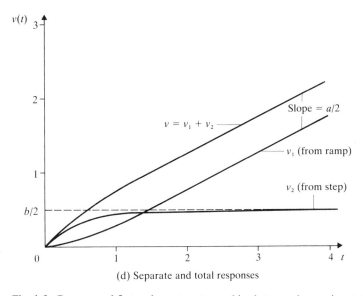

(d) Separate and total responses

Fig. 1.6 Response of first order system to combined step and ramp inputs ($a = b = 1$)

$G(s)$ is the transfer function relating the input force and output velocity and we may show this by a simple diagram (Fig. 1.6(b)). The Laplace transform of u is

$$U(s) = \frac{a}{s^2} + \frac{b}{s}$$

and

$$V(s) = G(s)U(s)$$

$$= \frac{a}{s^2(s+2)} + \frac{b}{s(s+2)}.$$

It is instructive to keep the two terms on the right-hand side separate as the first term will give us the response of the system to a ramp input $u = at$ and the second will give the response to the step input $u = b$. It will be seen that the total response of the unit is the sum of the two, a property of linear systems (Fig. 1.6(d)).

Inversion of $V(s)$ requires the use of partial fractions and Table 1.1. Expansions in partial fractions yield

$$V(s) = a\left\{\frac{0\cdot5}{s^2} - \frac{0\cdot25}{s} + \frac{0\cdot25}{s+2}\right\} + b\left\{\frac{0\cdot5}{s} - \frac{0\cdot5}{s+2}\right\}$$

and inversion gives

$$v(t) = a\{0\cdot5t - 0\cdot25 + 0\cdot25e^{-2t}\} + b\{0\cdot5 - 0\cdot5e^{-2t}\}$$

$$= v_1 + v_2, \text{ say.}$$

The initial value theorem shows, for $a = b = 1$

$$\lim_{t \to 0} v(t) = \lim_{s \to \infty} sV(s)$$

$$= \lim_{s \to \infty} \left\{\frac{1}{s(s+2)} + \frac{1}{s+2}\right\}$$

$$= 0,$$

and the final value theorem gives

$$\lim_{t \to \infty} v(t) = \lim_{s \to 0} sV(s)$$

$$= \lim_{s \to 0} \left\{\frac{1}{s(s+2)} + \frac{1}{s+2}\right\}$$

$$= \infty$$

These values can be checked by inspection above, as can the value of each contributing term as t becomes large.

Elementary matrix algebra

Matrix notation and algebra is invaluable when considering plants for which it is necessary to relate a number of inputs with a number of variables or outputs associated with the plant. As the number of variables increases so the computational power and time required for the handling of them increases. However, the basic features associated with matrix algebra can all be observed at a lower order at which hand computation enables familiarization to be established. As with the Laplace transforms the theorems of matrix algebra required for the work in the book will be stated with little development. However, those specific areas which are more closely tied with particular aspects of the work will be considered in more detail as they arise, e.g. in the transition matrix, eigenvalues, diagonalization. Our principal use of vectors and matrices is for the convenient solution of sets of linear first order differential equations.

Vectors

A vector is an ordered sequence of numbers, $x_1, x_2 \ldots x_n$. An n-dimensional column vector is denoted by

$$\mathbf{x} = \begin{bmatrix} x_1 \\ x_2 \\ \cdot \\ \cdot \\ \cdot \\ x_n \end{bmatrix}$$

and the row vector by $\mathbf{x} = [x, x_2 \ldots x_n]$. A vector with all its component zero is the null vector $\mathbf{0}$. The unit vector has one of its components equal to unity and the rest zero.

$$\mathbf{e}_i = \begin{bmatrix} 0 \\ 0 \\ \cdot \\ \cdot \\ \cdot \\ 1 \\ 0 \\ \cdot \\ \cdot \\ \cdot \\ 0 \end{bmatrix} \longleftarrow i^{\text{th}} \text{ component}$$

Multiplication of a vector by a scalar multiplies each component by the scalar value. Two vectors may only be equal if they are dimensionally the same and the sequence of numbers in one are equal to those in the other.

Matrices

A rectangular array of numbers, or elements, is called a matrix. Thus vectors are special cases of a matrix form. Each matrix element is characterized by a subscript so that the general matrix form is

$$\mathbf{A} = \begin{bmatrix} a_{11} & a_{12} & \cdots & a_{1m} \\ a_{21} & a_{22} & \cdots & a_{2m} \\ \cdot & & & \\ \cdot & & & \\ \cdot & & & \\ a_{n1} & a_{n2} & \cdots & a_{nm} \end{bmatrix}$$
$$= [a_{ij}]$$

where the first subscript indicates the row number and the second subscript indicates the column number. $\mathbf{A}$ is an $n \times m$ matrix; n is the number of rows, m is the number of columns. A square matrix of order n is formed when $m = n$. If all elements other than those on the diagonal a_{ii} are zero the matrix is a diagonal matrix. If these diagonal elements are also all unity then the matrix is a unity or identity matrix, $\mathbf{I}$. If all elements are zero the matrix is a null matrix.

The transpose of a matrix is formed by interchanging the rows and columns. Thus the transpose $\mathbf{A}^T$ is given by

$$\mathbf{A}^T = \begin{bmatrix} a_{11} & a_{21} & \cdots & a_{n1} \\ a_{12} & a_{22} & \cdots & a_{n2} \\ \cdot & & & \\ \cdot & & & \\ \cdot & & & \\ a_{1m} & a_{2m} & \cdots & a_{nm} \end{bmatrix}$$

and $\mathbf{A}^T$ is an $m \times n$ matrix.

If $\mathbf{A}^T = \mathbf{A}$, i.e. all elements a_{ij} of $\mathbf{A}$ are equal to those correspondingly positioned in $\mathbf{A}^T$, then $\mathbf{A}$ is a symmetric matrix. If $\mathbf{A}^T = -\mathbf{A}$, it is a skew symmetric matrix. Those two relationships can only apply of course to a square matrix.

Some matrix operations when written in general matrix notation exactly correspond to those of ordinary algebra but others do not. Two matrices can be added, or subtracted, if they have the same number of rows and columns:

$$\mathbf{A} \pm \mathbf{B} = \begin{bmatrix} a_{11} \pm b_{11} & a_{12} \pm b_{12} & \cdots & a_{1m} \pm b_{1m} \\ \cdot & & & \\ \cdot & & & \\ \cdot & & & \\ a_{n1} \pm b_{n1} & & \cdots & a_{nm} \pm b_{nm} \end{bmatrix}$$

Multiplication of a matrix by a scalar, say c, means that all elements a_{ij} are multiplied by c to give ca_{ij}. Multiplications of a matrix by a matrix is only possible when the matrices are 'conformable', i.e.

the number of rows in the second matrix must be equal to the number of columns in the first. If $\mathbf{A}$ is an $n \times m$ matrix and $\mathbf{B}$ is an $m \times p$ matrix then we can form the product $\mathbf{AB}$ by 'postmultiplying' $\mathbf{A}$ by $\mathbf{B}$ or 'premultiplying' $\mathbf{B}$ by $\mathbf{A}$ but the product $\mathbf{BA}$ does not exist. The product $\mathbf{AB}$ is defined by the matrix $\mathbf{C}$ whose elements c_{ij} are given by

$$c_{ij} = \sum_{k=1}^{m} a_{ik} b_{kj} \quad \text{for} \quad i = 1, \ldots n, \quad j = 1, \ldots p.$$

The matrix $\mathbf{C}$ is an $n \times p$ matrix. Note that in many cases $\mathbf{B}$ will be an $n \times 1$ matrix, i.e. a column vector and $\mathbf{A}$ may be a row vector $(1 \times m)$, possibly originating as a transpose of a column vector. If $\mathbf{A}$ and $\mathbf{B}$ are such that both products $\mathbf{AB}$ and $\mathbf{BA}$ may be formed and are also equal then $\mathbf{A}$ and $\mathbf{B}$ are said to 'commute'. The associative and distributive laws of algebra hold for matrix multiplication:

i.e. $(\mathbf{AB})\mathbf{C} = \mathbf{A}(\mathbf{BC}) = (\mathbf{AB})\mathbf{C}$

$\quad (\mathbf{A} + \mathbf{B})\mathbf{C} = \mathbf{AC} + \mathbf{BC}$

$\quad \mathbf{C}(\mathbf{A} + \mathbf{B}) = \mathbf{CA} + \mathbf{CB}.$

From the above it follows that powers of a matrix can only be formed if that matrix is square. The lth power of $\mathbf{A}$ is given by

$$\mathbf{A}^l = \mathbf{AA} \ldots \mathbf{A}$$
$$\quad 1 \ 2 \ldots l.$$

The transpose of a matrix is frequently required, e.g. to obtain the inverse of a matrix as below, and if this is itself comprised of a sum or product then

$$(\mathbf{A} + \mathbf{B})^T = \mathbf{A}^T + \mathbf{B}^T$$
$$(\mathbf{AB})^T = \mathbf{B}^T \mathbf{A}^T.$$

Matrix inversion. The inversion of a matrix involves the concept of division and this is less direct than the multiplication outlined above. Cancellation in the scalar manner is not possible and we proceed by post- or pre-multiplication. If the row and column containing the specific element a_{ij} is deleted from the square matrix $\mathbf{A}$, the determinant of the resulting matrix, which will be $(n-1) \times (n-1)$, is the 'minor' M_{ij} of the matrix $\mathbf{A}$. Multiplication of this determinant by $(-1)^{i+j}$ yields the 'co-factor' A_{ij} of the element a_{ij}, i.e.

$$A_{ij} = (-1)^{i+j} M_{ij}.$$

The determinant of $\mathbf{A}$, $|\mathbf{A}|$, can be expressed in terms of one row and its co-factors by

$$|\mathbf{A}| = a_{i1} A_{i1} + a_{i2} A_{i2} + \ldots a_{in} A_{in}$$
$$\quad = \sum_{j=1}^{n} a_{ij} A_{ij} \quad \text{for any } i.$$

Also the expansion may be similarly expressed in terms of the elements of a column and its co-factors

$$|\mathbf{A}| = \sum_{i=1}^{n} a_{ij}A_{ij} \quad \text{for any } j.$$

A determinant of a matrix with two equal rows, or columns, is zero and replacing $a_{i1}, a_{i2}, \ldots$ by $a_{r1}, a_{r2}, \ldots$ in the above row expansion but retaining the same cofactors gives

$$a_{r1}A_{i1} + a_{r2}A_{i2} + \ldots a_{rn}A_{in} = 0 \qquad (i \neq r)$$

since this is the expansion of a determinant in which the ith and r^{th} row are equal. Again the similar argument holds for the column based expansion. Thus in general we may write

$$\sum_{k=1}^{n} a_{jk}A_{ik} = \delta_{ji} |\mathbf{A}|$$

and

$$\sum_{k=1}^{n} a_{ki}A_{kj} = \delta_{ij} |\mathbf{A}|$$

where δ_{ji} is the Kronecker delta

$$\delta_{ji} = 1 \qquad i = j$$
$$\delta_{ji} = 0 \qquad i \neq j.$$

Another $n \times n$ matrix $\mathbf{B}$ is now defined where elements b_{ij} are the cofactors A_{ji} of matrix $\mathbf{A}$, i.e. it is the transpose of the cofactor matrix of $\mathbf{A}$. $\mathbf{B}$ is the adjoint matrix, written $adj\mathbf{A}$,

$$\mathbf{B} = adj\mathbf{A} = \begin{bmatrix} A_{11} & A_{21} & \ldots & A_{n1} \\ A_{12} & A_{22} & \ldots & A_{n2} \\ \cdot & & & \\ \cdot & & & \\ \cdot & & & \\ A_{1n} & A_{2n} & \ldots & A_{nn} \end{bmatrix}$$

$$= \begin{bmatrix} b_{11} & b_{12} & \ldots & b_{1n} \\ b_{21} & b_{22} & \ldots & b_{2n} \\ \cdot & & & \\ \cdot & & & \\ \cdot & & & \\ b_{n1} & b_{n2} & \ldots & b_{nn} \end{bmatrix}$$

For the product $\mathbf{A} \, adj\mathbf{A}$ the element positioned in the jth row and the ith column is, by the product rule,

$$\sum_{k=1}^{n} a_{jk}b_{ki} = \sum_{k=1}^{n} a_{jk}A_{ik}$$
$$= \delta_{ji} |\mathbf{A}|$$

from above. Thus full expansion of the product $\mathbf{A}$ $adj\mathbf{A}$ yields a diagonal matrix, each element of the diagonal, $i = j$, being $|\mathbf{A}|$, i.e.

$$\mathbf{A} \, adj\mathbf{A} = |\mathbf{A}|\mathbf{I}$$

and similarly it can be shown

$$\mathbf{A} \, adj\mathbf{A} = (adj \, \mathbf{A})\mathbf{A} = |\mathbf{A}|\mathbf{I}. \tag{1.2}$$

(Remember that $|\mathbf{A}|$ is a scalar quantity.)

All this is necessary in order to establish a usable expression for the inverse of a matrix. A square matrix $\mathbf{D}$ is the inverse of $\mathbf{A}$ if

$$\mathbf{DA} = \mathbf{AD} = \mathbf{I}.$$

$\mathbf{D}$ is represented by $\mathbf{A}^{-1}$ and exists if the determinant of $\mathbf{A}$ is not zero, i.e. $\mathbf{A}$ is nonsingular. Thus by this definition,

$$\mathbf{I} = \mathbf{A}^{-1}\mathbf{A} = \mathbf{A}\mathbf{A}^{-1} \qquad (cf. \ a \, . \, a^{-1} = 1)$$

and from equation (1.2) above

$$\mathbf{I} = \frac{\mathbf{A} \, adj\mathbf{A}}{|\mathbf{A}|}$$

also, so that

$$\mathbf{A}^{-1} = \frac{adj\mathbf{A}}{|\mathbf{A}|}.$$

Commonly useful properties of the inverse are

$$(\mathbf{AB})^{-1} = \mathbf{B}^{-1}\mathbf{A}^{-1}$$
$$(\mathbf{A}^{-1})^{T} = (\mathbf{A}^{T})^{-1}$$
$$(\mathbf{A}^{-1})^{-1} = \mathbf{A}.$$

Differentiation. Since our principle interest is in dynamics, differentiation with respect to time will feature strongly. The derivative of a matrix is the matrix whose elements are the derivatives of the elements of the original matrix. The integral of a matrix is similarly defined.

The derivative of the matrix product is given by

$$\frac{d}{dt}[\mathbf{A}(t)\mathbf{B}(t)] = \frac{d\mathbf{A}(t)}{dt}\mathbf{B}(t) + \mathbf{A}(t)\frac{d\mathbf{B}(t)}{dt}$$

and of the inverse by

$$\frac{d\mathbf{A}^{-1}(t)}{dt} = -\mathbf{A}^{-1}(t)\frac{d\mathbf{A}(t)}{dt}\mathbf{A}^{-1}(t).$$

Differentiation with respect to vectors occurs in optimization and will be covered as the need arises. The most commonly occurring form of a time derivative of a matrix will be that of the column vector, $\dot{\mathbf{x}}$.

Rank. If some complete rows or columns are omitted from a matrix then a reduced matrix called a submatrix of $\mathbf{A}$ is formed. Thus a number of submatrices can be formed from an initial matrix, and they may or may not be square. If, however, a matrix $\mathbf{A}$ contains at least one $r \times r$ square submatrix, the determinant of which is not zero (nonvanishing) while containing higher order square submatrices all of which do have zero determinants, then the matrix $\mathbf{A}$ is said to be of rank r. Thus a square $n \times n$ matrix $\mathbf{A}$ with $|\mathbf{A}|$ equal to zero will be of rank less than n. It is then called a singular matrix. If the rank is n, i.e. $|\mathbf{A}| \neq 0$, $\mathbf{A}$ is a nonsingular matrix.

Example   The values of two variables $\mathbf{x}$ are related by a set of algebraic relations to two outputs $\mathbf{y}$. The outputs are measured and two further variables $\mathbf{z}$ are evaluated from them for subsequent use in a control system.

The vector $\mathbf{y}$, $\begin{bmatrix} y_1 \\ y_2 \end{bmatrix}$, is given by $\mathbf{y} = (\mathbf{A} + \mathbf{B})\mathbf{x}$ where $\mathbf{x} = \begin{bmatrix} x_1 \\ x_2 \end{bmatrix}$ and matrix $\mathbf{A} = \begin{bmatrix} 1 & 1 \\ 1 & 0 \end{bmatrix}$, $\mathbf{B} = \begin{bmatrix} 0 & 1 \\ 2 & 1 \end{bmatrix}$. $\mathbf{z}$, $\begin{bmatrix} z_1 \\ z_2 \end{bmatrix}$, is related to $\mathbf{y}$ by the equation $\mathbf{z} = \mathbf{C}\mathbf{y}$ where $\mathbf{C} = \begin{bmatrix} 1 & 0 \\ 1 & 2 \end{bmatrix}$. Determine $\mathbf{D}$ in the expression $\mathbf{z} = \mathbf{D}\mathbf{x}$ and hence obtain an explicit expression for $\mathbf{x}$ in terms of $\mathbf{z}$.

By straight addition

$$\begin{bmatrix} y_1 \\ y_2 \end{bmatrix} = \begin{bmatrix} 1 & 1 \\ 1 & 0 \end{bmatrix}\begin{bmatrix} x_1 \\ x_2 \end{bmatrix} + \begin{bmatrix} 0 & 1 \\ 2 & 1 \end{bmatrix}\begin{bmatrix} x_1 \\ x_2 \end{bmatrix}$$

$$= \begin{bmatrix} 1 & 2 \\ 3 & 1 \end{bmatrix}\begin{bmatrix} x_1 \\ x_2 \end{bmatrix}.$$

Now $\mathbf{z} = \mathbf{C}\mathbf{y} = \mathbf{C}(\mathbf{A} + \mathbf{B})\mathbf{x} = \mathbf{D}\mathbf{x}$. Therefore by multiplication

$$\begin{bmatrix} z_1 \\ z_2 \end{bmatrix} = \begin{bmatrix} 1 & 0 \\ 1 & 2 \end{bmatrix}\begin{bmatrix} 1 & 2 \\ 3 & 1 \end{bmatrix}\begin{bmatrix} x_1 \\ x_2 \end{bmatrix}$$

$$= \begin{bmatrix} 1+0 & 2+0 \\ 1+6 & 2+2 \end{bmatrix}\begin{bmatrix} x_1 \\ x_2 \end{bmatrix}$$

$$= \begin{bmatrix} 1 & 2 \\ 7 & 4 \end{bmatrix}\begin{bmatrix} x_1 \\ x_2 \end{bmatrix}.$$

If $z = Dx$, premultiplying by D^{-1} yields $D^{-1}z = D^{-1}Dx = x$, i.e. by inversion

$$x = D^{-1}z = \frac{adjD}{|D|}z$$

$$= \frac{1}{(4-14)}\begin{bmatrix} 4 & -7 \\ -2 & 1 \end{bmatrix}^T z$$

$$= \frac{-1}{10}\begin{bmatrix} 4 & -2 \\ -7 & 1 \end{bmatrix}\begin{bmatrix} z_1 \\ z_2 \end{bmatrix}$$

Therefore

$$\begin{bmatrix} x_1 \\ x_2 \end{bmatrix} = \begin{bmatrix} -0.4 & 0.2 \\ 0.7 & -0.1 \end{bmatrix}\begin{bmatrix} z_1 \\ z_2 \end{bmatrix}.$$

Thus the use of vectors, matrices, addition, multiplication, transposing and inversion of matrices has been illustrated. Note the strict order of terms during the formation of matrix products. Further examples will occur in specific problems.

1.5 Summary

The relationship between control theory and its implementation has been outlined. The study involves the dynamic behaviour of processes and it is necessary to take measurements in order to effect control of the process. This control may be in either an open loop or a closed loop manner. In order to facilitate the manipulation and the solution of ordinary linear differential equations use is made of the Laplace transformation and matrix algebra.

Chapter 2
Modelling

2.1 The reasons for modelling

The nature of a model of a plant depends largely on the use for which it is required and the complexity and limits of the plant. A model means, in the context of a study of the dynamics and control of plant, a representation of the plant behaviour in terms of mathematical statements. Within this chapter the main concern will be the formation of these statements, usually as differential equations, and the arrangement of them into a form which enables us to use them in understanding the dynamics of plant and to which we can add and apply control theory. The combination of model and control theory will enable the behaviour of the plant under various conditions and controls to be investigated, and will help in the choice of controls for a given plant, and possibly lead to changes at the design stage of plant.

The 'representation of the plant behaviour in terms of mathematical statements' is open to various interpretations. The representation can be based on observed behaviour, it can be based on natural physical and chemical laws, it may be a combination of these two approaches. In addition it may be a very detailed representation or it may be only a very simple form indicating approximate behaviour or behaviour within a limited operating condition. Interaction with other plant may be important, the dynamic behaviour may be of more significance than steady state operation, we may know a lot about a process and its inputs or we may know very little, the plant may exist or it may be at the design stage – all of these factors are among those which determine the most suitable model. Many of these alternatives will be handled almost subconsciously during the modelling. Because of the many factors which need consideration there are many ways in which modelling can be classified and formalized. The divisions may be made in terms of nonlinear and linear, lumped and distributed parameters, deterministic and stochastic models. Emphasis may be placed on the plant design side or on the control side, on the use of physicochemical knowledge or on black box (experimental input–output) determination of plant characteristics. However, all of these classifications cut at some time or other across the others. Within this chapter deterministic modelling will be studied, that is the modelling of a system based on knowledge or presupposed knowledge of the process or plant. As indicated above this may in itself be subject to a variety of approaches. Initially, we may look at the factors which, consciously or otherwise, guide the development of a plant model.

2.2 General factors in deterministic modelling

Savas (1965) has discussed the principle factors in relation to the modelling of process plant, his ultimate aim being concerned with the computer control of plant. The factors he considered with some additions may be expressed concisely in a family tree manner. This is shown in Figs. 2.1 to 2.3; the combination of all three figures forming one approach to the problems of deterministic modelling. Each branch will be discussed in turn, it being borne in mind that (in practice) a number of stages will be considered simultaneously or at the exclusion of others. Other factors will be so straightforward in choice or immediately so obviously unsuitable that they may not be consciously considered at all, e.g. modelling on the molecular level for a steady liquid flow process. Variations in this pattern of classification are possible.

The basis of modelling

At the start of a modelling exercise it must be clear as to the purpose of the model, the extent (boundaries) of the model, which variables need consideration and the type of model that will be used. The decision on the type of model will depend to a large extent on a consideration of the first three of these points.

As the purpose of the model plays a major role in determining the form of the model, this must be decided initially and it must be recognized that the model developed specifically for one purpose may not be the best, or even a suitable form at all, for a further purpose. It will be necessary to use a simple form of model for a subsystem if we are concerned with the overall performances of a large system. Increase in overall plant complexity may require a reduction in the detail of modelling of specific units or subsystems. The new design of a plant will require that certain plant parameters which will be fixed in a control study of an existing plant are treated as variable quantities. Similarly an economic study may shift the emphasis again away from that of the predominantly dynamic study and model which may be sought for control uses. The economic aspect may play a subservient role in some studies but it will normally be incorporated in some respect, either directly or indirectly.

Because the process or plant under consideration generally interacts with others, boundaries for the model must be decided upon. The position of fringe units as within or outside the system boundaries, both upstream or downstream, must be decided or a limited time of interest may be determined as a temporal boundary. The position of a

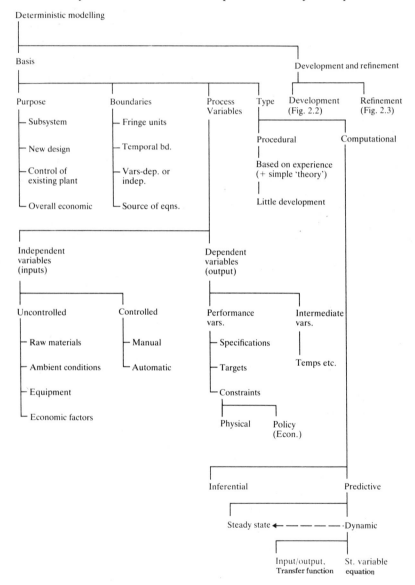

Fig. 2.1 One method of considering some major factors in deterministic modelling – a basis for the model

boundary will determine whether some variables are dependent or independent for the model within the boundaries. To some extent the boundaries which determine the system size will determine in addition both the complexity of the model and the source of the model equations. If the system is a small section of a plant then either special measurement or physical and chemical arguments will be needed to establish the model.

Before the model can be developed at all the necessary variables must be identified in order that the plant behaviour can be adequately described. These variables fall essentially into two categories, the independent variables, also known as input variables, and the dependent or output variables whose quality is dependent on the plant behaviour. Of the independent variables some will be beyond the capabilities or desirability of control. These uncontrolled inputs include changes in raw material, ambient conditions and possibly changes within the process equipment itself such as heat transfer surfaces fouling and reducing performance, or continuous catalyst decay. Economic factors will also be subject to variation, e.g. market demand and material costs. Those factors which can be controlled may be controlled either manually or automatically. At the one extreme we have manual process adjustment and at the other in large system studies there are the decisions which influence overall policy. Automatic control likewise may be at an elementary or complex level as will be shown in the later chapters.

The dependent variables may likewise be divided into intermediate variables which are used as intermediates in equation formation or which may be secondary process properties reflecting a dependent variable of major interest, and performance variables. The latter include product specifications, and production and quality targets. The dependent variables may be constrained (limited to a set range) by physical and economic considerations. The fourth branch to the 'basis' tree is the type of model to be used. Savas used the simple division of 'procedural' and 'computational'. The procedural model is hardly a model in the true sense of representing system behaviour but it is a set of instructions which are input to the plant in a predetermined way so that the plant output, or intermediate stages, follow a desired pattern. It may be based exclusively on empirical knowledge of plant behaviour and requires no direct mathematical relationship between inputs and outputs. An example is the positioning of a shaped driven template to a controller set point arm for control of pressure within a digester for wood-pulp production. Then the pressure within the digester and hence the 'cooking' process follow a preset pattern. Because of the absence of a mathematical basis extension is minimal, but a known input–output relationship is inferred. The computational model on the other hand is comprised of mathematical relationships between the variables and as such is capable of greater development and manipulation. The subdivision of this type is open to a considerable number of variations, e.g. based on nonlinearity and linearity and on the type of partial differential equations which arise. A more basic division is into inferential models, or parts of models, which relate unmeasured variables and measured variables and predictive models which relate the dependent variables of the plant to the independent variables, i.e. they enable a prediction to be made of the plant response to inputs. Both forms of equation will generally be present in a full model, the predictive form being dominant. These may be of steady state or dynamic form, the first being a special case of the more general dynamic equations with which we shall deal. In turn these equations may be expressed in the input–output, or transfer function, form or in terms of state variable equations. The latter will be expanded in section 2.7.

Development

The process of establishing the relationships between the variables is referred to as development. In order to obtain some measure of process performance a performance function is sought by which the effectiveness of control or plant design can be gauged. This simple criterion will include economic and quality considerations and will be subject to constraints in many cases. This function is a basis of optimization studies. The process equations form a very large part of many process control problems. The relationships may be established on an essentially theoretical basis but some empirical help or confirmation will generally be required. In other cases the mainstay will be empirical information and relationships. Each complement the other and a model is seldom purely one form or the other and each has its advantages and disadvantages. When considering either the theoretical or empirical model the level, or degree of detail, to which one needs to go must also be considered. The essential equations are the continuity equations of mass, momentum and energy and these may be written from the fine molecular and atomic level to the macroscopic, or lumped parameter model, with varying degrees of detail between them. In addition the equations may be written in terms of discrete entities, or age distributions and this gives the population balance models. We shall return to this below.

The theoretically based models use the basic physical and chemical rules taking in, as well as the continuity equations, reaction mechanisms (if known), diffusion theory etc. The dynamic equations will be ordinary differential equations or partial differential equations, these being reduced for steady state cases to algebraic and ordinary differential equations respectively. Major advantages of a fully theoretical model are its reliability and flexibility in allowing for major changes in plant and control, in predicting behaviour into a wide operating range, and in our ability to represent – at least to a first degree – the behaviour of new, unbuilt, plant. However, this ideal situation is in general seriously undermined by a lack of precise knowledge and the need to make assumptions which at some time need verification. Even then the extremely complex nature of many process models can make their solution impossible or economically unattractive. Similarly there is the conflict between advantages and disadvantages if one relies principally on empirical tests and measurements to determine the structure and equations of a model. The empirical results may be used to supplement theoretical work and may confirm the basic theory or show up deviations due to omitted or unknown factors. They may also be essential at an intermediate stage to confirm assumptions upon which further development may rest. It is tempting to use routine data in this way but some of the pitfalls which may arise are shown in Fig. 2.2. Some essential data may be missing, in error, or subject to hidden correlations which may lead to misleading conclusions. Unless special precautions are taken there will be no error estimate and the information may be on charts which require much labour in transposition for data analysis by computer. These hazards may be avoided in carefully executed tests on a 'blackbox' basis, the purpose being to relate inputs and outputs without prior knowledge of the process. Although this method of identification may yield fruitful relationships in growing numbers, the relationships established are particular to the unit under test, have validity probably in only a narrow operating range and require good test techniques and facilities. In the absence of other knowledge or in difficult-to-define systems this method is most important (Åström 1970; Åström and Eykhoff 1971).

Refinement

Figure 2.3 shows refinement as the taking of a model and verifying it, reducing it and/or improving it. The processes of development and refinement are thus closely related and refinement may be considered as a part of the development of the model to its final form. At each stage of reduction and improvement verification is required. Verification will include an error analysis which will show the limits of model accuracy and those measurements likely to cause largest errors. It may indicate that the initial computational method for evaluation of the model needs changing to a more accurate alternative. This analysis will ideally be supported and illustrated by simulation before results obtained are actually put into practice by construction or control of the actual plant. A feedback from this stage will lead to possible adjustments in the model itself. Apart from this straightforward follow up to the initial model development it may be found that, as mentioned above, the model is of a size or complexity which renders it unsuitable for use as it stands, or it may need restating in a form more suited for computation or linearizing prior to further use. If the complexity is too great then some reduction in size is necessary, which may be effected by neglecting some effects, making additional simplifications and approximations or by direct model reduction techniques. During reduction certain physical effects may be 'lost'. It is important that the correct variables and a physical understanding are retained and some partitioning of the model may be required.

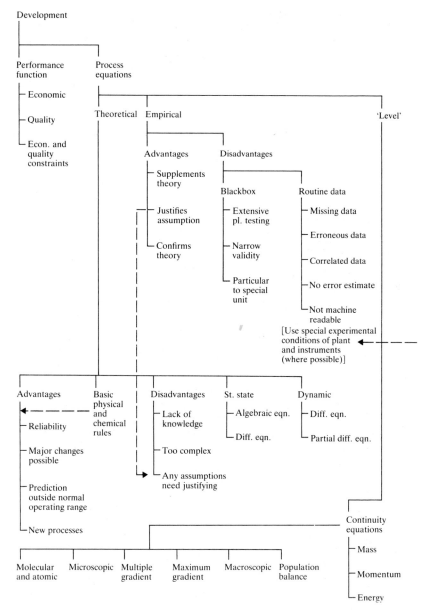

Fig. 2.2 A breakdown of the development considerations in process modelling

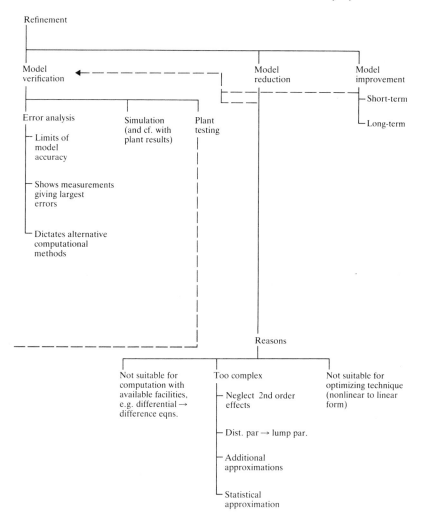

Fig. 2.3 A breakdown of the refinement considerations in process modelling

As distinct from verification and reduction, model improvement infers the correction of parameters and values in the short term and possibly a complete rephrasing of the model as more long-term operation is achieved. Thus any model will be subject to scrutiny continuously.

2.3 Continuity equations

In mechanical, electrical and process systems the underlying principle in modelling is that of conservation or continuity. This is true for both the steady state and dynamic cases. The three basic laws are those of the conservation of mass, momentum and energy. These may be expressed by the single equation:

Accumulation within the system
= Total inflow into system
− Total flow out of system
+ generation within system boundaries
− consumption within system boundaries   (2.1)

With mechanical systems the main equation is that of conservation of momentum and in process systems the conservation of mass and of energy feature more strongly. Equation (2.1) as a mass balance may be applied to the entire mass of the system or to any species within the system. Chemical reaction is taken care of by the generation and consumption terms. Similarly, in energy balances these terms allow us to handle exothermic and endothermic heats of reaction, the normal laws of reaction and stochiometry enabling us to determine changes in temperature.

Associated with these equations is the idea of a defined system and boundaries. Boundary specification is therefore important, whether we are looking at a lumped system or a small element within, say, a flowing fluid or mechanism. In dealing with the control of systems it is usual to consider dynamic behaviour so that from the statement above we shall end up with differential equations by considering the equation over a small time Δt and taking the limit as Δt tends to zero. From the time boundaries, e.g. $t = 0$ and $t = \infty$, and the conditions at the system boundaries, sufficient boundary conditions must be established to enable the solution of the differential equations to be fully effected – subject to adequate solution methods.

Most of the problems associated with modelling, the mathematical representation of behaviour, occur within the process type system as opposed to the more direct electromechanical type when we are talking of engineering systems. (Thinking in terms of sociological and organizational systems brings up even more problems.) Prior to looking specifically at process modelling some attention will be given to electromechanical systems. The use of fundamental laws naturally depends to a large extent on a knowledge of the field of science, engineering, etc. under consideration. Thus true modelling ability only comes from experience and an understanding of the true nature of a problem. It may be necessary either to form a team with varied experience to solve large problems or to rely on models produced by 'outside' sources. In all cases, however, a study of the dynamics and possible control of a system is only really effective when coupled with a sound knowledge of that system.

2.4 Electromechanical systems

Within this category we can include purely mechanical systems as well as systems of the servomotor, electromechanical type. The type of control system which may spring to mind is the fuel governor for a petrol engine, the flow of fuel being controlled by a feedback of the engine speed. By considering this as a purely mechanical system we are overlooking the reaction processes within the engine and the flow processes for the fuel and air. Thus we see that the distinction in this case is somewhat artificial and relies on the level to which the modelling is carried out or on just where we draw the system boundaries. Mechanical and electromechanical systems are more frequently of the servomechanism type than are continuous chemical processes where the problem has been more frequently that of regulation i.e. maintaining operation at a fixed level. The response of a mechanical system in following a change in command or desired operating point is usually much more rapid than for an equivalent process system change. Responses are completed within a matter of seconds or fractions of a second compared with the slow responses of processes which normally take a minimum of minutes but are frequently of the duration of hours or even days. The actual testing of mechanical systems does have the advantage that test times are much shorter but on the other hand more rapid response of instrumentation and control devices is also necessary. Thus in mechanical systems the emphasis has always been on the more positive and rapidly acting electrical and hydraulic controls so that low pressure pneumatic systems have been almost entirely restricted to the process field. The need for compactness associated with the smaller electrical and mechanical plant, e.g. in aircraft, has also eliminated the use of low pressure pneumatic systems in these systems.

Although all types of systems are subject to nonlinearities those present in the representation of mechanical systems are frequently of a discontinuous nature, including Coulomb friction, backlash and hard limits or saturation characteristics. Other nonlinearities, such as the square law of some governors, are more similar to the 'continuous' nonlinear behaviour met with in process modelling and which can be treated within a limited operational range by linearization techniques.

Electrical servos

The modelling of electrical system is based on the rules of electrical networks, e.g. Kirchoff's laws, but normally in control systems we shall require a combination both of the electrical rules and of those of mechanics. The electrical components may be of direct current, alternating current or mixed characteristics and of passive (resistance-capacitance-inductance) or active (amplifier) type. Although the input or output of electrical systems may be expressed in voltage or current terms many devices have an input or output which is a position or speed. That is, there is a combination of pure, static, electrical components with moving parts. Although this combination may be effected through a conventional form of rotating rotor motor or generator, it may be brought about also by movement transducers utilizing effects of varying capacitance, inductance or resistance. The field and variation in the design of such systems is wide (Neubert 1975; Gibson and Tuteur 1958) and in this chapter we shall restrict ourselves to looking at how our model equations arise in electromechanical servomechanisms, the archtype of which is the d.c. servo system. No attempt is made to discuss the details of electrical curcuit theory or amplifier construction which is well covered in electrical textbooks. However, this may be a requirement for the detailed design of systems and its relevance should not be overlooked. When considering servomechanisms as part of a large system such detailed study may be seen as the equivalence of the 'microscopic' description of process modelling which follows in section 2.5. The position and use of the servomechanism in dynamic systems generally is best illustrated by modelling it in a 'macroscopic' or 'lumped' sense.

Figure 2.4 illustrates the basic elements of an electrical servomechanism. The purpose of this device is to move an output shaft carrying an inertia load through an angle θ_o in response to a movement of the input shaft through an angle θ_i. Although nonlinearities may occur through saturation of the amplifier, electrical damping and nonlinear friction forces in the motor, backlash in gears, and hysteresis we shall assume, quite rightly in many cases, that the system behaves in a linear fashion and model it on that basis. The output shaft might be instrumental in changing, possibly via other power input stages, the position of a speed setting on a motor, a scanning aerial position, a process valve stem position, a variable tension in a mechanical system and so on. Note that at present we are concerned with establishing a set of equations, and hence their solution, for the system. Only after this has been done can we consider the dynamics and possible variations in the system in detail. As we proceed it will be observed that the form of equation obtained tends to be repeated, for electrical, mechanical, process and other systems. It is because of this analogy between various systems that we may proceed to a general treatment of dynamic systems. In most cases though one must still possess a good knowledge of the particular system to be modelled. The electric motor is armature controlled with constant field current I_f. To establish the required model, in this instance the overall relationship between the output θ_o and the input θ_i, we write down the equations for the relationship between successive parts of the system. The transducers to convert the input or demand signal θ_i and the output θ_o to voltages are shown as simple potential dividers. From some datum position of the input shaft we may write the input voltage V_i as proportional to the new shaft position, i.e.

$$V_i = k\theta_i$$

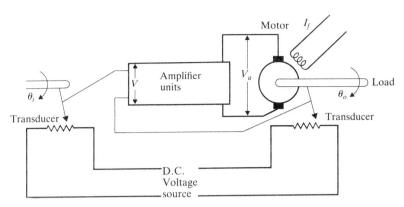

Fig. 2.4 Basic servomechanism

and for the output

$$V_o = k\theta_o.$$

The difference in the movement of the input and output shafts from their respective datum points is thus proportional to the voltage difference $V_i - V_o$, i.e.

$$V = V_i - V_o = k(\theta_i - \theta_o). \tag{2.2}$$

This voltage difference V is a measure of the error, the difference between the desired output shaft position as represented by the input shaft θ_i and the actual output shaft position. If the amplifier units are assumed to have linear proportional characteristics then

$$V_a = KV \tag{2.3}$$

and if the d.c. motor torque T is proportional to the armature voltage

$$T = K_m V_a. \tag{2.4}$$

This torque is applied to the output shaft and thus is the direct cause of changes in θ_o. If the load has moment of inertia J and is also damped (including the motor inertia and damping) the relationship between θ_o and T is

$$J\ddot{\theta}_o + \lambda\dot{\theta}_o = T \tag{2.5}$$

where λ is the damping coefficient. Even writing this last equation down may not be straightforward the first time and we might think in the following way. The gross torque from the motor is T; the torque required to overcome viscous forces which are proportional to the velocity of rotation $\dot{\theta}_o$ is $\lambda\dot{\theta}_o$; the net torque to produce acceleration in the inertia load is thus $T - \lambda\dot{\theta}_o$ and by basic laws of mechanics we ultimately arrive at $T - \lambda\dot{\theta}_o = J\ddot{\theta}_o$, equation (2.5). Thus even what appears to be a simple part of a model may need to be reasoned out in some detail, although in this case fairly trivially.

Equations (2.2) to (2.5) may now be combined to eliminate the intermediate variables V_i, V_o, V_a which we do not explicitly require, but which, nevertheless, through the structure of the system play a major part in establishing a useful relationship between our key variables θ_o and θ_i. Equations (2.2) to (2.5) are a model of the system. To express the output as a function of input we may use the Laplace transform with convenience and for movements about an equilibrium condition equation (2.5) becomes

$$Js^2\theta_o(s) + \lambda s\theta_o(s) = T(s) \tag{2.6}$$

so that

$$\theta_o(s) = \frac{T(s)}{Js^2 + \lambda s}$$

$$= \frac{K_m Kk(\theta_i(s) - \theta_o(s))}{Js^2 + \lambda s}$$

and

$$\frac{\theta_o(s)}{\theta_i(s)} = \frac{K_m Kk}{Js^2 + \lambda s + K_m Kk}. \tag{2.7}$$

The inversion of such forms is treated in detail in section 3.2 but if $0 < \lambda/(2\sqrt{K_m KkJ}) < 1$ and θ_i is a step input of magnitude θ_i then

$$\theta_o = \theta_i \left\{ 1 - \frac{e^{-\zeta\omega_n t}}{\sqrt{1-\zeta^2}} \cdot \sin\left[\omega_d t + \tan^{-1}\frac{\sqrt{1-\zeta^2}}{\zeta}\right] \right\} \tag{2.8}$$

where

$$\omega_n^2 = K_m Kk/J, \qquad \zeta = \lambda/(2J\omega_n)$$

and

$$\omega_d = \sqrt{1-\zeta^2} \cdot \omega_n.$$

This example has illustrated the consideration to be given in setting up a simple system model. Obviously there are a multitude of similar systems which one might consider. Notice, however, that the model may be a separate equation or a set of equations and whereas equations (2.2) to (2.5) are a general model of the system under the made assumptions, equation (2.8) is true only for a step input and hence is a much more restricted representation of the system and a particular input.

Let us now look briefly at other mechanical systems.

Simple mechanical systems

For the purpose of modelling dynamics of mechanical systems it is convenient to consider firstly those which are comprised essentially of rigid or simply extensible bodies but linked as a mechanism, and secondly those containing a fluid as an essential working part of the system. Included in the former, to be discussed first, are the mass-spring-damper arrangements, mechanical governors and generally bodies in which speed or position is important, e.g. engines, vehicles, powered tools. The process of modelling is as in the above section, we wish to establish the equations describing dynamic behaviour. Because the models of these systems are frequently discussed in terms of an equivalent mass-spring-damper arrangement let us look initially at just such an arrangement. When considering further examples later it will become apparent how representative the equations describing the performance of this simple system are. Once again though, the ability to establish the equations for the system will depend on the understanding in a qualitative way of the physics of the system.

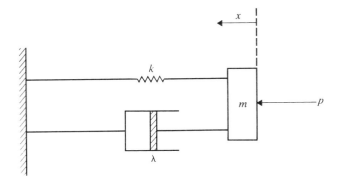

Fig. 2.5 A simple mass-spring-damper system

Figure 2.5 shows a body of mass m constrained by a spring of stiffness k and a dashpot with damping coefficient λ. It is subject to a force p. Initially the force p is zero and the spring takes up its natural length, neither extended nor compressed. The mass m is then at an equilibrium point and the displacement x is zero also. Consider now the mass displaced and acted upon by a force p. The forces in the x direction are

(i) from the spring $-kx$
(ii) from the damper $-\lambda\dot{x}$
(iii) from the force p

and the result of these is acceleration of the mass, $\ddot{x}$. Thus

$$m\ddot{x} = -kx - \lambda\dot{x} + p$$

or

$$\ddot{x} + \frac{\lambda}{m}\dot{x} + \frac{k}{m}x = \frac{p}{m}. \tag{2.9}$$

This may be expressed as

$$\ddot{x} + 2\zeta\omega_n\dot{x} + \omega_n^2 x = p/m$$
$$= p'\omega_n^2 \qquad (2.10)$$

where

$$\omega_n = \sqrt{\frac{k}{m}}$$

is the natural undamped frequency of the system,

$$p'k = p$$

and

$$\zeta = \frac{\lambda}{2m\omega_n} = \frac{\lambda}{2}\sqrt{\frac{1}{km}}.$$

is the damping factor or coefficient (e.g. Prentice 1970). The actual form of the response will depend on the force p, and if p is zero then equation (2.10) describes the behaviour of the system if released from a position x away from its natural equilibrium condition. Expressing (2.10) in terms of the Laplace transforms we see that

$$(s^2 + 2\zeta\omega_n s + \omega_n^2)X(s) = \omega_n^2 P'(s)$$

or

$$\frac{X(s)}{P'(s)} = \frac{\omega_n^2}{s^2 + 2\zeta\omega_n s + \omega_n^2} \qquad (2.11)$$

Immediately the similarity between equation (2.11) and equation (2.7) is apparent and the time solution is of equivalent form, section 3.2.

Example   Figure 2.6 shows a heavy duty mixer driven via a fluid coupling by an electric motor. The addition of fresh material to the partially full mixer trough places an impulsive torque change on the blades. Assume that changes in the motor speed are small enough to be neglected and that the mixer returns after each addition to its initial speed. What is the relationship between steady mixer and motor speeds and how does mixer speed vary with time following an addition of material?

If it is assumed that the motor speed remains constant, an assumption dependent on motor size, inertia and gearing in comparison with the stiffness of the drive and the mixer requirements, then the problem has two major components, the fluid drive and the mixer itself.

The fluid drive enables sudden surges on the driven side to be accommodated without placing sudden loads onto the motor. The

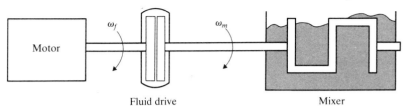

Fig. 2.6 Dynamic mechanical system

transmitted torque across the drive is

$$T_d = \lambda(\omega_f - \omega_m) \qquad (2.12)$$

where ω_f is the motor speed, ω_m the mixer (output) speed and λ is a coefficient for the component.

The input torque to the mixer has to overcome resistive forces of a frictional nature, and the inertial forces created by lifting and mixing the contents as well as by the mass of the blades themselves. This load torque T_m may thus be written

$$T_m = J\dot{\omega}_m + \xi\omega_m \qquad (2.13)$$

Where J is the total inertia and the resistive torque is taken as proportional to the speed ω_m. The magnitude of these factors would require measurement or estimation from the particular mixer or by scaling up approximately from experimental plant with a knowledge of the nature of the contents. (For details on torque measurements see, for example, Neubert (1975).)

With a rigid coupling between the output side of the drive and the mixer

$$T_d = T_m$$

and equation (2.12) and (2.13) combine to give

$$J\dot{\omega}_m + (\xi + \lambda)\omega_m = \lambda\omega_f$$

Under steady running condition $\dot{\omega}_m$ is zero and the steady running speed of the mixer is

$$\omega_m = \frac{\lambda}{\xi + \lambda} \cdot \omega_f.$$

Thus mixer speed is less than motor speed and the transmitted torque is

$$T_d = \frac{\lambda\xi}{\xi + \lambda} \cdot \omega_f.$$

If an increase in load torque δT is imposed then equation (2.13) becomes

$$T_m = J\dot{\omega}_m + \xi\omega_m + \delta T. \qquad (2.14)$$

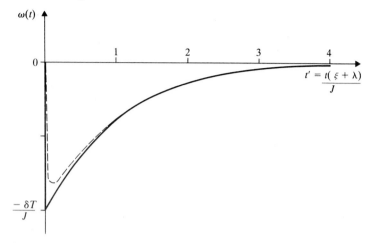

Fig. 2.7 Change in mixer speed, $\omega(t) = -\dfrac{\delta T}{J} e^{-t'}$, following an impulse δT

If this change is an impulse when the system is running at steady conditions then we may combine (2.12) and (2.14) so that

$$\lambda(\omega_f - \omega_m) = J\dot{\omega}_m + \xi\omega_m + \delta T(t)$$

and subtract the initial steady conditions at $\omega_m(0)$

$$\lambda(\omega_f - \omega_m(0)) = J\dot{\omega}_m(0) + \xi\omega_m(0)$$

to give

$$-\lambda\omega = J\dot{\omega} + \xi\omega + \delta T(t) \qquad (2.15)$$

where

$$\omega = \omega_m - \omega_m(0)$$

$$= \text{change in mixer speed.}$$

Now rearrange equation (2.15) and take the Laplace transforms so

$$(Js + \xi + \lambda)\omega(s) = -\delta T(s).$$

If $\delta T(t)$ is an impulse of magnitude δT then $\delta T(s)$ is δT and

$$\omega(s) = \frac{-\delta T}{Js + (\xi + \lambda)}$$

$$= \frac{-\delta T/J}{s + (\xi + \lambda)/J}$$

and the change in speed $\omega(t)$ is

$$\omega(t) = -\frac{\delta T}{J} \cdot e^{-t(\xi + \lambda)/J}.$$

Thus the addition causes a sudden decrease in speed followed by a recovery to the initial speed.

In practice the sharp speed change at $t = 0$ will not take place and the response to a sudden loading would be more like the dotted line in Fig. 2.7. The unit impulse is an idealized concept, not perfectly reproducible in practice.

Hydraulic and pneumatic systems

Within these systems a fluid plays a major part in the working of the system ranging from high pressure hydraulic systems, e.g. as used in presses, to low pressure pneumatic devices as used in process control systems. Within hydraulic systems it is most often suitable to neglect compressibility of the working fluid but in the higher pressure pneumatic systems this is not usually permissible. However, at the low pressure end compressibility may again be neglected. These factors obviously have to be considered in modelling the behaviour of any particular system and the effects of leakage may also require attention (Gibson and Tuteur 1958).

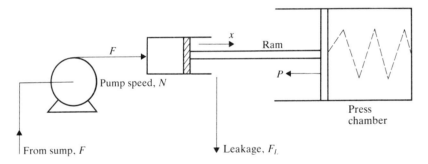

Fig. 2.8 Schematic of hydraulic press

Figure 2.8 is a simplified representation of a press used for the compression and baling of waste material. Hydraulic fluid is pumped from a reservoir into a cylinder and the press piston moves to compress the waste held in the baling chamber. Assume that once the ram and the material are in contact that the resistive force P is proportional to the ram (piston) movement x. The fluid is incompressible but there is a leakage flow F_L proportional to the pressure p in the cylinder of cross section area A. The flowrate F from the positive displacement pump is proportional to the speed N. If compression is slow inertia effects may be neglected and we have

flow $F = C . N$

useful flow $F - F_L = A\dot{x}$,   leakage $F_L = kp$

resistive force $P = Kx = pA$

and

$$\dot{x} = \frac{F - F_L}{A} = \frac{CN - kp}{A}$$

$$= \frac{CN - kKx}{A} A$$

i.e.

$$\dot{x} + \left(\frac{kK}{A^2}\right)x = \frac{CN}{A} \tag{2.16}$$

and the relationship between pump speed and ram movement is a first order equation. Although a linear equation results here, in practice the force P is likely to increase much more rapidly as compression progresses and a nonlinear relationship would replace $P = Kx$. Note that where rates, e.g. flowrate, speed of rotation, velocity, are used as variables, then a time derivative is automatically introduced. The relationship between cylinder pressure p and position x is non-dynamic.

In this section the intention has been not to show specific hardware devices in detail but to illustrate how from a knowledge of the action of individual components we may see how the system as a whole behaves. This requires that we model by way of basic laws of electricity and mechanics the interaction between these units. Within 'process' systems we are likely to encounter more difficulties if we wish to model in detail, but our simplest form of modelling will give equations very similar to the above. An alternative approach to dynamic models is given by MacFarlane (1970).

Example   An actuator-spoolvalve combination is connected to a damped inertial load, Fig. 2.9. Given the fundamental characteristics of the spoolvalve establish the relationship between a small movement in the spoolvalve position δx and the change in load position δy. (A detailed treatment of hydraulic systems is given by Gibson and Tuteur (1958).)

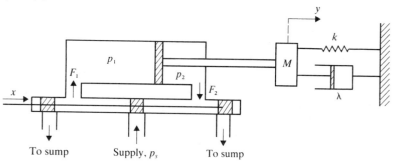

Fig. 2.9   Basic hydraulic system

By considering flow through the valve ports it may be shown that the valve load flow F_1 is a function of spoolvalve shift x and the load pressure p_L, where $p_L = p_1 - p_2$, the pressure across the actuator. In the absence of leakage in the actuator $F_1 = F_2$ and

$$F_1 = C \left\{ (x_o + x) \sqrt{\frac{p_s - p_L}{2}} - (x_o - x) \sqrt{\frac{p_s + p_L}{2}} \right\}$$

where C is a proportionality constant for the spoolvalve, x is the spool displacement from its central position, x_o is the 'underlap', a constant for the valve, and p_s is the supply pressure, assumed constant also. Thus F_1 is a function of the variables p_L and x alone;

$$F_1 = F_1(x, p_L)$$

and for small deviations about an equilibrium state and flow F_0

$$\delta F_1 = \frac{\partial F_1}{\partial x} \bigg|_{F_0} \delta x + \frac{\partial F_1}{\partial p_L} \bigg|_{F_0} \delta p_L.$$

For small deviations the partial differential terms may be considered constant, and it is convenient to express them as

$$\frac{\partial F_1}{\partial x} = A . K$$

so that

$$K = \frac{1}{A} \left(\frac{\partial F_1}{\partial x} \right)$$

and

$$\frac{\partial F_1}{\partial p_L} = -\frac{A^2 K}{f}$$

so that

$$f = -A^2 K \bigg/ \left(\frac{\partial F_1}{\partial p_L} \right),$$

where A is the effective cross-section area of the actuator, K is the valve gain constant and f is an equivalent stiffness. Then

$$\delta F_1 = AK \left(\delta x - \frac{A}{f} \delta p_L \right). \tag{2.17}$$

If the output shaft movement is δy then the change in flow δF_1 is related to δy by

$$\delta F_1 = A \frac{d(\delta y)}{dt}. \tag{2.18}$$

The change in output force applied to the load is

$$A \cdot \delta p_L = M \frac{d^2(\delta y)}{dt^2} + \lambda \frac{d(\delta y)}{dt} + k(\delta y) \qquad (2.19)$$

$$= \phi(\delta y).$$

From equations (2.17), (2.18) and (2.19) we may eliminate the intermediate variable δp_L:

$$\frac{d(\delta y)}{dt} = \frac{\delta F_1}{A}$$

$$= K\left(\delta x - \frac{A}{f} \delta p_L\right)$$

$$= K\left(\delta x - \frac{\phi(\delta y)}{f}\right).$$

Rearranging we have

$$\frac{d(\delta y)}{dt} + \frac{K}{f} \phi(\delta y) = K\delta x. \qquad (2.20)$$

As we are considering changes from an equilibrium point we can conveniently introduce the Laplace transform in equation (2.20) to give

$$\left\{ s + \frac{K}{f}(Ms^2 + \lambda s + k) \right\} \delta y(s) = K\delta x(s)$$

or the transfer function

$$\frac{\delta y(s)}{\delta x(s)} = \frac{f/M}{s^2 + (f + \lambda K)/MK \cdot s + k/M}. \qquad (2.21)$$

It is seen that this has a standard form of second order representation. If the dependence of the flow on the load pressure is small, i.e.

$$F_1 = F_1(x),$$

then

$$\frac{\partial F_1}{\partial p_L} = 0$$

and

$$\frac{\delta y(s)}{\delta x(s)} = \frac{K}{s}. \qquad (2.22)$$

That is the actuator and spoolvalve act as a pure integrator. In practice δy cannot continue to increase, and there may be feedback linkages also from the actuator output to the spoolvalve position.

The response of first and second order systems, equations (2.22) and (2.21), to different forms of input are evaluated in Chapter 3.

Note that inertia forces and friction in the spoolvalve itself have been neglected.

2.5 Process system modelling

In section 2.2 the 'level' of modelling was introduced and this is particularly important in the study of large or complex systems and, in general, chemical or physical 'processes' fall within one or both of these categories. As the level, or degree of detail, into which modelling is carried increases, so naturally does the overall size and complexity of the model itself for any given system. As already mentioned under 'model refinement' this can lead to computational difficulties. Also, the quantity of information contained in a detailed study may be in excess of that required for a given purpose. Thus the 'purpose' of the model, we again affirm, affects the modelling itself. A very complete treatment of modelling at various degrees of detail for process units is given by Himmelblau and Bischoff (1968), including representation in the alternative forms of vector notation, cylindrical coordinates etc. Of principal interest in the larger process models, as distinct from the detailed examination of specific complex mechanisms such as, e.g. simultaneous heat and mass transfer in a drying paper sheet, are the maximum gradient description and macroscopic description, more generally known as distributed parameter and lumped parameter descriptions. However, where diffusion is also important then a fuller description as given by the multiple gradient level is required. It is instructive to consider the change in model structure as we move from the most detailed representation to the less detailed, but most frequently used, forms of modelling. The ease of application also increases as the degree of complexity decreases so that the general pattern is:

Level of model	Molecular	Micro-scopic	Multiple gradient	Maximum gradient	Macro-scopic
Degree of detail ← ———————————————— Increasing ———————————					
General extent ——————————————————Increasing———————————————→ of use					

The basic principle of continuity extends throughout the range and the laws of diffusion etc. where required enter at the more detailed levels only. The following sections are based on mass transfer and mass conservation but development for momentum and energy conservation balances is similar (Himmelblau and Biscoff, 1968).

Molecular description

At this level each molecule is considered and the properties summed to give overall properties. Although of use in very special cases such as low pressure gases the use in dynamics and control of processes is not sufficient to warrant discussion here.

Microscopic description

The system is now considered as a continuum, and equations of mass, momentum and energy conservation are applied at a point in the system. In the case of the mass balance, considered here, reaction and diffusion terms are included. To extend from the point equations to cover the whole system, integration is required up to the system boundaries. The diffusion terms, reaction and flow terms may all be functions of space (and time) so that the problem is still formidable. However, with increased computational power such a model now helps in the interpretation of experimental results and in design and control applications, (Hartley and Richards 1974). There is, however, still a conflict in that as our ability increases to look at larger systems, it is generally not possible, nor necessarily desirable, to look at subsystems and sub-processes in increased detail at the same time. Thus, we are still left with the choice of studying either in detail for specific processes or in a wider sense the interactions of larger plants and possibly organizations.

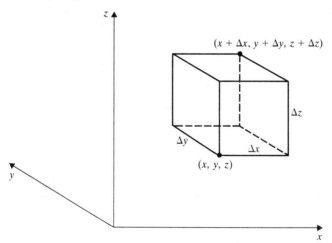

Fig. 2.10 Element of continuum

Consider a small volume treated at co-ordinates x, y, z as in Fig. 2.10. If the continuum contains a number of components, e.g. a mixture of soluble salts in a solvent, then the continuity equation may be used for each in turn as well as for the continuum as a whole. To reduce the use of subscripts consider just one component of the mixture, so that unless otherwise specified all lower-case notation refers to that component.

Bulk flow will have velocity, and hence mass flow, components in directions x, y, z. There are concentration gradients through each face of the elemental volume as well. In addition the possibility of generation or loss of our chosen component by reaction within the volume is also included. The net effect of all these will be accumulation or loss within the elemental volume over a given element of time, Δt.

If ρ is the mass density of the considered component, i.e. mass of component per unit volume, and v_i is the fluid mass bulk velocity component in direction $i = x, y, z$, then the net transport by bulk flow into the volume in time Δt can be expressed as the sum of the net flow in each of the three directions:

net flow in x direction $\Delta t\{(\rho v_x \, \Delta y \, \Delta z)\,|_x - (\rho v_x \, \Delta y \, \Delta z)\,|_{x+\Delta x}\}$

net flow in y direction $\Delta t\{(\rho v_y \, \Delta x \, \Delta z)\,|_y - (\rho v_y \, \Delta x \, \Delta z)\,|_{y+\Delta y}\}$

net flow in z direction $\Delta t\{(\rho v_z \, \Delta y \, \Delta x)\,|_z - (\rho v_z \, \Delta y \, \Delta x)\,|_{z+\Delta z}\}$.

If the component transport by diffusion is j_i, the mass flux of the component per unit area per unit time due to concentration gradients, then the net transport by diffusion in the x direction is

$$\Delta t\{(j_x \, \Delta y \, \Delta z)\,|_x - (j_x \, \Delta y \, \Delta z)\,|_{x+\Delta x}\}$$

and similarly for the y and z directions.

The generation or loss within the volume due to an average reaction rate r where r is the component production rate per unit volume per unit time, is

$$\Delta t r (\Delta x \, \Delta y \, \Delta z).$$

The sum of these three sets of terms will give the net accumulation within the volume, $\Delta x \, \Delta y \, \Delta z$. This is the difference between the mass of the component there before and after Δt, i.e.

$$(\rho \, \Delta x \, \Delta y \, \Delta z)_{t+\Delta t} - (\rho \, \Delta x \, \Delta y \, \Delta z)_t.$$

Summing the expressions and dividing through by $\Delta t \, \Delta x \, \Delta y \, \Delta z$ yields

$$\frac{\rho_{t+\Delta t} - \rho_t}{\Delta t} = \frac{\rho v_x\,|_x - \rho v_x\,|_{x+\Delta x}}{\Delta x} + \frac{\rho v_y\,|_y - \rho v_y\,|_{y+\Delta y}}{\Delta y}$$

$$+ \frac{\rho v_z\,|_z - \rho v_z\,|_{z+\Delta z}}{\Delta z} + \frac{j_x\,|_x - j_x\,|_{x+\Delta x}}{\Delta x}$$

$$+ \frac{j_y\,|_y - j_y\,|_{y+\Delta y}}{\Delta y} + \frac{j_z\,|_z - j_z\,|_{z+\Delta z}}{\Delta z} + r$$

The 'point' mass balance is obtained by letting Δx, Δy, Δz, and Δt tend to the limit so that term by term we have:

$$\frac{\partial \rho}{\partial t} = -\frac{\partial(\rho v_x)}{\partial x} - \frac{\partial(\rho v_y)}{\partial y} - \frac{\partial(\rho v_z)}{\partial z} - \frac{\partial j_x}{\partial x} - \frac{\partial j_y}{\partial y} - \frac{\partial j_z}{\partial z} + r \qquad (2.23)$$

where the partial differential is given, by definition, by

$$\frac{\partial f(x, y, z, \ldots, t)}{\partial x} = \lim_{\Delta x \to 0} \left[\frac{f(x + \Delta x, y, z, \ldots, t) - f(x, y, z, \ldots, t)}{\Delta x} \right].$$

If the bulk flow velocity is V then equation (2.23) may be more concisely written in the vector form

$$\frac{\partial \rho}{\partial t} = -\nabla(\rho V) - \nabla j + r \qquad (2.24)$$

where ∇ is the gradient operator $\dfrac{\partial}{\partial x} + \dfrac{\partial}{\partial y} + \dfrac{\partial}{\partial z}$.

This is the simplest expression for the general case. Note that in this expression the component mass density, velocity, the diffusion term and reaction are all position and time dependent without any loss of generality. Utilization of equation (2.24) requires, apart from a knowledge of the diffusion law giving j, an explicit knowledge of how these properties change with time and position throughout the system. This will be absent except for extremely well defined cases since turbulence, shape, local temperature variations etc. all have to be taken into account before the general point equation can be integrated to give the dynamics of the system.

Multiple gradient description

Because of the difficulties of full interpretation of the microscopic description equations for mass, momentum and energy, they are slightly modified to give the 'multiple gradient' description. The 'point' nature and full variability of the properties are replaced by empirically determined coefficients. Thus the diffusion coefficient used to determine j may be experimentally determined by laboratory apparatus and although determinations lead to values specifically related to the apparatus of determination it is known by experience that coefficients are obtained of wide practical use. We no longer attempt to integrate over intricate local turbulence fields, and velocities V are given by observed 'effective' values, which are time average values. Obviously the effective coefficients determined in this way will depend on flow conditions, turbulence and mixing and may be expressed in the form of some correlation or dependent relationship.

The main criterion is that the possibility of diffusion terms, either true diffusion or terms giving a diffusion-like effect of mass transfer, in at least one direction is retained. These terms are also referred to generally as 'dispersion' terms which in view of their uncertain nature is a better term than 'diffusion' at the multiple gradient levels. This level of modelling has been used in laminar flow, turbulent flow, two phase system and reaction system studies.

Maximum gradient description

A considerable simplification arises if the 'dispersion' terms are dropped, i.e. if the effects of these terms are not significant when compared to the bulk flow terms. If also the character of the process is dominated by flow in one direction only then further simplification is possible, although accompanied by a reduction in the detailed description of the process or unit. Now only the largest component of the gradient of the dependent variable is retained, i.e. we have a 'mximum gradient' description. This is the equation type usually referred to as the 'plug flow' model and it is the simplest form of non-lumped representation, the simplest form of the distributed parameter system model. Now it may be necessary to introduce an additional term, say $f(x, t)$ to allow for net transport into the system through the 'side' surfaces. This will be enlarged upon in an example.

Equation (2.23) or (2.24), the mass balance, thus reduces to the form:

$$\frac{\partial \rho}{\partial t} = -\frac{\partial \rho v_x}{\partial x} + r + f. \tag{2.25}$$

Properties of the system are now average values for the cross-section perpendicular to the direction of flow retained. They are considered to be functions of x only.

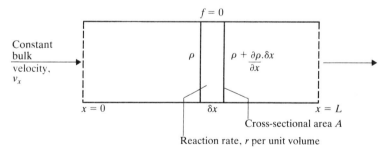

Fig. 2.11 Maximum gradient representation

Flow through a tubular reactor, Fig. 2.11, illustrates equation (2.25). In such a case $f(x, t)$ will usually be zero when considering the mass balance but for a heat balance on the system it will represent the heat transfer terms through the reactor wall.

A mass balance on the element in time Δt gives:

$$A \, \delta x \cdot \frac{\Delta \rho}{\Delta t} = A \cdot v_x \rho - A v_x \left(\rho + \frac{\partial \rho}{\partial x} \cdot \delta x \right) + r A \, \delta x$$

i.e. as $\Delta t \rightarrow 0$

$$\frac{\partial \rho}{\partial t} = -v_x \frac{\partial \rho}{\partial x} + r.$$

Macroscopic description

At this stage no spatial gradients are retained and all dependent variables are average values taken over the full system volume V. All detail within the system, or subsystem, is suppressed in the model and the basic mass balance equation for the chosen component becomes

$$\frac{\partial \rho}{\partial t} = r + \frac{f}{V} \qquad (2.26)$$

where ρ and r are the same as before and f is the net flow of the component into the system through its boundaries. The initial partial differential equation involving four independent variables, x, y, z, t has now been reduced to an ordinary differential equation in the one independent variable t. Such a form is amenable to far simpler and conclusive analytical treatment than the partial differential equations, especially when more than one equation is involved simultaneously. This simplest form of dynamic model is obtained when considering continuous flow through a perfectly stirred vessel (Fig. 2.12).

$$V \frac{d\rho}{dt} = Vr + Q(\rho_i - \rho)$$

i.e.

$$\frac{d\rho}{dt} = r + \frac{f}{V}$$

where $f = Q(\rho_i - \rho)$, the net flow of component into the vessel.

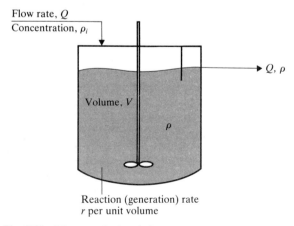

Flow rate, Q
Concentration, ρ_i

Q, ρ

Volume, V

ρ

Reaction (generation) rate
r per unit volume

Fig. 2.12 Macroscopic description

Attention has been focussed on the mass balance equations but momentum and energy equations may be developed also. In fact the energy and mass equations will require solving as a whole especially where reaction occurs. A change in composition is simultaneous with a change in the energy equations because of the heat of reaction. Thus temperatures, compositions and possibly other properties of interest, are coupled together by the reaction equation. This reaction equation will be of a nonlinear form, temperature and composition dependent in the vast majority of cases.

Equation (2.26) and the previous equations require stipulated boundary conditions, including initial conditions, for full solution. Common conditions are specified properties, fluxes and reaction rates at the system boundaries in the case of both mass balance and heat balance considerations. So far no distinction has been made between linearity or otherwise in the model equations and this is covered below in section 2.7. Note that steady state equations are obtained from the dynamic equations by equating the time derivatives to zero.

Example Using the maximum gradient model basis what is the response of the exit temperature T_L to a step change in steam temperature in the jacket of the heater, T_j, shown in Fig. 2.13? Assume that the dynamics of the jacket are fast, that the temperature is uniform on the steam side, and that the heat capacity of the tube wall is negligible.

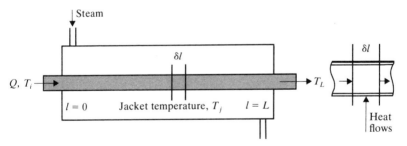

Fig. 2.13 Heat transfer problem, maximum gradient model

The principal concern is with the temperature within the inner tube. The assumptions have given us a lumped (macroscopic) model for the jacket of the system with uniform temperature throughout. The provision of a step change in this temperature infers that the dynamics of the jacket volume are fast compared with those of heat transfer to the inner fluid. It infers also that the flow of steam is sufficient to compensate rapidly for the increased heat transfer to the material of the exchanger and to the fluid. The complexity of the model required and the difficulty of solution increases as we recognize appreciable heat capacity of the central tube and jacket and if the uniform temperature jacket volume is replaced by a fluid in cocurrent, countercurrent or cross flow (Silindir 1970).

Neglecting the heat capacity of the exchanger tube and shell itself the heat balance, continuity equation (2.1), for the inner fluid may be written taking an element of tube δl long,

accumulation of heat within the element = net inflow of heat.

There is no generation term and the net inflow consists of heat entering and leaving with the fluid plus heat transferred through the tube wall.

If the steady volumetric flow = Q

density = ρ

specific heat = c

internal diameter of tube = D

heat transfer coefficient
referred to internal diameter = h

fluid temperature = T

then in time δt

$$\delta l \frac{\pi D^2}{4} \rho c \frac{\partial T}{\partial t} \cdot \delta t = Q\rho c T \, \delta t - Q\rho c \left(T + \frac{\partial T}{\partial l} \delta l \right) \delta t + hD\pi \, \delta l (T_j - T) \, \delta t$$

$$= -Q\rho c \frac{\partial T}{\partial l} \cdot \delta l \cdot \delta t + h\pi D \, \delta l (T_j - T) \, \delta t$$

i.e.

$$\frac{\pi D^2}{4} \rho c \frac{\partial T}{\partial t} = -Q\rho c \frac{\partial T}{\partial l} + h\pi D (T_j - T)$$

which is seen to be of the form of equation (2.25) developed for mass transfer. Combine the constant parameters of the system to give

$$\tau \frac{\partial T}{\partial t} + \alpha \frac{\partial T}{\partial l} = (T_j - T) \tag{2.27}$$

where

$$\tau = \frac{D\rho c}{4h}, \qquad \alpha = \frac{Q\rho c}{h\pi D}$$

and T is a function of both l and t, the position and time.

Even making the original simplifying assumptions it is still necessary to solve this partial differential equation. We may use the Laplace transformation and by further simplifying the initial conditions we may obtain an analytical solution. The use of the Laplace transform in this case converts the partial differential equation to an ordinary differential equation as the first step in the solution.

Transforming equation (2.27) from the time domain,

$$\tau(sT(s) - T(0)) + \alpha\frac{dT(s)}{dl} = T_j(s) - T(s)$$

where $T(s)$ is the transform of T. Note that we now have an equation in which $T(s)$ is a function of the independent variable l. Also $T(0)$ the initial temperature distribution on the tube side of the exchanger is l dependent. Before we can proceed to express $T(s)$ as a function of l, i.e. solve this equation in $T(s)$, we need to know $T(0)$ itself. However, the problem is much simplified if we take as our initial conditions,

$$T_j = T = T_i \quad \text{at} \quad t = 0 \quad \text{for all } l.$$

Such a situation might arise where initially there is no steam flow in the jacket and losses are neglected. Return now to equation (2.27) and replace the temperature T by $T_i + \theta$ where $\theta(t, l)$ is the change in tube temperature and T_j by $T_i + T_j'$ where T_j' is the change in the shell side temperature, i.e.

$$\tau\frac{\partial(T_i + \theta)}{\partial t} + \alpha\frac{\partial(T_i + \theta)}{\partial l} = (T_i + T_j' - T_i - \theta)$$

or

$$\tau\frac{\partial\theta}{\partial t} + \alpha\frac{\partial\theta}{\partial l} = T_j' - \theta \tag{2.28}$$

Equation (2.28) now expresses the dynamics of the heat exchange system in terms of deviations from the defined, special case, initial conditions.

Rearranging equation (2.28) and taking the Laplace transform gives

$$(\tau s + 1)\theta(s) + \alpha\frac{d\theta(s)}{dl} = T_j'(s) \tag{2.29}$$

as now $\theta(0) = 0$ for all l. We also know that $\theta(l = 0)$ is zero for all t if the inlet temperature is unchanged and this gives a boundary condition for the solution of equation (2.29). $T_j'(s)$ is by definition and assumptions constant for all l also. It may be left in this form or put into a specific form (e.g. $1/s$) without affecting this part of the solution.

Solving equation (2.29) we obtain from the standard first order equation methods,

$$\theta(s) = \frac{T_j'(s) \cdot}{(1 + \tau s)}(1 - e^{-(\tau s + 1)l/\alpha}).$$

Restricting our attention to the exit temperature, i.e. at $l = L$

$$\theta_L(s) = \frac{T_j'(s)}{(1 + \tau s)}(1 - e^{-(\tau s + 1)L/\alpha})$$

and if the change in T_j is a step change of magnitude T'_j then

$$T'_j(s) = \frac{T'_j}{s}$$

so that

$$\theta_L(s) = \frac{T'_j}{s(1+\tau s)}(1 - e^{-(\tau s + 1)L/\alpha}). \tag{2.30}$$

Before trying to obtain the inverse of this expression in order to express the fluid exit temperature as a function of time it is useful to break it down into two terms,

$$\theta_L(s) = \frac{T'_j}{s(1+\tau s)} - \frac{T'_j}{s(1+\tau s)} e^{-(1+\tau s)L/\alpha}.$$

It is easier to use standard forms of transform pairs to invert these terms than if we look at the expression as a whole. Although the example has illustrated a method of solution for the maximum gradient model it has left a problem of its own, the inversion of terms which have exponential functions of s. These do not normally yield straightforward functions in time and an alternative formulation and numerical solution method is required. Special attention is paid to the problem by Gould (1969) and Douglas (1972). However, where the exponential is of the simple form $e^{-\beta s}$ the shift or translated function theorem (section 1.4) may be used. Note that this simple form only arises because of the assumptions made in the simplified model of the exchanger, e.g. neglecting wall heat capacity.

Solution of equation (2.30) now proceeds on a term by term basis. The quadratic term may be expressed in partial fractions

$$\frac{1}{s(1+\tau s)} = \frac{1}{s} - \frac{\tau}{1+\tau s}$$

and

$$\theta_L(t) = T'_j(1 - e^{-t/\tau}) - T'_j e^{-L/\alpha}(U(t - t_0) - e^{-(t-t_0)/\tau})$$

where $t_0 = \dfrac{\tau L}{\alpha}$ and $U(t - t_0)$ and $e^{-(t-t_0)/\tau}$ are both zero for $t < t_0$. This expression for $\theta_L(t)$ may be rewritten

$$\theta_L(t) = T'_j(1 - e^{-t/\tau}) - T'_j e^{-L/\alpha}U(t - t_0)(1 - e^{-(t-t_0)/\tau})$$

with

$$U(t - t_0) = 0, \ t < t_0$$
$$1, \ t \geqslant t_0.$$

The delayed terms in the solution show the presence of the distributed element in the system. Because of the lumping of the jacket volume and the neglect of the wall resistance some effect is felt immediately if the jacket temperature changes (Fig. 2.14). Thus we see that the model fits the assumptions and will represent the system behaviour closely if the assumptions are valid. The sharpness of the response is due to the absence of a dispersion term in the model. The

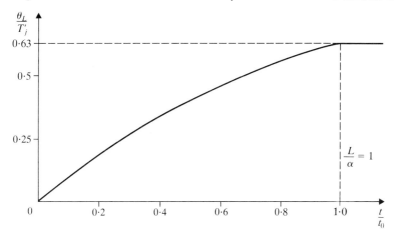

Fig. 2.14 Response of heat exchange system (Fig. 2.13) to step input:

$$\frac{\theta_L}{T_i'} = (1 - e^{-t/t_0 \cdot L/\alpha}) - e^{-L/\alpha} U(t - t_0)(1 - e^{-(t/t_0 - 1)L/\alpha})$$

use of different values for α, L etc. will change the absolute shape of the curve but will still give rise to a discontinuity when the delayed terms first become apparent. Remember that at time $t = 0$ the exchanger is filled along its length with fluid of uniform temperature and the initial temperature in tube and shell is the same.

Compare this solution with that obtained for a lumped parameter, macroscopic, model in the next example.

Example Consider that the tubular exchanger is replaced by a stirred but jacketed tank. The total heat transfer area is A and the overall heat transfer coefficient is H (Fig. 2.15). Assuming a uniform jacket temperature the continuity equation now reads, with $T = T_{out}$, the exit temperature,

$$\rho c V \frac{dT}{dt} = \rho c Q(T_i - T) + AH(T_j - T) \qquad (2.31)$$

In a steady state $\dfrac{dT}{dt} = 0$, and using the subscript$_o$ for the steady state condition

$$0 = \rho c Q(T_{io} - T_o) + AH(T_{jo} - T_o). \qquad (2.32)$$

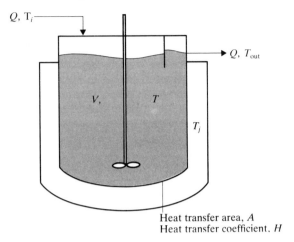

Fig. 2.15 Heat transfer problem, macroscopic model

If we subtract equation (2.32) from (2.31) and assume that T_i remains constant, i.e. $T_{io} = T_i$
then

$$\rho c V \frac{dT}{dt} = \rho c Q(T_o - T) + AH\{(T_j - T_{jo}) - (T - T_o)\}.$$

Redefine

T = change in temperature = $T - T_o$

T_j = change in jacket temperature = $T_j - T_{jo}$,

so we may write

$$\rho c V \frac{dT}{dt} = -\rho c Q T + AH(T_j - T)$$

i.e.

$$\tau \frac{dT}{dt} + T = \alpha T_j$$

where τ, the time constant of the system is

$$\tau = \frac{\rho c V}{(\rho c Q + AH)} \quad \text{and} \quad \alpha = \frac{AH}{\rho c Q + AH}.$$

The solution of this equation for a step input may be obtained by way of illustration via Laplace transforms. If T_j is again the magnitude

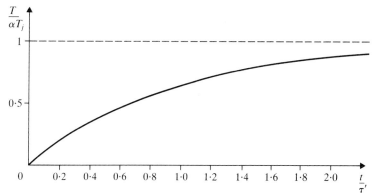

Fig. 2.16 Response of lumped heat exchange system (Fig. 2.15)

$$\frac{T}{\alpha T_j} = 1 - e^{-t/\tau}$$

of the step in jacket temperature

$$(\tau s + 1) T(s) = \frac{\alpha T_j}{s}$$

$$T(s) = \alpha T_j \cdot \frac{1}{s(1 + \tau s)}$$

$$= \alpha T_j \left\{ \frac{1}{s} - \frac{\tau}{1 + \tau s} \right\}$$

$$T(t) = \alpha T_j (1 - e^{-t/\tau})$$

This is a smooth curve with no abrupt changes and the solution of the equation is also a degree easier (Fig. 2.16).

2.6 Population balance models

The transport phenomena based models described above may be complimented by population balance, or age distribution, models. In general the transport phenomena models have been most widely used, especially in the field of control studies. However, population balance models and the empirical work associated with them, are capable of giving considerable information on the most suitable structure to use as a basis for further modelling, especially in the cases of non-ideal behaviour such as channelling and dead spaces. For example, they indicate the flow pattern and non-ideality of mixers which may enable a unit to be divided, for modelling purposes, into a lumped parameter perfectly mixed vessel in series or in parallel with a plug flow section. Thus from an analytical origin they may be used in the synthesis of plant models. Their use in this way will be indicated in the following sections.

Age distribution functions

The 'age' of an element of fluid, or solid, within a vessel at time t is the time that has elapsed between the time of entry of the element and the given time t. An age distribution function $I(t)$ is defined such that the fraction of fluid elements having ages between t and $t+\Delta t$ is given by $I(t)\,\Delta t$. $I(t)$ has the units 'fraction of age per unit time' and the sum of all such fractions is unity, i.e.

$$\int_0^\infty I(t)\,dt = 1, \tag{2.33}$$

$t=0$ is an arbitrary initial time.

The fraction of vessel contents younger than an age τ is $\int_0^\tau I(t)\,dt$ and the fraction older is $\int_\tau^\infty I(t)\,dt$. The sum of the two must be unity, equation (2.33). If fluid flows into a vessel of volume V at a volumetric rate Q then in time Δt the flow in is $Q\,\Delta t$. If none of this leaves the vessel then as $\Delta t \to 0$ the fraction in the vessel with age t as $t \to 0$ is

$$I(t)_{t\to 0}\,\Delta t = \frac{Q\,\Delta t}{V}$$

so

$$I(t)_{t\to 0} = \frac{Q}{V} = \frac{1}{\bar{t}} \tag{2.34}$$

where $\bar{t}$ is the mean residence time, V/Q.

The age distribution function thus has the form of Fig. 2.17. The age distribution function of the exit stream, or the 'residence time

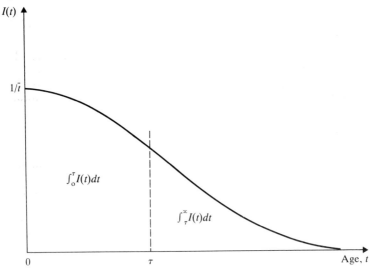

Fig. 2.17 Internal age distribution function, $I(t)$

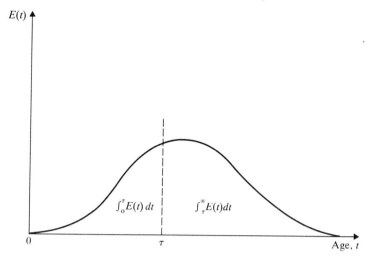

Fig. 2.18   Residence time distribution function, $E(t)$

distribution', $E(t)$, is defined such that the fraction of fluid elements in the exit stream having ages between t and $t + \Delta t$ is $E(t)\,\Delta t$. It is called the 'residence time distribution' since it gives the age distribution reached in that fluid leaving the vessel at time t before leaving the vessel. It need not be, and in only the very special case of perfect mixing is it equal to the (internal) age distribution function. The fraction of ages sums to unity,

$$\int_0^\infty E(t)\,dt = 1 \tag{2.35}$$

and $E(t)$ has also the units of fraction of ages per unit time. The fraction in the exit stream of age less than τ is $\int_0^\tau E(t)\,dt$ and the fraction of older age is $\int_\tau^\infty E(t)\,dt$; again these two sum to unity. The mean residence time $\bar{t}$ is given by the usual first moment equation

$$\bar{t} = \int_0^\infty tE(t)\,dt = V/Q \tag{2.36}$$

Note that $E(t)_{t=0}$ is zero and in a real vessel there is a minimum value for a residence time which is greater than zero.

These age distribution functions are related to each other and to a third function, the 'intensity function', $\Lambda(t)$. This is defined as the fraction of the fluid in the vessel that is of age t that will leave between the time t and $t + \Delta t$. Worthwhile relationships can only be obtained for vessels of constant content volume. Then the continuity equation and unsteady state age population balance may be used to establish (Himmelblau and Bischoff, 1968) the relationships for a constant volume V and flow rate Q,

$$E(t) = -\bar{t}\frac{dI(t)}{dt} \tag{2.37}$$

$$\Lambda(t) = \frac{1}{\bar{t}}\frac{E(t)}{I(t)} = -\frac{d}{dt}\ln[\bar{t}I(t)]. \tag{2.38}$$

Thus the same basic information is contained in all three functions because of their one-to-one relationships but some aspects of non-ideal flow are shown up more easily by a particular function.

It has been common practice to use the mean residence time t to define a dimensionless time base, $\theta = t/\bar{t}$. Then $\bar{t}\,d\theta = dt$ and $E(\theta) = \bar{t}E(t)$, $I(\theta) = \bar{t}I(t)$, $\Lambda(\theta) = \bar{t}\Lambda(t)$ and equations (2.37) and (2.38) become

$$E(\theta) = -\frac{d}{d\theta}I(\theta) \tag{2.39}$$

$$\Lambda(\theta) = \frac{E(\theta)}{I(\theta)} = -\frac{d}{d\theta}\ln I(\theta). \tag{2.40}$$

Determination and interpretation of age distribution functions

The responses of flow processes to a step input and impulse input of traces are known as F and C curves respectively (Danckwerts, 1953). They may be plotted to a dimensionless time base, θ, and the use of inert tracers means that the process itself is unaffected and the concentration of tracers in the output steam, the F and C curves is a measure of the residence time and age distribution function. Typical responses are shown in Figs. 2.19 and 2.20.

In the case of the step input, c_0 is the tracer concentration in the inlet stream and hence, for a constant volume vessel, it is also the final concentration in the outlet. For the impulse input which can only be approximated to in practice, c_0 is the concentration that the tracer would have if taken as evenly distributed throughout the vessel. Thus the C curve starts at zero and ends at zero when all the tracer has left the vessel. The c_0 may be determined from a knowledge of the flow and vessel volume or from the area under the C curve,

$$c_0 = \int_0^\infty C(\theta)\,d\theta. \tag{2.41}$$

For a closed vessel, one in which we consider there to be no back diffusion at the vessel inlet and outlet, the C curve and the residence time distribution E and the F curve and the internal age distribution I are directly related. Then

$$E(\theta) = C(\theta) \tag{2.42}$$

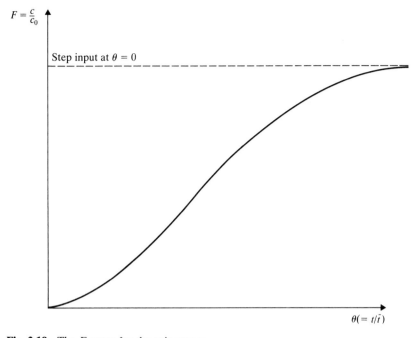

Fig. 2.19 The F curve for the exit stream

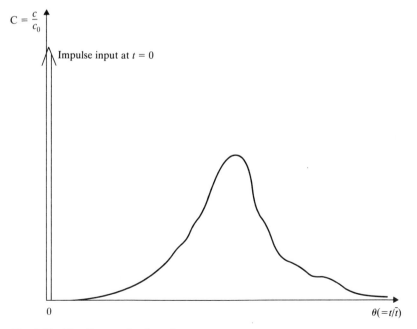

Fig. 2.20 The C curve for the exit stream

and

$$I(\theta) = 1 - F(\theta). \tag{2.43}$$

Other relationships directly follow, e.g.

$$E(\theta) = \frac{dF(\theta)}{d\theta} = C(\theta)$$

The C curve is a recording of the age fraction of fluid elements that enters at $t = 0$ and leaves at $t = t$, i.e. the residence time distribution. The internal age distribution can be expressed in terms of the F curve by equating the amount of tracer remaining in the vessel to the total which enters but does not leave by time t or θ.

Remembering that these results are giving information about flow through the vessel one would expect a representation of the two extremes of plug flow and perfect mixing to be possible with intermediate patterns for 'partial mixing'. In addition intermediate conditions of incomplete mixing may include poor use of the total volume or may result from the presence of dead space which contains virtually stagnant fluid and which reduces the effective vessel volume. A principal cause of such behaviour is poor agitation or poor vessel design. Contrasting with this is the situation where some fluid may flow into and then almost immediately out of the vessel without mixing with the bulk fluid (Fig. 2.21). This could be caused by poor connection positioning and made worse by poor stirring. How then do these effects show up and how do they enable the vessel to be adequately modelled?

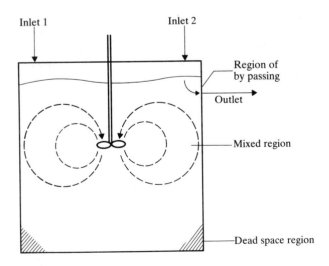

Fig. 2.21 Occurrence of bypassing and dead space

For perfect plug flow there is no mixing within the vessel and all fluid elements have the same residence time $\bar{t}$. This we might expect from a tubular reactor with high flowrates and low diameter-to-length ratio. Thus the F curve itself is a step function but delayed relative to the input by the time $\bar{t}$ (Fig. 2.22a), i.e.

$$F(\theta) = U(t - \bar{t})$$
$$= U(\theta - 1) \tag{2.44}$$

where

$$U(t - \bar{t}) = 0, \, t < \bar{t}$$
$$1, \, t \geq \bar{t}$$

hence

$$I(\theta) = 1 - U(\theta - 1). \tag{2.45}$$

$E(\theta)$ may be obtained by differentiation of $F(\theta)$, or $I(\theta)$, or by direct consideration of the C curve, which will be a delayed impulse, to give

$$E(\theta) = \delta(\theta - 1) \tag{2.46}$$

where $\delta(\theta)$ is the Dirac delta function, having unit area, but occurring over an infinitely short time and hence having instantaneous infinite magnitude. The intensity function is obtained directly from equation (2.38) and yields

$$\Lambda(\theta) = \delta(\theta - 1). \tag{2.47}$$

These functions are shown in Fig. 2.22 together with curves for intermediate stages of mixing, but with no dead space, and for perfect mixing.

As the degree of mixing increases from plug flow to perfect the peak of the residence time moves to the left away from the mean residence time until the perfect mixing pattern is established. Considering the perfect mixing case, e.g. a perfectly stirred vessel for mixing components, and using the step input again,

$$V \frac{dc}{dt} = Qc_0 - Qc \quad \text{(continuity equation)}$$

giving the solution

$$\frac{c}{c_0} = 1 - e^{-t/\bar{t}} \quad (\bar{t} = V/Q)$$

By definition this is $F(\theta)$, i.e.

$$F(\theta) = 1 - e^{-\theta}$$

Consequently

$$I(\theta) = e^{-\theta} \tag{2.48}$$
$$E(\theta) = e^{-\theta} \tag{2.49}$$
$$\Lambda(\theta) = 1. \tag{2.50}$$

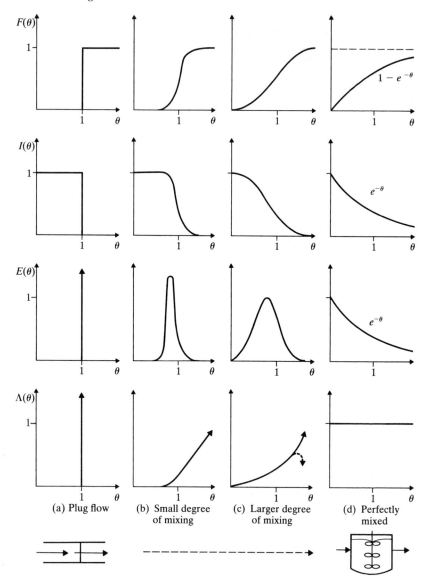

Fig. 2.22 Age distribution functions

Thus we see that it is possible to tell in these cases where a tracer technique can be used whether we are faced with plug flow, perfect mixing or some near approximation to them or if the vessel does not easily fall into either of these brackets.

If a part of the vessel retains stagnant or near stagnant fluid then this means that the fluid passing through will have a mean residence time less than if all the vessel were 'active'. If we know the true volume and hence can calculate the mean residence time based on this, then the difference between this and the 'observed' residence time will become apparent. Even if this is not so dead space will still show up as the residence time distribution curve will have a long tail, i.e. the proportion of liquid with long residence times will be increased. As $E(\theta)$ gets smaller, the experimental determination and its graphical results may be cut short, say at θ_1 (Fig. 2.23, curve *b*). The apparent

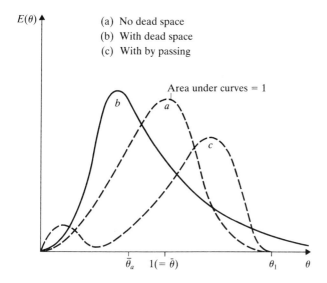

Fig. 2.23   Effect of dead space and bypassing on residence time distribution

mean dimensionless residence time is then

$$\bar{\theta}_a = \int_0^{\theta_1} \theta E(\theta)\, d\theta \quad \theta_1 > 2\bar{\theta}.$$

True mean

$$\bar{\theta} = \int_0^\infty \theta E(\theta)\, d\theta$$

$$= \int_0^{\theta_1} \theta E(\theta)\, d\theta + \int_{\theta_1}^\infty \theta E(\theta)\, d\theta$$

$$= \bar{\theta}_a + \int_{\theta_1}^\infty \theta E(\theta)\, d\theta.$$

Thus, depending on the size of the second integral, i.e. the length of the 'tail', $\bar{\theta}_a$ will be correspondingly smaller than the true $\bar{\theta}$. $\bar{\theta}_a$ is the mean residence time based on the truncated experiment. Using the total flow rate Q, the apparent mean residence time, t_a, is

$$\bar{t}_a = \frac{V_{active}}{Q} = \frac{V - V_{dead}}{Q}$$
$$= \bar{t}\bar{\theta}_a$$

where $\bar{t}$ is the real residence time evaluated from known V and Q, therefore

$$V_{dead}/V = 1 - \bar{\theta}_a.$$

Hence the proportion of dead space may be estimated and if the true volume is known a better volume for simulation, $V - V_{dead}$ may be found for use in a transport phenomena type model. Bypassing of some of the flow Q is shown up by the double lump feature of Fig. 23, curve *c*. Some of the fluid Q_b is now passing through rapidly and this shows up in the internal age distribution figure (Fig. 2.24), where, because there is a quantity of fluid present with a short age compared with the remainder, there is a sudden drop in the $I(\theta)$ curve. Because the remaining fluid rate $Q - Q_b$ is flowing through essentially the whole volume the second main peak is moved to the right giving a longer effective mean residence time. In practice the humps of Fig. 23, curve *c* and the drop in Fig. 2.24 may not be so clear but these effects are

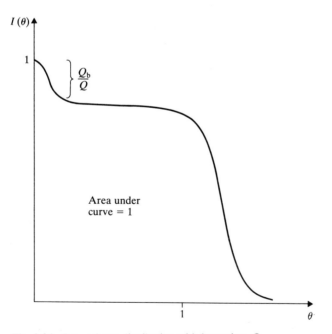

Fig. 2.24 Internal age distribution with bypassing, Q_b.

generally well shown in the intensity curve $\Lambda(\theta)$. With dead space the tail of this curve will decrease and then increase again. With bypassing there will be an earlier hump in the $\Lambda(\theta)$ curve. Thus these 'stagnancy' effects can be detected.

When these defects or other less easily observed non-idealities are found to exist the model for the vessel may be made up as a 'combined model' with regions of perfect mixing, dead space etc. (Levenspiel and Bischoff, 1963).

Combined models

An alternative determination of the active volume and active flow is described by Cholette and Cloutier and gives probably the simplest form of combined model. However, if no clearly defined deficiencies exist then a good approximation of this form will not necessarily be obtained and the method is limited. Within a general combined model flows are considered to link various regions. Various combinations in series and parallel may be postulated and the exact meanings of dead space and specification of the number of required parameters discussed (Levenspiel, 1962; Adler and Hovorka 1961) but the Cholette and Cloutier stirred tank provides a good basic example. In this a stirred tank of volume V is represented by an active perfectly mixed region V_a, a completely dead region V_d with no flow or transfer to other regions and a bypassing flow Q_b (Fig. 2.25).

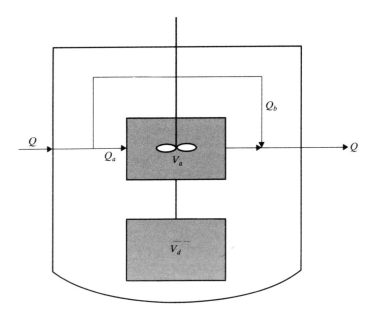

Fig. 2.25 Cholette and Cloutier's model for a stirred tank (Cholette and Cloutier, 1959)

By carrying out the experimental determination of a single F curve Cholette and Cloutier were able to determine the ratio of V_a/V and of Q_a/Q and to show the effect of agitation in these values.

Age distribution studies have been used in a number of applications (e.g. Cholette and Cloutier, 1959; Hull and Rosenberg 1960; Parent and Ray 1964) but their possible uses in the study of control stem from the hope that they enable a description suitable for mathematical simulation, i.e. a model, to be established and they also highlight the importance of the proper design and location of measurement elements. If a system possesses dead space then the true representation is no longer a simple linear model and the system 'gain' in terms of an input change effect on the short-term properties of the system will be higher than expected. The controller might be designed or controller settings adjusted to allow for this.

For a full description of population balance models involving multidimensional distributions in terms of fluid element and entity properties one should see Himmelblau and Bischoff (1968).

Example   A suspension is made up by adding solids to a fluid in a mixing tank. Some of the fluid bypasses the tank in plug flow as in Fig. 2.26 in order that it may be heated. The two streams join together again as shown. How would you estimate by a single tracer experiment with measurement at A and B only the division of flow between the two paths and what is the transfer function relating the tracer concentration at B to that at A?

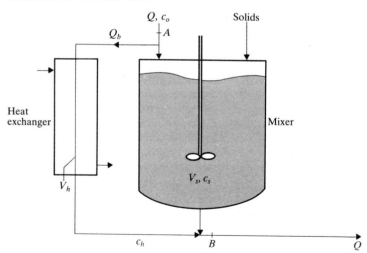

Fig. 2.26   Mixing with bypass

Assume that the pipe lengths between A and B and the mixer are negligible. Total flow is Q, bypass flow is Q_b. The heat exchanger and mixer vessels have volumes V_h and V_s respectively of the process fluid. Consider a step input of tracer at concentration c_o into the inlet flow to give an F curve.

For the mixer, assuming perfect mixing and following the addition of tracer

$$V_s \frac{dc_s}{dt} = (Q - Q_b)c_o - (Q - Q_b)c_s. \tag{2.51}$$

At $t = 0$, $c_s = 0$ and the solution of the first order equation is

$$c_s = c_o(1 - e^{-(Q - Q_b)t/V_s}) \tag{2.52}$$

The tracer flows in plug flow through the exchanger with a delay from A to B equal to its mean residence time, V_h/Q_b. Therefore the tracer concentration c_h is given by

$$c_h = c_o U(t - t_o) \tag{2.53}$$

where

$$U(t - t_o) = 0, \, t < t_o$$
$$1, \, t \geq t_o$$

and

$$t_o = \frac{V_h}{Q_b}.$$

The tracer concentration at B in response to the step input is the weighted sum of equation (2.52) and (2.53),

$$Qc_B = (Q - Q_b)c_o(1 - e^{-(Q - Q_b)t/V_s}) + Q_b c_o U(t - t_o)$$

or

$$\frac{c_B}{c_o} = \frac{Q - Q_b}{Q}(1 - e^{-(Q - Q_b)t/V_s}) + \frac{Q_b}{Q} U(t - t_o). \tag{2.54}$$

[To express this on a normalized time base the total volume $V_h + V_s$ should be used to determine the overall mean residence time. This is unnecessary for our purposes.]

To establish the transfer function between input c_o and output c_B it is possible to work instead from the basic equations separately rather than from equation (2.54). The need is for $G(s) = \dfrac{C_B(s)}{C_o(s)}$, i.e. the transfer function for the general case.

From equation (2.51)

$$(V_s s + (Q - Q_b))C(s) = (Q - Q_b)C_o(s)$$

so

$$\frac{C_s(s)}{C_o(s)} = \frac{Q-Q_b}{V_s s + (Q-Q_b)}$$

$$= \frac{1}{1 + V_s/(Q-Q_b) \cdot s}$$

where $V_s/(Q-Q_b)$ is the time constant of the stirred vessel.

From the argument preceding equation (2.53), for the heater pure flow delay

$$\frac{C_h(s)}{C_o(s)} = e^{-st_o}.$$

The third relationship required is the weighting

$$c_B = \frac{Q-Q_b}{Q} \cdot c_s + \frac{Q_b}{Q} c_h$$

i.e.

$$C_B(s) = \frac{Q-Q_b}{Q} C_s(s) + \frac{Q_b}{Q} C_h(s).$$

Combination of these gives the overall transfer function

$$\frac{C_b(s)}{C_o(s)} = \frac{Q-Q_b}{Q} \cdot \frac{1}{1 + V_s/(Q-Q_b) \cdot s} + \frac{Q_b}{Q} e^{-st_o} \qquad (2.55)$$

If c_o is a step input then this yields equation (2.54) on inversion. This stepwise procedure may be illustrated by a block diagram. Such block diagrams will feature widely in subsequent chapters.

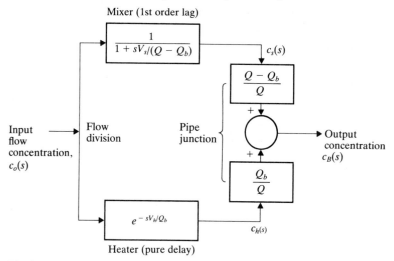

Fig. 2.27   Block diagram representation for tracer test

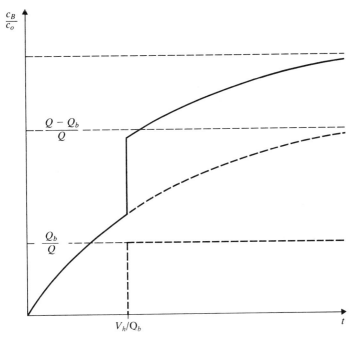

Fig. 2.28 Step response of plug flow and mixing in parallel paths

The step response of the system is shown in Fig. 2.28. From an experimental determination of c_B/c_o the amount of bypass Q_b/Q can be determined. In turn, V_h the heater volume can also subsequently be found.

2.7 Linearization, transfer functions and state variables

Linear systems and transfer functions have been introduced briefly in Chapter 1. These, with the concept of state variables, form the basis of the mathematical representation of many systems for control studies.

Linearization

The fundamental property of linear systems and the property on which they may be defined is that of superposition. If a given input, say u, into a system results in an output x and a further input v results in an output y when u and v are applied one at a time only, then the system is said to be linear if an input $u+v$ results in the output $x+y$. Note that this is true both for the ultimate steady state and for all dynamic conditions through which the system passes from the time of input being applied to the time a steady state, if any, is achieved. Thus if the input is doubled in magnitude $u + u$, the output is also doubled in magnitude and so on. It is for this reason that we can refer to a 'unit' input with regularity in the study of linear systems and still retain full generality. If this superposition does not hold then the system is nonlinear. Linearity is a property of the system, not of the inputs, and superposition holds for both constant and time-varying inputs.

The ability to represent the dynamic behaviour of a system by linear equations, i.e. equations to which superposition applies, has distinct advantages, of which not the least is that the solution of linear equations (as general analytical results) is comparatively simple in many cases. The use of linear matrix algebra in particular has shown particular advantages in the study of control systems. Thus, although it is increasingly possible to solve nonlinear equations to yield numerical results by computer, the solution of linear equations is still easier and may lead to the very useful general results.

Unfortunately the actual presence of real linear systems is extremely limited and carried to the extremes of their performance one would be safe in saying probably no real full-range linear systems exist. For example, even the simple mechanical case of a spring which is considered to obey Hooke's law, i.e. that the extension (output) is proportional to the force (input) becomes nonlinear as it becomes more and more extended and it obviously has a finite length. The dynamic laws like Newton's second law (accleration is proportion to force for a given mass) do not break down under normal conditions but the force itself may be modified by the results of acceleration, i.e. a frictional or windage force proportional to the square of the speed may develop, and because of this inherent feedback the overall system equations become nonlinear. Specific nonlinearities such as saturation of an amplifier, or the on-off action of a switch may be treated by separate methods (Chapter 8). These problems become even more acute in process systems where, for example, reactions introduce a high degree of nonlinearity. In the face of such difficulties when can a linear representation of a system, i.e. a linear differential equation be used?

For a general discussion attention will be restricted to lumped systems, i.e. those which can be represented by ordinary differential equations, our macroscopic model.

Above it is seen that even simple systems may develop nonlinear behaviour when they are subject to large inputs. When inputs are limited the system at least approximates to linear behaviour, and in fact this is true in most cases if the inputs are taken small enough, provided the system is stable, i.e. the result of a disturbance does not continue to grow. In regulation control theory where a main concern is keeping plant outputs constant in the face of comparatively small deviations it is apparent that linearization of the system representation can be beneficial. Also in such cases it is possible to consider deviations about some desired steady-operating conditions. A change in input, desired or otherwise, then causes a change in output. Working in terms of these changes it is obvious that zero change in input will produce no change in the output, so that preceding a change our initial steady conditions are zero. Similarly we may consider changes in other variables in the system, not necessarily actual system 'outputs', and we shall refer to these variables later as state variables. A further advantage in using changes in variables as our working basis is that it involves numerical accuracy in computation. If a change of 1·5 per cent is considered in the input from a steady state of 100 per cent then, working in absolute values the input is 101·5 but working in the perturbation alone the input is 1·5, which may be 'scaled up' to the 100 per cent level itself. By using the change itself we avoid the differencing of large numbers and the subsequent loss of accuracy. In using the linear approximations we rely on the system staying near its initial state. Initially we shall use a single input–single output case to illustrate the linearization of a nonlinear system.

Consider first when the output X is given by some nonlinear algebraic function of the input U, e.g. $X = aU^2 + bU$. (Note that in this case doubling the input does not double the output.) Denote the general function relating the output to the input and which is determined by the process as

$$X = f(U).$$

Both X and U may be functions of time t, $X(t)$ and $U(t)$, i.e.

$$X(t) = f(U(t)).$$

The input U is represented by being comprised of a nominal (steady state) value U_o and a deviation, $U - U_o$. For a deviation about the value $f(U_o)$, $f(U)$ may be expanded in the standard Taylor series,

$$f(U) = f(U_o) + \frac{\partial f(U_o)}{\partial U} \cdot (U - U_o) + \frac{\partial^2 f(U_o)}{\partial U^2} \cdot \frac{(U - U_o)^2}{2!} + \dots \quad (2.56)$$

where $\dfrac{\partial f(U_o)}{\partial U}$ is the rate at which $f(U)$ changes with U, evaluated at the nominal value U_o. For small deviations $(U - U_o)$ the higher terms in the expansion rapidly decrease and the expansion reduces to

$$f(U) \simeq f(U_o) + \frac{\partial f(U_o)}{\partial U} \cdot (U - U_o)$$

so that

$$X \simeq f(U_o) + \frac{\partial f(U_o)}{\partial U} \cdot (U - U_o). \quad (2.57)$$

The nominal value of X, X_o, is given under steady conditions by the input U_o, i.e. $X_o = f(U_o)$. The change, $X - X_o$, is thus given from equation (2.57) by

$$X - X_o = f(U) - f(U_o)$$

$$\simeq \frac{\partial f(U_o)}{\partial U} \cdot (U - U_o).$$

Since our major interest is in the deviation, define the deviations as $x = X - X_o$, $u = U - U_o$, then

$$x = \frac{\partial f(U_o)}{\partial U} \cdot u \tag{2.58}$$

where x and u are now the deviations.
Now

$$\frac{\partial f(U)}{\partial U} = \frac{\partial f(U)}{\partial (U_o + u)}$$

$$= \frac{\partial f(U)}{\partial u}$$

and as $x = f(U) - f(U_o)$
so

$$\frac{\partial f(U)}{\partial u} = \frac{\partial}{\partial u} \{x + f(U_o)\}$$

$$= \frac{\partial x}{\partial u}$$

and

$$\frac{\partial f(U_o)}{\partial u} = \frac{\partial x}{\partial u} \bigg|_{U_o, X_o}$$

since $f(U_o)$ is evaluated at U_o and is independent of u. Equation (2.58) thus becomes

$$x = \frac{\partial x}{\partial u} \bigg|_{U_o, X_o} \cdot u \tag{2.59}$$

where the subscripts U_o, X_o indicate evaluation of the derivative at these co-ordinates.

This is now a linear equation, $\dfrac{\partial x}{\partial u} \left(= \dfrac{\partial f}{\partial U} \right)$ is evaluated at the nominal condition and becomes a constant coefficient, and hence the change in output x is proportional to the change in input u (Fig. 2.29). The coefficient in the linearized equation is dependent upon the nominal condition, but once established it is not changed for small deviations about that condition. The effect of the nominal condition is shown by looking at two nominal points satisfying the relationship

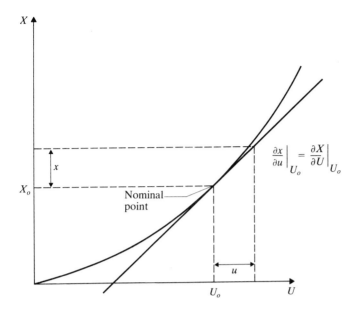

Fig. 2.29 Linearization of nonlinear simple function

$X = U^2 + 2U$. We see that

$$\frac{\partial x}{\partial u} = \frac{\partial X}{\partial U} = 2U + 2.$$

For $X_o = 0$, $U_o = 0$

$$\frac{\partial x}{\partial u} = 2$$

and

$$x = 2u.$$

For $X_o = 3$, $U_o = 1$

$$\frac{\partial x}{\partial u} = 4$$

and

$$x = 4u.$$

This illustrates in particular that linearization of a nonlinear function is only suitable when we are concerned with small deviations from a nominal condition.

The initial conditions corresponding to X_o, U_o are $x = 0$, $u = 0$. Now extend this argument to the dynamics of a simple system where the rate of change of X with time, dX/dt is dependent on the input U and on the value of X itself, i.e.

$$\frac{dX(t)}{dt} = f(X(t), U(t)) \tag{2.60}$$

Remembering that f, X and U are functions of t write this more simply as

$$\frac{dX}{dt} = f(X, U)$$

where again f may be a nonlinear function of X and U. We now see that to express the function $f(X, U)$ in terms of a nominal state X_o, U_o we need a more general form of the Taylor expansion;

$$f(X, U) = f(X_o, U_o) + \frac{\partial f(X_o, U_o)}{\partial X} \cdot (X - X_o) + \frac{\partial^2 f(X_o, U_o)}{\partial X^2}$$

$$\cdot \frac{(X - X_o)^2}{2!} + \ldots + \frac{\partial f(X_o, U_o)}{\partial U} \cdot (U - U_o)$$

$$+ \frac{\partial^2 f(X_o, U_o)}{\partial U^2} \cdot \frac{(U - U_o)^2}{2!} + \ldots$$

We choose our nominal condition X_o, U_o as an equilibrium (steady state) point, i.e. when $dX/dt = 0$. Depending on the form of $f(X, U)$ this may give us more than one steady state pair X_o, U_o. However, by choosing the nominal condition we see that

$$f(X_0, U_o) = 0$$

and hence on truncation of the series

$$f(X, U) \simeq \frac{\partial f(X_o, U_o)}{\partial X} \cdot (X - X_o) + \frac{\partial f(X_o, U_o)}{\partial U} \cdot (U - U_o)$$

$$= \frac{\partial f(X_o, U_o)}{\partial X} \cdot x + \frac{\partial f(X_o, U_o)}{\partial U} \cdot u.$$

At the equilibrium point, the deviations x and u are both zero. As the partial derivatives are evaluated at a specific X, U they may be written as 'constants' A, B.
Then

$$\frac{dx}{dt} = \frac{dX}{dt}$$

$$= f(X, U)$$

$$\simeq Ax + Bu. \tag{2.61}$$

However, in the general form A and B may be functions of time $A(t)$, $B(t)$. By assuming that they are time invariant, i.e. the partial derivatives of $f(X, U)$ are time invariant, the situation is further simplified. In making this assumption we are assuming that the governing physical conditions upon which the relationship may be established, or which it represents, no not change with time, or at least change only very slowly compared with the changes x resulting directly from a change u. For example, in a heat transfer model A and B will depend on heat transfer coefficients. It is assumed that these coefficients are independent of time if an equation of the form (2.61) is used and if the coefficients themselves are not considered as variables. Then the equation

$$\frac{dx}{dt} = Ax + Bu \tag{2.62}$$

is linear (in x and u) and stationary (A and B are not functions of time). The general solution of the ordinary linear differential equation is given by the sum of its complementary function and the particular integral

$$x(t) = e^{At} \cdot x(0) + \int_0^t e^{A(t-\tau)} Bu(\tau) \, d\tau \tag{2.63}$$

where x(0) is the initial value of x at t = 0, u is a function of time, and the second term on the right-hand side is a convolution integral. This equation will be reviewed when the multivariable case is considered.

In this section on linearization of a simple relationship it has been convenient to use upper case notation for absolute values (X, U) and the lower case for the deviations from these, (x, u). In the remainder of the text we shall reserve the upper case to denote a transformed variable, e.g. X(s), a constant, or a matrix. Variables x and u will be denoted by lower case only, and it will be apparent from the context whether a nominal or deviation value is to be used. If necessary a subscript, e.g. x_o, will be used to indicate a nominal condition about which perturbations occur.

Transfer functions and convolution integral

In section 1.4 the transfer function was introduced via the Laplace transform. It was noted that for the input–output relationships to be expressed by the transfer function

$$\frac{X(s)}{U(s)} = G(s)$$

the initial conditions on x(t) must be zero, i.e. x(0) is zero in equation (2.63). Taking the Laplace transform of equation (2.63) with x(0) = 0, i.e. of the convolution integral alone,

$$X(s) = \mathscr{L}\left[\int_0^t e^{A(t-\tau)} Bu(\tau) \, d\tau \right]$$

$$= \int_0^\infty e^{-st} \int_0^t e^{A(t-\tau)} Bu(\tau) \, d\tau \, dt.$$

Since this relation holds only for all $\tau < t$ and the convolution integral is zero for $\tau > t$ we can write

$$X(s) = \int_0^\infty e^{-st} \int_0^\infty e^{A(t-\tau)} Bu(\tau) \, d\tau \, dt.$$

Introducing the new variable $t' = t - \tau$ and reordering the integration yields

$$X(s) = \int_0^\infty e^{At'} \cdot e^{-s(t'+\tau)} \, dt' \int_0^\infty Bu(\tau) \, d\tau$$

$$= \int_0^\infty e^{-st'} e^{At'} \, dt' \int_0^\infty Bu(\tau) e^{-s\tau} \, d\tau$$

$$= \mathscr{L}[e^{At'}] \cdot \mathscr{L}[Bu(\tau)]$$

$$= \frac{1}{s-A} \cdot BU(s)$$

i.e.

$$\frac{X(s)}{U(s)} = \frac{B}{s-A} \tag{2.64}$$

By considering the initial dynamic equation (2.61) and transforming this it can be seen that

$$(s - A)X(s) = BU(s)$$

directly, i.e. equation (2.64) is confirmed.

This is an example of the occurrence of the general convolution integral, $\int_0^t f_1(t-\tau)f_2(\tau) \, d\tau$, with a specific $f_1(t-\tau)$. If the input is the delta function $\delta(t)$ with Laplace transform $U(s)$ equals unity, Table 1.1, then the inverse, $g(t)$, of the transfer function $G(s)$ itself is obviously the response to the impulse input because with this input $X(s)$ is equal to $G(s)$. This impulse function response $g(t)$ is also known as the weighting function of the system. Thus for a general input $u(t)$ the system response is the convolution integral of the weighting function $g(t)$ and the input $u(t)$ denoted frequently as

$$x(t) = g(t) * u(t)$$

$$= \int_0^t g(t-\tau)u(\tau) \, d\tau$$

$$= \int_0^t g(t)u(t-\tau) \, d\tau.$$

We see that this is in agreement with our earlier statement on the transform of the convolution integral and from our table of transforms

$$x(t) = \mathscr{L}^{-1}[X(s)] = \mathscr{L}^{-1}[G(s)U(s)]$$

$$= \mathscr{L}^{-1}\left[\mathscr{L}\int_0^t g(t-\tau)u(\tau)\,d\tau\right]$$

$$= \int_0^t g(t-\tau)u(\tau)\,d\tau.$$

Within this section attention has been focussed on a first order differential equation and this enables us to use standard forms of solution. Linear systems may obviously be, and generally are, of higher order, e.g. if they involve inertial forces. The transfer function then involves higher powers of s. Alternatively systems of higher order differential equations may be reduced to a set of first order differential equations. Any linear system may be represented in this way.

Consider the general second order equation representing a system with inertia and damping,

$$\ddot{x} + a\dot{x} + bx = u \tag{2.65}$$

where $\dot{x} = d/dt$ etc. The transfer function may be obtained directly since

$$(s^2 + as + b)X(s) = U(s)$$

if $x(0) = \dot{x}(0) = 0$.
Thus

$$G(s) = \frac{X(s)}{U(s)} = \frac{1}{s^2 + as + b}.$$

Alternatively equation (2.65) may be reduced to two first order equations.

Let $x_1 = x$ and $x_2 = \dot{x}$ and

then

$$\dot{x} = \dot{x}_1 = x_2$$

and

$$\ddot{x} = \dot{x}_2 = -a\dot{x}_1 - bx_1 + u \qquad \text{from equation (2.65)}$$

$$= -ax_2 - bx_1 + u.$$

These equations may be rearranged

$$\dot{x}_1 = x_2 \tag{2.66}$$
$$\dot{x}_2 = -bx_1 - ax_2 + u$$

and written in vector form (see Chapter 1)

$$\begin{bmatrix} \dot{x}_1 \\ \dot{x}_2 \end{bmatrix} = \begin{bmatrix} 0 & 1 \\ -b & -a \end{bmatrix} \begin{bmatrix} x_1 \\ x_2 \end{bmatrix} + \begin{bmatrix} 0 \\ 1 \end{bmatrix} u$$

or

$$\dot{\mathbf{x}} = \frac{d\mathbf{x}}{dt} = \mathbf{A}\mathbf{x} + \mathbf{b}u. \tag{2.67}$$

This method may be extended to higher order equations similarly.

Alternatively sets of simultaneous first order equations may arise directly without passing through a higher order representation and may be similarly represented in this vector-matrix form. The x_1 and x_2 are examples of 'state variables' of the system.

State variables

The more recent approach to the dynamics of systems is to consider the behaviour of the 'state' of the system as distinct from considering only particular outputs. The state is determined by all its 'state variables'. The state variables of a system are the smallest set of variables which contain sufficient information for all future states of the system to be determined. In addition the equations relating the state variables and also the future inputs to the system must be known. If the state variables are defined in this way then the effects of all prior inputs are combined within their values and to determine their values at some time t after an initial time, say t_0, only their values at t_0 are required and not their previous history. This smallest number of variables required to define the state is the 'order' of the system, n, and we talk of an nth order state space, in which the trajectories are traced by the state variables with increase in time. The simplest multivariable representation is for the two dimensional state and this state-space is referred to as the 'phase plane' (see Chapter 8).

In choosing the state variables it is convenient to choose variables of direct physical significance where possible, and in particular those which can be directly observed and measured as it is then easier to use them in the construction of control laws and in the implementation of them. In process systems in particular this may prove difficult. The actual output of the plant, although dependent on the state vector need not be and in complex plant is usually not, identical to the 'chosen state' of the plant. Thus a relationship is also required between the state vector and the output vector. Although the number of independent state variables required to adequately describe a system will be fixed the choice of the state variables is not unique.

As the problem is now one of considering a number of relationships between inputs, outputs and state variables of the system, a number of simultaneous equations are required to do this. As indicated above high order differential equations may be reduced to first order equations by the definition of suitable state variables, so that a set of first order differential equations may be used to represent general system behaviour. These may contain nonlinear equations but they may still be written in vector-matrix form. The time derivative of each state variable is expressed as a function of all state variables (although some may be absent, i.e. have zero coefficients in any particular relationship) and system inputs. Thus for the state variables x_i, $i = 1, \ldots, n$, and inputs u_i, $i = 1, \ldots, m$,

$$\dot{x}_i = f_i(x_1, \ldots, x_i, \ldots, x_n; u_1, \ldots, u_m)$$

or for all state variables the function vector $\mathbf{f}$ is used so that

$$\dot{\mathbf{x}} = \mathbf{f}(\mathbf{x}, \mathbf{u}). \tag{2.68}$$

If the system parameters vary with time then $\mathbf{f}$ will contain explicit functions of time. Our main concern at present is with time-invariant linear systems so that f_i is a linear function of the x_i and u_i and the coefficients are constant, i.e.

$$\dot{x}_i = a_{i1}x_1 + a_{i2}x_2 + \ldots + b_{i1}u_1 + b_{i2}u_2 + \ldots b_{im}u_m$$
$$= \sum_{j=1}^{n} a_{ij}x_j + \sum_{j=1}^{m} b_{ij}u_j.$$

Equations of this form may be written for all state variables to yield

$$\begin{bmatrix} \dot{x}_1 \\ \cdot \\ \cdot \\ \cdot \\ \dot{x}_n \end{bmatrix} = \begin{bmatrix} a_{11} & a_{12} & \ldots & a_{1n} \\ \cdot & & & \\ \cdot & & & \\ \cdot & & & \\ a_{n1} & a_{n2} & \ldots & a_{nn} \end{bmatrix} \begin{bmatrix} x_1 \\ \cdot \\ \cdot \\ \cdot \\ x_n \end{bmatrix} + \begin{bmatrix} b_{11} & b_{12} & \ldots & b_{1m} \\ \cdot & & & \\ \cdot & & & \\ \cdot & & & \\ b_{n1} & b_{n2} & \ldots & b_{nm} \end{bmatrix} \begin{bmatrix} u_1 \\ \cdot \\ \cdot \\ \cdot \\ u_m \end{bmatrix}$$

i.e.

$$\dot{\mathbf{x}} = \mathbf{A}\mathbf{x} + \mathbf{B}\mathbf{u} \tag{2.69}$$

where $\mathbf{x}$ and $\mathbf{u}$ are the state vector and input (or control) vector respectively and $\mathbf{A}$ and $\mathbf{B}$ are constant matrices for the time-invariant system. The outputs may not be identical with the state variables but will be linear combinations of them if the state variables are correctly chosen. Thus output y_i, $i = 1, \ldots, p$ $(p \leqslant n)$ is given by

$$y_i = c_{i1}x_1 + c_{i2}x_2 + \ldots + c_{in}x_n$$
$$= \sum_{j=1}^{n} c_{ij}x_j$$

or for all y_i

$$\mathbf{y} = \mathbf{C}\mathbf{x} \tag{2.70}$$

where $\mathbf{y}$ is the output vector and $\mathbf{C}$ the output constant matrix.

For an n-dimensional (n^{th} order) state vector, an m-dimensional input vector and a p-dimensional output vector, the 'system matrix' $\mathbf{A}$ is of order $n \times n$, the 'distribution matrix' $\mathbf{B}$ is of order $n \times m$ and the 'output or measurement matrix' $\mathbf{C}$ is of order $p \times n$.

Before moving to a more general consideration of the form of the solution of equation (2.69) compare this model of system behaviour with that for the single input–single output case, equations (2.61) and

(2.62). It is apparent that equation (2.62) is just the first order case of equation (2.69) with the state variable being identified to the output, i.e. c_{ij} for $i=j=1$ is unity in equation (2.70), and with a single input. Thus we might expect other similarities to emerge. Consider the form of the matrices **A** and **B**. The linear equations (2.69) may be linear by direct derivation or may result from the linearization of a nonlinear description or set of relationships. In the latter case we linearize the problem by considering small changes about a chosen state trajectory. This was illustrated in the single input–single output case and the procedure is similar, although a little more involved, in the multivariable case. Here one has to assume that all state variables remain sufficiently close to the nominal state and then we apply the Taylor series expansion to each state variable, remembering that it is a function of all others and of the input(s). (See note in section 2.7 above on notation.)

The general dynamic system description may be written in the form of equation (2.68) where **f** is now assumed to be a general, possibly nonlinear, form. The nominal state trajectory is defined by the vector $\mathbf{x}_0(t)$ with input vector $\mathbf{u}_0(t)$ so that

$$\dot{\mathbf{x}}_0(t) = \mathbf{f}(\mathbf{x}_0, \mathbf{u}_0).$$

Note that the nominal state is not necessarily time invariant. If **x** and/or **u** are slightly away from their nominal values, i.e. the deviations are $\mathbf{x}-\mathbf{x}_0$ and $\mathbf{u}-\mathbf{u}_0$ then

$$\dot{\mathbf{x}} = \mathbf{f}(\mathbf{x}, \mathbf{u})$$

and the rate of change of the deviation of **x** is given by

$$\frac{d(\mathbf{x}-\mathbf{x}_0)}{dt} = \dot{\mathbf{x}} - \dot{\mathbf{x}}_0$$

$$= \mathbf{f}(\mathbf{x}, \mathbf{u}) - \mathbf{f}(\mathbf{x}_0, \mathbf{u}_0).$$

It is assumed that the deviations are such that the form of the functions **f** are unchanged. It is easiest at this stage to consider one state variable

$$\dot{x}_i = f_i(\mathbf{x}, \mathbf{u})$$

$$= f_i(x_1, \ldots, x_n; u_1, \ldots, u_m)$$

and the Taylor series for $\dot{x}_i(x_1, \ldots, x_n; u_1, \ldots, u_m)$ in terms of deviations about its nominal trajectory $\dot{x}_{i,0}(x_{i,0}, \ldots, x_{n,0}; u_{1,0}, \ldots, u_{m,0})$. The series expansion is

$$f_i(\mathbf{x}, \mathbf{u}) = f_i(\mathbf{x}_0, \mathbf{u}_0) + \frac{\partial f_i(\mathbf{x}_0, \mathbf{u}_0)}{\partial x_1} \cdot (x_1 - x_{1,0}) + \ldots \frac{\partial f_i(\mathbf{x}_0, \mathbf{u}_0)}{\partial x_n} \cdot (x_n - x_{n,0})$$

$$+ \frac{\partial f_i(\mathbf{x}_0, \mathbf{u}_0)}{\partial u_1} \cdot (u_1 - u_{1,0}) + \ldots \frac{\partial f_i(\mathbf{x}_0, \mathbf{u}_0)}{\partial u_m} \cdot (u_m - u_{m,0})$$

$$+\frac{\partial^2 f_i(\mathbf{x}_0, \mathbf{u}_0)}{\partial x_1^2} \cdot (x_1 - x_{1,0})^2 + \ldots + \frac{\partial^2 f_i(\mathbf{x}_0, \mathbf{u}_0)}{\partial u_1^2} \cdot (u_1 - u_{1,0})$$

$$+ \ldots \frac{\partial^2 f_i(\ldots)}{\partial x_1 \partial x_2} \ldots \tag{2.71}$$

where the derivatives $\dfrac{\partial f_i(\mathbf{x}_0, \mathbf{u}_0)}{\partial x_j}$, $\dfrac{\partial f_i(\mathbf{x}_0, \mathbf{u}_0)}{\partial u_j}$ are the derivatives of the function $f_i(\mathbf{x}, \mathbf{u})$ evaluated along the nominal trajectory. On the assumption that both $\mathbf{x}$ and $\mathbf{u}$ remain 'sufficiently close' to the nominal state- and input-vectors respectively, then equation (2.71) can be truncated to the first order approximation. Writing $f_{i,0}$ for $f_i(\mathbf{x}_0, \mathbf{u}_0)$,

$$\frac{d(x_i - x_{i,0})}{dt} = f_i(\mathbf{x}, \mathbf{u}) - f_i(\mathbf{x}_0, \mathbf{u}_0)$$

$$\simeq \frac{\partial f_{i,0}}{\partial x_1} \cdot (x_1 - x_{1,0}) + \ldots \frac{\partial f_{i,0}}{\partial x_n} \cdot (x_n - x_{n,0})$$

$$+ \frac{\partial f_{i,0}}{\partial u_1} \cdot (u_1 - u_{1,0}) + \ldots \frac{\partial f_{i,0}}{\partial u_m} (u_m - u_{m,0}). \tag{2.72}$$

Now n equations of this form, one for each state variable x_i, describe the system behaviour. Comparison of the full set of equations (2.72, all i) with equation (2.69) shows that for the linearized system the system matrix $\mathbf{A}$ is comprised of the partial derivatives $\dfrac{\partial f_{i,0}}{\partial x_j}$ as its elements, and the distribution matrix $\mathbf{B}$ has $\dfrac{\partial f_{i,0}}{\partial u_j}$ as its elements. These 'Jacobian' matrices therefore are

$$\mathbf{A} = \begin{bmatrix} \dfrac{\partial f_{1,0}}{\partial x_1} & \cdots & \cdots & \dfrac{\partial f_{1,0}}{\partial x_n} \\ \dfrac{\partial f_{2,0}}{\partial x_1} & \dfrac{\partial f_{2,0}}{\partial x_2} & \cdots & \dfrac{\partial f_{2,0}}{\partial x_n} \\ \cdot & & & \\ \cdot & & & \\ \cdot & & & \\ \dfrac{\partial f_{n,0}}{\partial x_1} & \dfrac{\partial f_{n,0}}{\partial x_2} & \cdots & \dfrac{\partial f_{n,0}}{\partial x_n} \end{bmatrix} \quad \mathbf{B} = \begin{bmatrix} \dfrac{\partial f_{1,0}}{\partial u_1} & \cdots & \dfrac{\partial f_{1,0}}{\partial u_m} \\ \dfrac{\partial f_{2,0}}{\partial u_1} & \cdots & \dfrac{\partial f_{2,0}}{\partial u_m} \\ \cdot & & \\ \cdot & & \\ \cdot & & \\ \dfrac{\partial f_{n,0}}{\partial u_1} & \cdots & \dfrac{\partial f_{n,0}}{\partial u_m} \end{bmatrix} \tag{2.73}$$

If the diagonal terms of the $\mathbf{A}$ matrix are dominant then the dynamics of the specific state variables $x_1, x_2, \ldots,$ are most strongly influenced by their own values. Redefining our state variables as the deviations $x_i - x_{i,0}$, $u_i - u_{i,0}$ and calling these x_i and u_i respectively then equation

(2.72) for all $i = 1, \ldots, n$ becomes the linearized equation

$$\dot{\mathbf{x}} = \mathbf{A}\mathbf{x} + \mathbf{B}\mathbf{u} \tag{2.74}$$

with $\mathbf{A}$ and $\mathbf{B}$ defined by equation (2.73).

In its general form equation (2.74) has time variant elements in $\mathbf{A}$ and $\mathbf{B}$. In the regulator problem $\mathbf{x}_0$ and $\mathbf{u}_0$ will refer to the equilibrium steady state and $\dot{\mathbf{x}} = f(\mathbf{x}_0, \mathbf{u}_0) = \mathbf{0}$. If the perturbations from the state are small then the linearization may be performed and, since the operation remains in the vicinity of the single desired state, the elements of $\mathbf{A}$ and $\mathbf{B}$ will be constant, time invariant. In the servo problem it may be necessary, for large and rapid excursions, to use time variant terms within the system matrix and in batch processes $\mathbf{A}$ will also be time variant but our normal requirements will be that $\mathbf{A}$ and $\mathbf{B}$ are time invariant, i.e. constant matrices.

It will be seen that the transfer function and vector-matrix notation are complementary in both being used in the description of linear plant behaviour. The solution of these dynamic equations will be investigated in Chapter 3 where the handling of models established on the concepts of this chapter will be discussed.

Example   The key variables describing the quality of a process product X_1, X_2 are given in terms of themselves and a controlling process input U by the equations

$$\dot{X}_1 = -X_1^2 + 0.5 X_2$$
$$\dot{X}_2 = X_1 - 3X_2 + U.$$

Derive the linear vector-matrix equation relating changes about a nominal steady state $(X_{1,0}, X_{2,0})$.

Following equations (2.71) to (2.74) write

$$\dot{X}_1 = f_1(\mathbf{X}, U)$$
$$\dot{X}_2 = f_2(\mathbf{X}, U)$$

where

$$\mathbf{X} = \begin{bmatrix} X_1 \\ X_2 \end{bmatrix},$$

$$f_1(\mathbf{X}, U) \simeq (\mathbf{X}_0, U_0) + \frac{\partial f_{1,0}}{\partial X_1} \cdot (X_1 - X_{1,0}) + \frac{\partial f_{1,0}}{\partial X_2} \cdot (X_2 - X_{2,0})$$

$$+ \frac{\partial f_{1,0}}{\partial U} \cdot (U - U_0).$$

Using lower case letters for perturbations

$$f_1(\mathbf{X}, U) \simeq f_{1,0} + \frac{\partial f_{1,0}}{\partial x_1} \cdot x_1 + \frac{\partial f_{1,0}}{\partial x_2} \cdot x_2 + \frac{\partial f_{1,0}}{\partial u} \cdot u$$

and

$$\dot{x}_1 = \dot{X}_1 - \dot{X}_{1,0} = f_1(\mathbf{X}, U) - f_1(\mathbf{X}_0, U_0)$$

$$= \frac{\partial f_{1,0}}{\partial x_1} \cdot x_1 + \frac{\partial f_{1,0}}{\partial x_2} \cdot x_2 + \frac{\partial f_{1,0}}{\partial u} \cdot u$$

$$= -2X_{1,0} \cdot x_1 + 0 \cdot 5 \cdot x_2.$$

Similarly

$$\dot{x}_2 = 1 \cdot x_1 - 3 \cdot x_2 + 1 \cdot u$$

so that

$$\dot{\mathbf{x}} = \mathbf{Ax} + \mathbf{Bu}$$

or

$$\begin{bmatrix} \dot{x}_1 \\ \dot{x}_2 \end{bmatrix} = \begin{bmatrix} -2X_{1,0} & 0 \cdot 5 \\ 1 & -3 \end{bmatrix} \begin{bmatrix} x_1 \\ x_2 \end{bmatrix} + \begin{bmatrix} 0 \\ 1 \end{bmatrix} u.$$

Note that if we had started with all linear equations without constant terms then the perturbation equations would have been identical in form to the initial equation, e.g. as $\dot{x}_2$ and $\dot{X}_2$. In this example nominal steady conditions are $X_{1,0} = X_{2,0} = U_0$ or $X_{1,0} = X_{2,0} = 0 \cdot 5$, $U_0 = 1$.

2.8 Summary. The general pattern of dynamic system representation

At this stage it is worthwhile recapping briefly by looking again at the dynamic system representation as a whole in general terms. The following chapters will look at specific details and the more analytical aspects together with examples.

To study the dynamics and control of a system it is necessary to establish a model either by consideration on physical and chemical grounds or by plant tests. At this stage attention is restricted to the deterministic models which will be comprised of (nonlinear) differential (ordinary or partial) equations. To aid in determining the form of these the mechanics, transport phenomena or age and population distributions of the system must be sought and expressed in the form of the mathematical equations which then comprise the model. If these equations have, or can be suitably reduced to, a linear form then either the Laplace transform, linear matrix algebra, or a combination of these can be used in solving the equations. These equations and their solutions can be combined with, and used to determined, controllers for the system. The aim of these controllers is to improve in some way the steady state and/or dynamic behaviour of the plant.

Chapter 3
Basic system dynamics

3.1 Graphical representation

In Chapters 1 and 2 the concepts of an input–output relationship using Laplace transforms and transfer functions were introduced and the use of the vector-matrix notation was shown to be a complementary, or even a more general, form of representation which included the single input or single output case. The most general illustrative form of representation of the relationships established from modelling is the block diagram. An alternative form is the signal flow graph. Both of these representations were originally used for the input–output model but their use has been extended with the development of control theory so that multivariable vector-matrix models are illustrated in this way also.

For the input–output representation the block diagram and signal flow graphs (or chart) usually show the transformed relationships, although they may of course be equally used for the equations in the time domain (and in D-operator notation). As it is the more common usage, attention will be concentrated on the transformed relationships. This will infer a linear time invariant relationship and the discussion will be based on the lumped parameter representation also.

Block diagrams

A block diagram shows essentially two relationships; that between the input and response of a particular unit or sub-unit and secondly the flow of information or signals between the various elements making up the system as a whole. These elements may not be true physical sub-units of the systems but may represent convenient analytical units within the overall model. By block diagram algebra the configuration of the block diagram may be changed whilst retaining the same overall input–output relationship.

If the relationship between an output $x(t)$ and input $u(t)$ is

$$x(t) = g(u(t)) \tag{3.1}$$

then the transfer function representation, where valid, may be written

$$\frac{X(s)}{U(s)} = G(s). \tag{3.2}$$

In the subsequent discussion of linear systems, linearization and the introduction of perturbation variables where required will be assumed to have been introduced as described in Chapter 2. The block diagram may have one of the forms of Fig. 3.1.

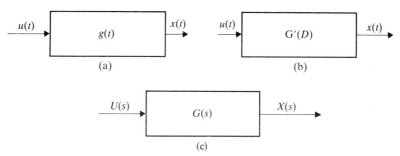

Fig. 3.1 Alternative bases for block diagrams

Figure 3.1(a) and (c) are those more commonly used and most attention will be given to (c), the Laplace transform notation. The arrows indicate the direction of signal flow. At times this may appear to conflict at first sight with the physics of the system since the block diagram does not necessarily represent the physical structure of the system. The summation of signals (of equal dimensions and identical units) is shown by a circle, with the sign to be attributed to the signal entering the summation being shown either just within the circle or alongside the arrow head. When a signal is required at more than one summation point or element the line is divided at a branch point. Each elemental block will then have only one input and one output, (Fig. 3.2).

Most control systems have a block diagram which includes a feedback path. By feedback is meant the use of the response of the system to modify the input to the system and this is the basis of feedback control. A system employing feedback of knowledge in this way is a closed loop system (Fig. 3.3(a)). In the absence of any feedback the system is open loop. For a complex plant the system may be closed loop with respect to one input–output relationship and open loop with respect to others.The feedback path may itself contain an element with a transfer function to modify the output of the system before utilizing it in the forward path through the main transfer function of the system (Fig. 3.3(b)). For example x may be a temperature and the input r a voltage so that in the feedback path there is a conversion from, say, °C to mV. If r is a reference value signifying the value which it is required

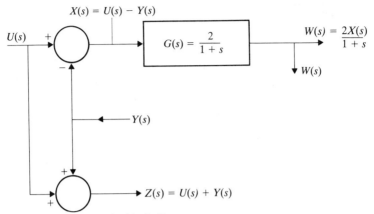

Fig. 3.2   Summations by block diagram

that the output follows then $r-x$, or its equivalent after x has been subject to some transformation, $H(s)$, is the error signal, e.

As well as inputs like r, additional inputs or disturbance may occur partway through the system. Also there may be feedback to an intermediate point so that the block diagram becomes more involved. This feedback may be caused by a physical effect such as friction. In a linear system a number of external disturbances may be lumped together by following the rules of block diagram algebra. This is part of the general pattern of block diagram reduction.

Before considering block diagram reduction let us summarize the path that is taken to the establishment of a block diagram representation. This is outlined in Fig. 3.4 for a 'zero' mass first order damped system. From a knowledge of the system its linearized equations are established, step (a). These are then transformed (b) and the block diagram element drawn (c). From a series of such relationships a full diagram is established. This may have some of the characteristics of that of Fig. 3.5. Block diagram reduction is essential if the system model produces a large number of relationships which give rise to a complicated arrangement as shown in Fig. 3.5. The subsequent analysis is simplified by the reduction in loops and by the better nesting of

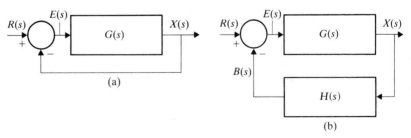

Fig. 3.3   Basic closed loop diagrams

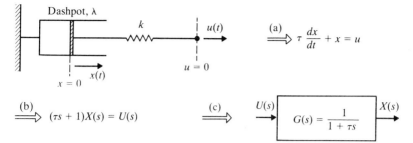

Fig. 3.4 Relationship between physical system and block diagram

loops within each other. The reduced number of blocks represent transfer functions though of increasing complexity as is shown by the stage-wise reduction of Fig. 3.5.

There are two basic rules: the forward transfer function product must be the same between chosen points within which the reduction is taking place, and the transfer function product around a closed loop must remain the same. The later rule can be looked on as a special case of the first rule, the two chosen points being coincident.

Looking specifically at Fig. 3.5, the external disturbances may be lumped by replacing $Z_1(s)$ by $Z_1(s) + Z_2(s)/G_2(s)$ and removing $Z_2(s)$ from its shown position. The sum effect on the signal entering $G_3(s)$ is unchanged. Similarly any number of external disturbances may be lumped and the 'point of entry' of $Z_1(s)$ and $Z_2(s)$ could be moved back to sum with $R(s)$ after passing through suitably established lumped blocks. As drawn we infer that the prime consideration is to determine the overall response, or transfer function relating $X(s)$ to $R(s)$. If the response to $Z_1(s)$ or $Z_2(s)$ is of prime importance it is advisable to redraw the diagram, e.g. showing the main path now between $Z_1(s)$ and $X(s)$ (Fig. 3.6). In this form it is easier to determine the overall transfer function $X(s)/Z_1(s)$.

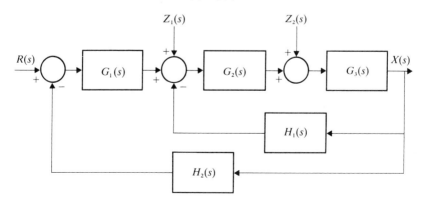

Fig. 3.5 System with multiple feedback paths and external disturbances

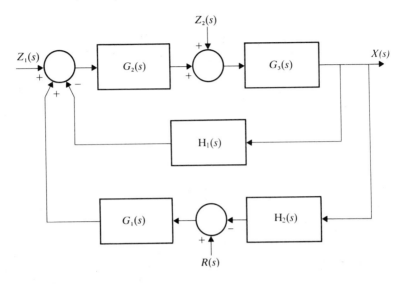

Fig. 3.6   Alternative form of Fig. 3.5

Because of the linearity laws the response to all inputs $R(s)$, $Z_1(s)$, $Z_2(s)$ together is obtained by determing each transfer function in turn $X(s)/R(s)$, $X(s)/Z_1(s)$, $X(s)/Z_2(s)$ assuming respectively that the other two inputs are zero, and then summing the effects by

$$X(s) = \frac{X(s)}{R(s)} \bigg|_{Z_1=Z_2=0} . R(s) + \frac{X(s)}{Z_1(s)} \bigg|_{R=Z_2=0} . Z_1(s)$$
$$+ \frac{X(s)}{Z_2(s)} \bigg|_{R=Z_1=0} . Z_2(s).$$

To determine $X(s)/R(s)$ for $Z_1(s) = Z_2(s) = 0$ the stages in Fig. 3.7 are

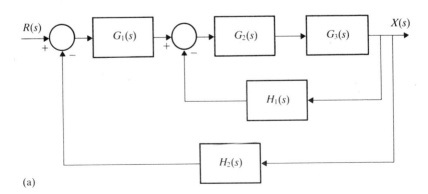

(a)

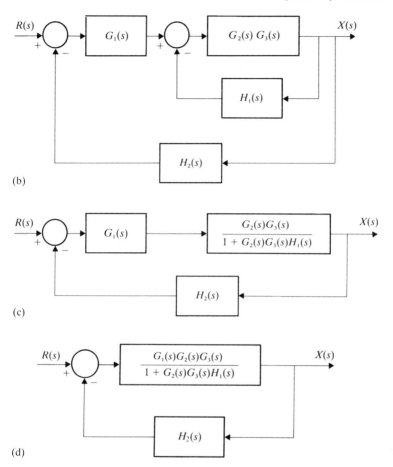

(b)

(c)

(d)

This stage may be reduced to (e) to retain the feed back form or to (f) finally

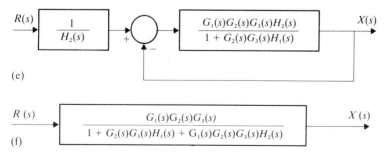

(e)

(f)

Fig. 3.7 Reduction of block diagram to give (i) unity feedback system (*e*) (ii) overall transfer function (*f*)

gone through. This figure shows the key rules in block diagram reduction, i.e. in relevant transfer function manipulation. The key rules of block diagram reduction used here, and which guide one to other manipulations of algebra are shown in Fig. 3.8.

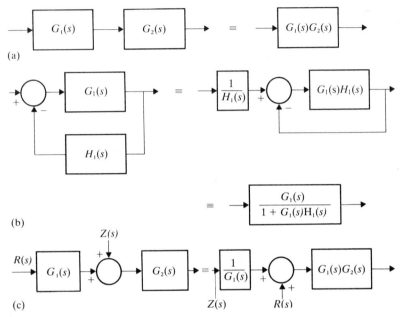

Fig. 3.8　Some basic block diagram algebra

The dynamics of systems with feedback and the corresponding block diagram will be further examined in Chapter 4. Note though that such input–output systems can be represented by diagrams like Fig. 3.7(f) so that our discussions on systems in pure input–output terms do not necessarily exclude those with feedback paths although these will be the subject of separate consideration later.

Block diagram representation of multivariable systems

The block diagrams in the above section may be considered as special cases of the general multivariable representation. The linear multivariable system (see section 3.4) may be represented by its Laplace transformed vectors and transfer function matrices (or transfer matrices). Taking the simplest form the transformed output vector $\mathbf{X}(s)$ may be related to the input $\mathbf{U}(s)$ simply by

$$\mathbf{X}(s) = \mathbf{G}(s)\mathbf{U}(s) \qquad (3.3)$$

where $\mathbf{G}(s)$ is a matrix, each element of which $G_{ij}(s)$ is a transfer function relating the ith output x_i to the jth input u_j. For the single input-single output case this reduces to the transfer function $G(s)$.

With scalar transfer functions the order of multiplication, when reducing the block diagrams for example, was not important. However, with manipulation of transfer matrices the order is important. If a further vector $\mathbf{Y}(s)$ is given by

$$\mathbf{Y}(s) = \mathbf{C}(s)\mathbf{X}(s) \qquad (3.4)$$

then

$$\mathbf{Y}(s) = \mathbf{C}(s)\mathbf{G}(s)\mathbf{U}(s) \qquad (3.5)$$

and the equivalent block diagram reduction is as Fig. 3.9. Note the order of $\mathbf{G}(s)$ and $\mathbf{C}(s)$ in each case as determined by equations (3.3) to

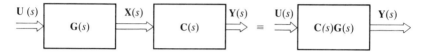

Fig. 3.9 Manipulation of transfer function matrices

(3.5). The banded arrow shows the multivariable nature of the problem. At a summation point the order of all vectors being summed must be equal.

As well as block diagrams being used in this way the first order differential equation may be illustrated diagrammatically in the time domain. If the system equation is

$$\dot{\mathbf{x}} = \mathbf{A}\mathbf{x} + \mathbf{B}\mathbf{u} \qquad (3.6)$$

then this may be shown as in Fig. 3.10(a). If we put this equation in the time domain into the s-domain, i.e. use Laplace transforms, then we

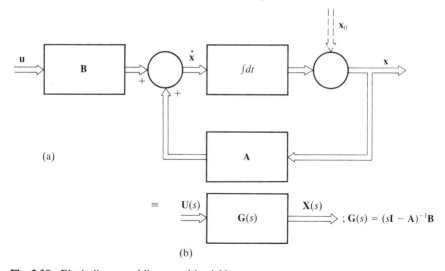

Fig. 3.10 Block diagram of linear multivariable system

have

$$sX(s) = AX(s) + BU(s)$$

or

$$(sI - A)X(s) = BU(s).$$

Premultiplying both sides by $(sI - A)^{-1}$ gives us

$$X(s) = (sI - A)^{-1}BU(s)$$
$$= G(s)U(s) \tag{3.7}$$

Signal flow graphs

Signal flow graphs offer an alternative graphical representation for complex systems and have a considerable associated algebra (Mason 1953; Naslin 1965). This book will use them far less than the block diagram form. The simultaneous linear equations describing the system are expressed in their Laplace transform notation and the diagram consists of a number of 'nodes' representing a system variable or signal and these are connected by 'branches' which are the signal multipliers (or transfer functions). This gain between nodes is also known as the transmittance. As in the block diagram, flow of information is in one direction and indicated by an arrow, but the relationships between variables may be obtained without a reduction equivalent to that required by the block diagram.

Features of the signal flow graph are illustrated in Fig. 3.11 which is equivalent to the block diagrams of Fig. 3.5. The nodes $R(s)$, $Z_1(s)$, $Z_2(s)$ are input nodes or sources. $X(s)$ is the output node or sink and the three other nodes with branches both going into and leaving them are mixed nodes. The node immediately following $G_3(s)$ could have been labelled $X(s)$ and the final unity gain branch omitted. As with block diagrams, different arrangements of the equation lead to different graphs so that neither represent a unique figure for a given system.

If required, a signal flow graph can be reduced so that it has only input and output nodes, directly relating independent and dependent

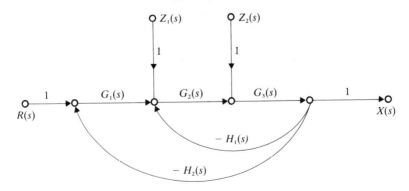

Fig. 3.11   Signal flow graph equivalent to Fig. 3.5

variables. The methods are indicated in Fig. 3.12 which may be compared with Fig. 3.8.

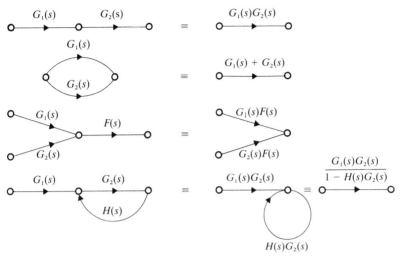

Fig. 3.12 Rules for reduction of signal flow graphs

The overall transfer function $G(s)$ between any input node i and any output node j may also be obtained using Mason's rule (Mason, 1953, 1956) without resource to the graph-reduction using the above rules. This states that

$$G_{ij}(s) = \frac{1}{\Delta} \sum G_k(s) \, \Delta_k \qquad (3.8)$$

where $G_k(s)$ = gain or transfer function of kth forward path connecting node i to node j,

Δ = determinant of graph

= 1 − sum of all different loop gains + sum of gain products of all possible combinations of two nontouching loops − sum of gain products of all possible combinations of three non-touching loops + . . .

$$= 1 - \sum_a L_a + \sum_{b,c} L_b L_c - \sum_{d,e,f} L_d L_e L_f + \ldots$$

where

$\sum L_a$ = sum of all different loop gains or transfer functions,

Δ_k = the term derived from Δ by keeping only those terms which are fully isolated (i.e. no common branch or node) from path k, called the minor of the kth path.

Like the block diagram the signal flow graph may be used for multivariable systems either using the transformed variables and carefully observing the rules of matrix algebra, or else directly in the time domain. Thus the equivalent to Fig. 3.10 is Fig. 3.13.

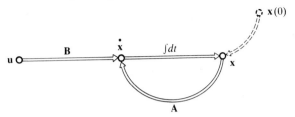

Fig. 3.13   Signal flow graph in the time domain for $\dot{x} = Ax + Bu$

In the event of initial conditions for **x** being non-zero an additional branch entering the node **x** with the initial conditions is required. For full details of signal graph manipulation the reader is referred to Naslin.

The above rule conveys little without an example.

Example   Consider the example in Fig. 3.14 which represents a system where there is interaction between two inputs and two outputs, see also Fig. 6.17. Inspection of Fig. 3.14 shows that there are the three closed loops in the direction of the arrows,

    (i) $E_1 M_1 C_1 E_1$
    (ii) $E_2 M_2 C_2 E_2$
and (iii) $E_1 M_1 C_2 E_2 M_2 C_1 E_1$ (drawn in more heavily)

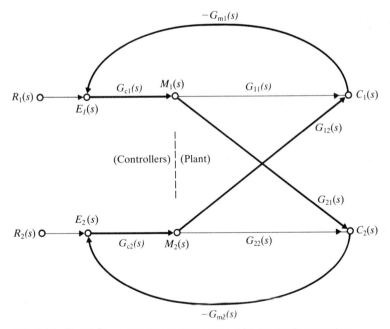

Fig. 3.14   Signal flow graph for control system with feedback controllers

Loops (i) and (ii) are the only non-touching (either by branch or node) loops so that the determinant Δ of the system graph will contain the sum of the loops gains of (i) (ii) and (iii), plus the gain product of loops (i) and (ii). Thus

$$\Delta = 1 + G_{c1}G_{11}G_{m1} + G_{c2}G_{22}G_{m2} - G_{c1}G_{21}G_{m2}G_{c2}G_{12}G_{m1}$$
$$+ G_{c1}G_{11}G_{m1}G_{c2}G_{22}G_{m2}.$$

(a) Between R_1 (or E_1) there is only one forward, i.e. open, path to C_2 and for this pair of nodes, therefore,

$$G_1 = G_{c1}G_{21}$$

and Δ_k, from its description above is simply unity as all three loops touch the path,

$$\Delta_1 = 1.$$

The transfer function relation $R_1(s)$ to $C_2(s)$ is then

$$\frac{C_2(s)}{R_1(s)} = \frac{G_1\,\Delta_1}{\Delta}$$
$$= \frac{G_{c1}G_{21}}{\Delta}.$$

(b) Between $R_2(s)$ and $C_2(s)$ there are two open paths in the direction of the arrows;

$$G_1 = -G_{c2}G_{12}G_{m1}G_{c1}G_{21}$$
$$G_2 = G_{c2}G_{22}$$

and

$$\Delta_1 = 1$$
$$\Delta_2 = 1 + G_{c1}G_{11}G_{m1}$$

so that

$$\frac{C_2(s)}{R_2(s)} = \frac{G_1\,\Delta_1}{\Delta} + \frac{G_2\,\Delta_2}{\Delta}$$
$$= \frac{-G_{c2}G_{12}G_{m1}G_{c1}G_{21} + G_{c2}G_{22}(1 + G_{c1}G_{11}G_{m1})}{\Delta}.$$

With experience this procedure may prove more rapid than the earlier block diagram reduction, although possibly liable to careless error. It has however generally proved less popular than block diagram reduction methods, which can be more easily interpreted in distinct physical steps.

3.2 Single input–single output systems

Although emphasis has shifted from the study of single input–single output systems to the general multivariable case, low order single input–single output systems still illustrate simply the type of behaviour which even those of higher order and complexity approximate to in many cases. In this section therefore the transient and frequency response of these systems will be discussed. The general first and second order differential equations will be used and specific examples given. The equations will be general in the sense also that the source of their formation, e.g. whether or not they represent systems containing feedback elements and whether they represent mechanical, electrical or process systems, will not feature in them. The specific significance and problems of feedback and control will be covered in the following chapters.

The dynamics of a system are assessed normally in terms of the response to either impulse, step or ramp inputs or in terms of the frequency response, i.e. the response of the system to a constant amplitude sinusoidal input. In addition to those set inputs we may consider stochastic inputs, those which vary in a random manner or contain a random element. An introduction only to these will be given in Chapter 11.

As we shall be looking at linear systems we may consider unit inputs. The general system behaviour of Fig. 3.15 is expressed in terms of its transfer function

$$X(s) = G(s)U(s).$$

The system transfer function changes with the system but we can standardize on the inputs.

Fig. 3.15 General single input–single output system

Unit impulse

Although a true impulse may only be approximated to in practice by a short time duration pulse this approximation may be quite adequate if the duration of the pulse input is small compared with the time of response of the system. The unit impulse may be defined as an input of zero time duration but having an area under its time curve of unity. This can be seen in Fig. 3.16 as $\varepsilon \to 0$ and we can write the definition as

$$\delta(t - t_0) = 0 \qquad t \neq t_0$$
$$\infty \qquad t = t_0$$

and

$$\int_{t_1}^{t_2} \delta(t - t_0)\, dt = 0 \qquad t_1 > t_0 \quad \text{or} \quad t_2 < t_0$$

$$1 \qquad t_1 < t_0 < t_2.$$

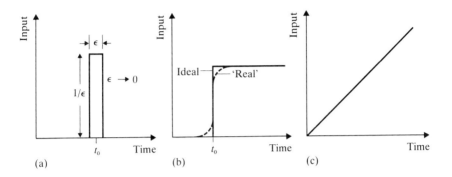

Fig. 3.16 (*a*) Impulse input as $\varepsilon \to 0$; (*b*) Step input; (*c*) Ramp input

The Laplace transform of the unit impulse is unity, i.e. $U(s) = 1$. This follows by direct application of the Laplace transform definition but if the impulse is delayed, $\delta(t - t_0)$, then the inverse has the value of unity only for $t \geqslant t_0$ and is zero for $t < t_0$. Thus for an impulse at $t = 0$ the output $X(s)$ is given by

$$X(s) = G(s) \tag{3.9}$$

and

$$x(t) = g(t),$$

the weighting function or impulse-response function.

Unit step

The unit step is shown in Fig. 3.16(b) and is defined by

$$U(t - t_0) = 0 \qquad t < t_0$$

$$1 \qquad t \geqslant t_0.$$

Physical step inputs are usually slightly rounded but may be approximated to, similarly to the impulse, and could represent sudden failure of an electric motor, breakage of a mechanical component, a switching action etc. If t_0 equals zero, i.e. we count time from the instant of input, the Laplace transform of $u(t)$ is $1/s$ so that

$$X(s) = \frac{1}{s} G(s). \tag{3.10}$$

Ramp input

A ramp input Fig. 3.16(c) is less severe than the step and it is more easily applied in practice. Real step inputs may be represented by a sharp ramp followed by a steady value. The transform of the ramp input $r(t) = at$, is a/s^2 so that

$$X(s) = \frac{a}{s^2} \cdot G(s). \tag{3.11}$$

The frequency response to sinusoidal inputs will be considered after we have looked at the response of simple systems to the above inputs.

First order system

The general first order linear dynamic system equation is

$$\tau \frac{dx(t)}{dt} + x(t) = ku(t). \tag{3.12}$$

k is the steady state gain of the system, the proportional factor between x and u in the steady state. The coefficient τ is assumed to be time invariant. The transfer function between output x and input u is

$$\frac{X(s)}{U(s)} = \frac{k}{(\tau s + 1)}. \tag{3.13}$$

[By scaling x suitably by division of (3.12) by k this could always be

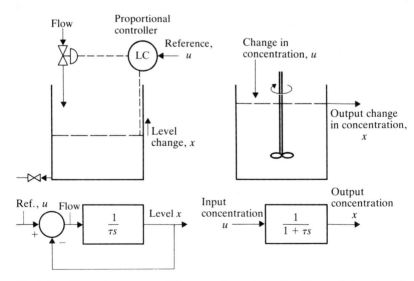

Fig. 3.17 Alternative sources of the same form of equation (first order), $\tau\dot{x} + x = ku$, or $X(s) = k\dfrac{U(s)}{1 + \tau s}$

written

$$\tau \frac{dx'}{dt} + x' = u$$

where

$$x' = x/k,$$

and this may represent any form of physical system which is first order, linear and time invariant.]

To illustrate that the systems in this chapter may be comprised of either forward transfer elements alone or include feedback, equation (3.13) equally represents both systems of Fig. 3.17. The first is level control in a vessel, assuming the controller and valve have very fast dynamics, and the second is the change in concentration from a mixing tank following a change in one of the input variables. Thus it is the overall transfer function which expresses the dynamics of the system and the systems described by any one form of equation can vary widely (see Chapter 2). Thus analytical solutions of equations like (3.13) and those that follow have a widespread applicability.

Impulse response

For a unit impulse input equation (3.13) becomes, for $t \geqslant 0$,

$$X(s) = \frac{k}{\tau s + 1}$$

and

$$x(t) = \frac{k}{\tau} e^{-t/\tau}. \tag{3.14}$$

The response is shown in Fig. 3.18 for τ positive. The system having no 'inertia' term is instantly displaced by the impulse to a value k/τ and then returns on a decaying exponential to its original state at infinite time. Because of the fact that in theory it takes this infinite time to reach its original state it is convenient to classify the system by its 'time constant', τ. Putting $t = \tau$ in equation (3.14) gives

$$x(\tau) = \frac{k}{\tau} e^{-1} = 0 \cdot 368 \frac{k}{\tau}$$

In time τ the first order system returns to only $0 \cdot 368$ of its maximum displacement, i.e. $63 \cdot 2$ per cent of its return path. Substituting $t = 3\tau$ shows that it has returned by 95 per cent at this time and at $t = 5\tau$ it has returned by 99 per cent , i.e. it is at its initial value to within 1 per cent of the total initial displacements. Thus τ may be determined by experiment in this way. Also a plot of $\log x(t)$ against time t will enable τ to be determined from its slope and its intercept will give the steady

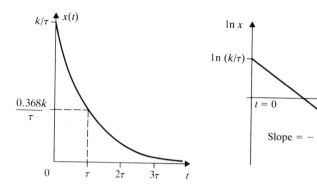

Fig. 3.18   First order impulse response

state gain k;

$$\ln x(t) = \ln \left(\frac{k}{\tau}\right) - \frac{t}{\tau}.$$

Step response

For a step input at $t = 0$ the dynamic equation becomes

$$X(s) = \frac{1}{s} \cdot \frac{k}{\tau s + 1} \tag{3.15}$$

as

$$U(s) = \frac{1}{s}.$$

Implementation of partial fractions gives us

$$X(s) = k\left(\frac{1}{s} - \frac{\tau}{\tau s + 1}\right)$$

and on inversion using Table 1.1

$$x(t) = k(1 - e^{-t/\tau}). \tag{3.16}$$

As with the impulse response put $t = \tau$ and then

$$x(\tau) = k(1 - e^{-1})$$
$$= 0{\cdot}632k.$$

so that after a time equal to the time constant of the system the response has reached 63·2 per cent of its final value, assuming τ is positive. τ may be determined by experiment using this fact, or by taking the slope of x against t at $t = 0$, or by a log plot of $(k - x(t))$ against time. Again, within 5τ the response has reached to within 1 per cent of its final value and is only about 2 per cent off after 4τ has elapsed.

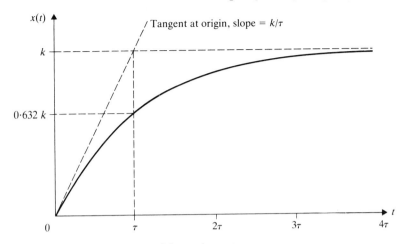

Fig. 3.19 Transient response of first order system

The impulse is the derivative of the step function and, as expected for a linear time-invariant system, the output or impulse response can be seen from equations (3.14) and (3.16) to be the derivative of the step response. The unit step input is itself the derivative of the ramp input $u(t) = t$ and a similar relationship is thus expected between the step and ramp responses as exists between the impulse and step responses.

Ramp input

For a ramp input of unit slope,

$$u(t) = t$$

and

$$U(s) = \frac{1}{s^2}.$$

Thus

$$X(s) = k\left(\frac{1}{s^2} \cdot \frac{1}{\tau s + 1}\right)$$

$$= k\left(\frac{1}{s^2} - \frac{\tau}{s} + \frac{\tau^2}{\tau s + 1}\right).$$

Inversion yields

$$x(t) = k(t - \tau + \tau e^{-t/\tau}). \tag{3.17}$$

(Note that $\dot{x}(t) = k(1 - e^{-t/\tau})$, equal to the step response.) This response is shown in Fig. 3.20.

As t continues to increase the third term of equation (3.17)

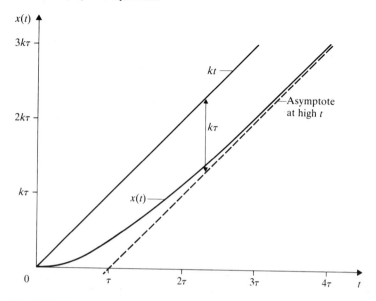

Fig. 3.20   Ramp response of first order system

becomes negligible and the output tends to a straight line

$$x(t) \simeq k(t - \tau)$$
$$= ku(t) - k\tau,$$

i.e. if $k = 1$ there is a constant difference equal to the time constant of the system, between the output and the input. Note that the smaller the time constant the more rapid is the response of the system and the quicker is its approach to a 'steady state value'.

Second order system

Although it is possible to represent the second order linear system equation by two first order equations and thus by the vector matrix notation to obtain a solution (section 3.4), this is neither necessary nor desirable in many instances. Second order equations can be solved comparatively easily and occur in systems containing 'inertia'. This may be true mechanical inertia or created by such effects as hold up. The common characteristic is that the inertia systems dynamics may all be represented by an equation of the form

$$\frac{d^2x}{dt^2} + 2\zeta\omega_n\frac{dx}{dt} + \omega_n^2 x = \omega_n^2 u(t). \tag{3.18}$$

This equation has been used as it is one of the most commonly met second order equations, e.g. it is that for describing a vibrating mass in mechanics. The notation of equation (3.18) is that normally used and it has become almost standard notation. The general transfer function for this second order system is

$$\frac{X(s)}{U(s)} = \frac{\omega_n^2}{s^2 + 2\zeta\omega_n s + \omega_n^2}. \tag{3.19}$$

As with the first order case this may represent either a direct response alone or may include feedback elements. The ζ is the damping ratio and ω_n the undamped natural frequency of the system. (As ω_n has the units of time^{-1} the inverse of ω_n is sometimes used to give similarity with the idea of the time constant.) The nature of the response depends on the damping ratio ζ which defines the degree of damping in the system, i.e. a measure of the presence and extent of oscillation in the system response. A low value of ζ, $\zeta < 1$, results in an oscillatory response when the system is disturbed and the system is then underdamped. For $\zeta = 1$, the system is critically damped and for $\zeta > 1$ the system is said to be overdamped and no oscillations are present in the output. The nature of ζ will become more obvious as the impulse and step responses are considered for the various cases of $\zeta < 1$, $\zeta = 1$, $\zeta > 1$.

Impulse response

For a unit impulse input,

$$X(s) = \frac{\omega_n^2}{s^2 + 2\zeta\omega_n s + \omega_n^2} \tag{3.20}$$

and the inversion of the transformed variable and function depends on the magnitude of ζ. The inverse may be obtained from tables or using the Heaviside inversion theorems (e.g. Friedly, 1972). For $0 \leq \zeta < 1$, i.e. underdamped

$$x(t) = \frac{\omega_n}{\sqrt{1-\zeta^2}} e^{-\zeta\omega_n t} \sin(\omega_n\sqrt{1-\zeta^2} \cdot t) \tag{3.21}$$

$\zeta = 1$, critically damped,

$$x(t) = \omega_n^2 t e^{-\omega_n t} \tag{3.22}$$

$\zeta > 1$, overdamped,

$$x(t) = \frac{\omega_n}{2\sqrt{\zeta^2-1}} \{\exp[-(\zeta-\sqrt{\zeta^2-1})\omega_n t] - \exp[-(\zeta+\sqrt{\zeta^2-1})\omega_n t]\}. \tag{3.23}$$

Whether the system is underdamped or overdamped it always returns to the initial state after an impulse input. For the overdamped and critically damped case it does this without 'overshooting' (Fig. 3.21). When it does oscillate, equation (3.21), for the underdamped case we can see that the frequency of oscillation is determined by the factor

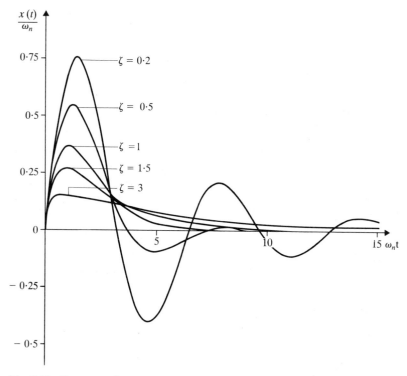

Fig. 3.21 Response of second order system to unit impulse

$\omega_n\sqrt{1-\zeta^2}$ and this is the damped natural frequency of the system, ω_d. It is only applicable of course for $\zeta < 1$. The time base for the graphs of $x(t)$ is usually taken as $\omega_n t$, the dimension of time being normalized in this way and the general solution may be represented by $x(t)/\omega_n$ plotted against $\omega_n t$.

Step response

For the unit step input $U(s) = 1/s$,

$$X(s) = \frac{\omega_n^2}{s(s^2 + 2\zeta\omega_n s + \omega_n^2)}. \tag{3.24}$$

For the different values of ζ the inversion to give $x(t)$ may be obtained from tables but an insight to the different forms of solution is obtained by looking at the factors of the denominator of equation (3.24). For $0 < \zeta < 1$, underdamped,

$$X(s) = \frac{1}{s} - \frac{s + \zeta\omega_n}{(s + \zeta\omega_n)^2 + (1 - \zeta^2)\omega_n^2} - \frac{\zeta\omega_n}{(s + \zeta\omega_n)^2 + (1 - \zeta^2)\omega_n^2}$$

and

$$x(t) = 1 - e^{-\zeta \omega_n t}\left(\cos \omega_d t + \frac{\zeta}{\sqrt{1-\zeta^2}} \cdot \sin \omega_d t\right)$$

$$= 1 - \frac{e^{-\zeta \omega_n t}}{\sqrt{1-\zeta^2}} \sin\left(\omega_d t + \tan^{-1}\frac{\sqrt{1-\zeta^2}}{\zeta}\right) \tag{3.25}$$

where

$$\omega_d = \sqrt{1-\zeta^2} \cdot \omega_n.$$

If $\zeta = 0$, i.e. no damping, then

$$x(t) = 1 - \sin(\omega_n t + \pi/2)$$

$$= 1 - \cos \omega_n t \tag{3.26}$$

so that the system continues to oscillate if undamped at the frequency ω_n for ever increasing t.

If $\zeta = 1$, critically damped,

$$X(s) = \frac{\omega_n^2}{s(s+\omega_n)^2}$$

and

$$x(t) = 1 - e^{-\omega_n t}(1 + \omega_n t). \tag{3.27}$$

If $\zeta > 1$, overdamped,

$$X(s) = \frac{\omega_n^2}{s(s+\zeta\omega_n + \omega_n\sqrt{\zeta^2-1})(s+\zeta\omega_n - \omega_n\sqrt{\zeta^2-1})}$$

and

$$x(t) = 1 + \frac{\omega_n}{2\sqrt{\zeta^2-1}}\left\{\frac{\exp[-(\zeta+\sqrt{\zeta^2-1})\omega_n t]}{\zeta+\sqrt{\zeta^2-1}}\right.$$

$$\left. - \frac{\exp[-(\zeta-\sqrt{\zeta^2-1})\omega_n t]}{\zeta-\sqrt{\zeta^2-1}}\right\}. \tag{3.28}$$

Figure 3.22 shows the effect of varying ζ when the system is subject to a step input. As might be expected the oscillatory behaviour corresponds to that in the cases of impulse input, i.e. oscillatory behaviour is a characteristic of the system rather than of the input.

We see that as the damping increases the response becomes similar to that for a first order system. This is because one of the exponential terms of equation (3.28) decays much more rapidly than the other. It is possible therefore at higher time to represent heavily damped second order systems by an approximate first order equation. For smaller values of t distinct differences are still present, e.g. dx/dt is zero at $t = 0$ for the second order case but not for the first.

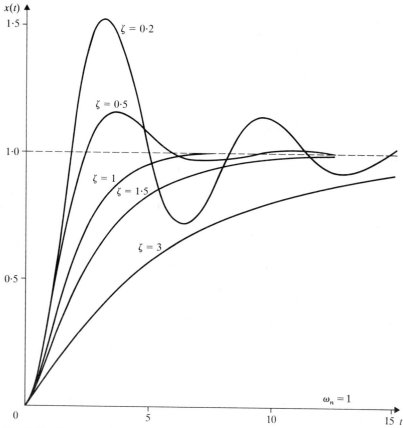

Fig. 3.22 Response of second order system to unit step imput

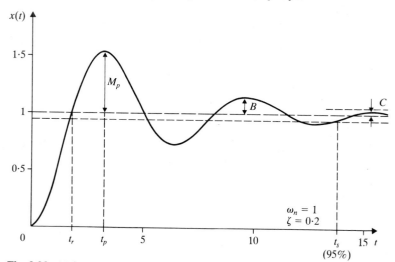

Fig. 3.23 Main features of the transient response (underdamped, second order system)

Because of the differences in overshoot, rate of rise from the zero position when disturbed and the oscillatory tail of second order responses it is desirable to use standard definitions to describe the system behaviour. In design work it may be necessary to compromise high initial rates of change with low overshoot and this is particularly so when we consider control. The essential features of the second order time or transient responses are shown in Fig. 3.23. The 'rise time' t_r is the time required for the response to first reach its final steady value. Putting $x(t) = 1$ in equation (3.25) t_r is given by

$$t_r = \frac{1}{\omega_d} \tan^{-1}\left(\frac{-\sqrt{1-\zeta^2}}{\zeta}\right). \tag{3.29}$$

The 'overshoot' M_p is defined as

$$\frac{the\ maximum\ amount\ by\ which\ the\ first\ peak\ exceeds\ the\ final\ value}{final\ value}.$$

It is normally expressed as a percentage. Differentiation of equation (3.25) with respect to time and equating this derivative to zero gives the time, the 'peak time' t_p, at which this occurs,

$$t_p = \frac{\pi}{\omega_d}. \tag{3.30}$$

This is actually the first root in an infinite number of roots. Substituting this back into equation (3.25) gives

$$M_p = \exp\left[-\frac{\zeta\pi}{\sqrt{1-\zeta^2}}\right]. \tag{3.31}$$

The 'decay ratio' is the ratio of any two successive overshoots on the same side of the steady state value and is given by (see Fig. 3.23),

$$\frac{B}{M_p} = \frac{C}{B} = \ldots = \exp\left[\frac{-2\pi\zeta}{\sqrt{1-\zeta^2}}\right]. \tag{3.32}$$

This observation may be used to determine the value of the damping ratio ζ, since by rearrangement of equation (3.32),

$$\zeta = \frac{\ln(B/C)}{2\pi\sqrt{\{1 + \ln(B/C)/2\pi\}}}. \tag{3.33}$$

Because of the sustained oscillations a 'settling time', t_s, is defined as the time after which the response stays within a certain region of the steady state value. This region is usually ±2 per cent or 5 per cent of the steady value. As the response curve intersection with the limit region is difficult to determine it is sufficient here to state that for the 2 per cent criterion the settling time is approximately four times the time constant $\frac{1}{\zeta\omega_n}$ and for the 5 per cent criterion it is approximately three times the value. Taking the settling time as a definition of the transient time we see that by varying ω_n this settling time may be changed independently of the damping and overshoot, which are dependent only on ζ.

The ramp response will be discussed in terms of the second order control system (Chapter 4). Higher order systems will also be discussed later although it may be mentioned here that the denominator may always be factored into first and second order terms and so the response will be made up of those terms which we have already considered for first order and second order systems.

It may be noted here also that alternative definitions to those used above may be found, e.g. for the rise time (e.g. Takahashi, Rabins and Auslander 1972).

Having looked at the response of first and second order systems to step, impulse and ramp inputs we can now turn to the response of the system when the input is continually varying in the form of a sinusoidal forcing function.

3.3 Frequency response

Frequency transfer function

When a system reaches a complexity where modelling is difficult or impossible on analytical grounds then the system behaviour must be determined by experiment. The experiments may have to take place during normal operation when the plant must be disturbed as little as possible and when other noise, unwanted disturbances, must be eliminated from the results. Frequency response methods may then be used and this is particularly so in process measurements where they have been used extensively, (Young 1965; Oldenburger 1956). To gain benefit from the frequency response results it is necessary to know what they mean in terms of the transient behaviour, i.e. the behaviour in the face of other disturbances like the step input. The feature of most importance in this respect is that of stability, discussed in section 5.1 more fully.

By the frequency response of a linear system we mean the output of the system when the input is a steady sinusoidal function of constant amplitude and frequency, and any initial transients set up at the beginning of the test have died down. The output then is also of sinusoidal form but with a different amplitude and is normally out of phase with the input. The frequency of the output is the same as that of the input. The ratio of the output to input amplitudes is the gain and the related phase between the output and input is the phase angle. Both of these depend on the frequency used and the complete frequency response is determined by tests over a range of frequency. The results are presented, in various ways, with frequency as the independent variable.

In addition to the convenience of the method for processes, however complex, the support for the frequency response method has been largely on account of the facts that the complete frequency response tells, in principle, all that is required about the process for control purposes, it can be measured with considerable accuracy, complex results can be handled graphically with comparative ease and deductions about the stability of the system can be made readily. In view of these factors let us consider how we can use the transfer function method to give the frequency response of those systems for which we do know the equations. The characteristics of the frequency response can then be seen.

First consider the transfer function $G(s)$ and assume that it can be written as the ratio of two polynomials,

$$
\begin{aligned}
G(s) &= \frac{p(s)}{q(s)} \\
&= \frac{k(s-z_1)(s-z_2)\ldots(s-z_m)}{(s-p_1)(s-p_2)\ldots(s-p_n)}.
\end{aligned}
\tag{3.34}
$$

For engineering systems the order of $p(s)$ will be lower than that of $q(s)$. For example, if

$$
G(s) = \frac{s+1}{s^2+2s+2}
$$

this may be factored to give

$$
G(s) = \frac{s+1}{(s+1+j)(s+1-j)}
$$

where $j = \sqrt{-1}$. In this case

$$
k = 1,
$$

$$
z_1 = -1,
$$

$$
p_1 = -1+j,
$$

$$
p_2 = -1-j.
$$

z_1 is a 'zero' of $G(s)$, i.e. it is the value of s that causes the transfer function to vanish if the substitution is made $s = z_1$. Similarly, if s is put equal to p_1, etc. in the transfer function expression one of the bracketed factors becomes equal to zero, e.g. $(s-p_2)=0$ if s is put equal to p_2. The transfer function $G(s)$ then becomes infinite and the value of s causing this is a 'pole' of the transfer function. Thus a transfer function may be written in terms of its poles and zeros as in equation (3.34). If $G(s)$ has distinct poles then it may be written by the partial fraction expansion as

$$G(s) = \frac{a_1}{s - p_1} + \frac{a_2}{s - p_2} + \ldots \frac{a_n}{s - p_n}. \tag{3.35}$$

For an impulse input, $U(s) = 1$, then the response is, by piecewise inversion,

$$x(t) = a_1 e^{p_1 t} + a_2 e^{p_2 t} + \ldots a_n e^{p_n t}. \tag{3.36}$$

Provided $p_1 \ldots, p_n$ all have negative real parts the value of $x(t)$ decays; otherwise for positive real parts of any of the poles p_i the value of $x(t)$ increases with time without limit, and in an exponential manner. Thus an unstable behaviour results in the later case. This instability is unaffected by the input forcing function $u(t)$ or by the values of the zeros in the numerator and it is a quality of the system itself. The zeros affect only the values and polarities of the partial fraction expansion coefficients a_i of the transfer function. Let us proceed now with the frequency response assuming that we have a stable system, as described above, with transfer function $G(s)$ and poles p_i with negative real parts.

If input $u(t) = U \sin \omega t$ (or we could choose a cosine function)

$$X(s) = G(s) U(s)$$

$$= \frac{p(s)}{q(s)} \cdot \frac{\omega U}{s^2 + \omega^2}. \tag{3.37}$$

Expanding this in partial fractions leads to

$$X(s) = \frac{A_1}{s + j\omega} + \frac{A_2}{s - j\omega} + \frac{b_1}{s - p_1} + \frac{b_2}{s - p_2} + \ldots \frac{b_n}{s - p_n} \tag{3.38}$$

where A_1, A_2 and $b_1 \ldots b_n$ are constants.

Inversion of the Laplace transforms gives the time response of $x(t)$ for a sinusoidal input,

$$x(t) = A_1 e^{-j\omega t} + A_2 e^{j\omega t} + b_1 e^{p_1 t} + \ldots b_n e^{p_n t}. \tag{3.39}$$

For a stable system, with time increasing the terms $b_1 e^{p_1 t} \ldots b_n e^{p_n t}$ decrease to zero as all $p_i < 0$ and

$$x(t)_{\text{steady state}} = A_1 e^{-j\omega t} + A_2 e^{j\omega t}. \tag{3.40}$$

(This will be so also if repeated factors occur in $G(s)$ giving rise to terms in $x(t)$ of te^{pt}.) (Note that the notations $(s + p_1)$ and $(s - p_1)$ have been used about equally in the literature to mean the same thing. The powers of equation (3.39) will then be shown as negative and positive respectively.)

Coefficients A_1 and A_2 may now be calculated. Multiply equations

(3.37) and (3.38) by the factor $s + j\omega$ and then let $s = -j\omega$ to give

$$A_1 = G(s) \cdot \left. \frac{\omega U(s + j\omega)}{s^2 + \omega^2} \right|_{s = -j\omega}$$

$$= \left. \frac{G(s)\omega \cdot U}{(s - j\omega)} \right|_{s = -j\omega}$$

$$= \frac{G(-j\omega) \cdot U}{-2j} \qquad (3.41)$$

where

$$G(s)|_{s = -\zeta\omega} = G(-j\omega).$$

Similarly

$$A_2 = \frac{G(j\omega)U}{2j}$$

$$= \text{conjugate of } A_1.$$

Thus from equation (3.40)

$$x(t)_{s.s} = U \left\{ \frac{G(j\omega)e^{j\omega t} - G(-j\omega)e^{-j\omega t}}{2j} \right\}.$$

Substitution of $s = j\omega$ in $G(s)$ yields some complex number $G(j\omega)$ which may be written in the standard terms of its magnitude $|G(j\omega)|$ and angle ϕ. The angle ϕ, or $\underline{/G(j\omega)}$ is given by $\phi = \tan^{-1}$ (imaginary part of $G(j\omega)$/real part of $G(j\omega)$), Fig. 3.24. In the complex plane the value $G(j\omega)$ is represented by the vector from the origin to the point $G(j\omega)$ which has a real and an imaginary part,

$$G(j\omega) = \text{real part of } G(j\omega) + \text{imaginary part of } G(j\omega)$$

$$= |G(j\omega)| \cos\phi + j |G(j\omega)| \sin\phi$$

$$= |G(j\omega)| (\cos\phi + j \sin\phi).$$

But

$$e^{j\phi} = \cos\phi + j \sin\phi \qquad \text{(De Moivre's theorem)}$$

so that

$$G(j\omega) = |G(j\omega)| \cdot e^{j\phi}. \qquad (3.42)$$

For a detailed revision of the complex number algebra a standard mathematics text may be consulted, e.g. (Kreyszig 1968). It is seen also from Fig. 3.24 that for any $G(j\omega)$,

$$|G(j\omega)| = \sqrt{\{(\text{real part})^2 + (\text{imaginary part})^2\}} \qquad (3.43)$$

$$\underline{/G(j\omega)} = \tan^{-1} (\text{imaginary part/real part}). \qquad (3.44)$$

Similarly

$$G(-j\omega) = |G(-j\omega)| e^{-j\phi}$$

$$= |G(j\omega)| e^{-j\phi}$$

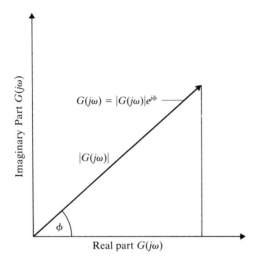

Fig. 3.24 $G(j\omega)$ in the complex plane

and

$$x(t)_{\text{s.s}} = U\left\{\frac{|G(j\omega)|\, e^{j(\omega t + \phi)} - |G(j\omega)|\, e^{-j(\omega t + \phi)}}{2j}\right\}$$

$$= \frac{U}{2j} \cdot |G(j\omega)| \cdot (e^{j(\omega t + \phi)} - e^{-j(\omega t + \phi)}).$$

For a general angle θ

$$e^{j\theta} - e^{-j\theta} = \cos\theta + j\sin\theta - (\cos\theta - j\sin\theta)$$
$$= 2j\sin\theta.$$

Thus

$$e^{j(\omega t + \phi)} - e^{-j(\omega t + \phi)} = 2j\sin(\omega t + \phi)$$

and

$$x(t)_{\text{s.s.}} = U \cdot |G(j\omega)| \sin(\omega t + \phi)$$
$$= U\,|G(j\omega)|\sin(\omega t + \underline{/G(j\omega)}). \tag{3.45}$$

Thus the steady state output is a sine wave of the same frequency ω as the input but out of phase by the angle ϕ. The amplitude of the output is $U \cdot |G(j\omega)|$ which is also frequency dependent. For sinusoidal inputs if we denote the amplitude of the output $x(t)_{\text{s.s.}}$ by $|X(j\omega)|$ and the amplitude of the input by $|U(j\omega)|$ then

$$|X(j\omega)| = |U(j\omega)|\,|G(j\omega)|$$

and

$$|G(j\omega)| = \left|\frac{X(j\omega)}{U(j\omega)}\right| = \frac{output\ amplitude}{input\ amplitude}.$$

The phase shift ϕ in equation (3.45) is caused by the system and may be written

$$\phi = \underline{/G(j\omega)}$$

$$= \underline{/\frac{X(j\omega)}{U(j\omega)}}.$$

Combining these the frequency transfer function $G(j\omega)$ is written as

$$G(j\omega) = \frac{X(j\omega)}{U(j\omega)}$$

$$= |G(j\omega)|\ \underline{/G(j\omega)}$$

$$= \left|\frac{X(j\omega)}{U(j\omega)}\right|\ \underline{/\frac{X(j\omega)}{U(j\omega)}}. \qquad (3.46)$$

Compare this with the normal transfer function

$$G(s) = \frac{X(s)}{U(s)}$$

i.e. to obtain the frequency transfer function of a system we just replace s by $j\omega$ in the normal transfer function. This then gives for each value of input frequency ω an amplitude and phase angle for the output sine waves. If the phase angle ϕ is positive the output leads the input and if ϕ is negative the output is behind the input and there is a phase lag. The setting up and use of the frequency transfer function is illustrated by considering standard first and second order systems as examples.

Example Frequency response of first and second order systems. For the first order system of equation (3.13), with unity gain $k = 1$, $G(s) = 1/(\tau s + 1)$. To determine the frequency response (using an input of $u(t) = \sin \omega t$) substitute $s = j\omega$ in $G(s)$,

$$G(j\omega) = \frac{1}{\tau j\omega + 1}$$

$$= \frac{1 - \tau\omega j}{\tau^2\omega^2 + 1}.$$

The amplitude

$$|G(j\omega)| = \frac{1}{\sqrt{(\tau^2\omega^2 + 1)}}$$

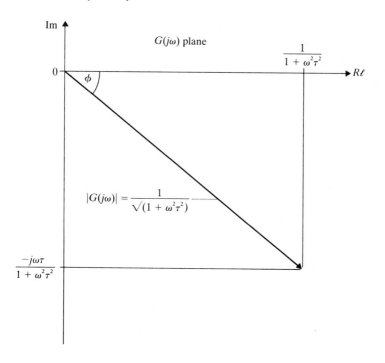

Fig. 3.25   The complex plane for the first order system

and the phase angle ϕ is $-\tan^{-1} \tau\omega$.

Thus for the sinusoidal input we substitute for $|G(j\omega)|$ and ϕ in equation (3.45) to yield

$$x(t) = |U(j\omega)| \cdot |G(j\omega)| \cdot \sin(\omega t + \phi)$$

$$= 1 \cdot \frac{1}{\sqrt{(1+\tau^2\omega^2)}} \cdot \sin(\omega t - \tan^{-1} \tau\omega)$$

$$= |X(j\omega)| \cdot \sin(\omega t + \phi) \tag{3.47}$$

As $\omega \to 0$, $|X(j\omega)| \to 1$, $\phi \to 0$

and as $\omega \to \infty$, $|X(j\omega)| \to 0$, $\phi \to -\pi/2$ rad, $-90°$.

For the unit gain ($k = 1$) first order system at low frequency the output approaches the input in magnitude and phase but as the frequency is increased the phase lag rises to a maximum of $\pi/2$ rads and the gain decreases towards zero.

Taking the standard second order system equation (3.18) and replacing s by $j\omega$ in the transfer function equation (3.19) to form the frequency transfer function,

$$G(j\omega) = \frac{\omega_n^2}{\omega_n^2 - \omega^2 + 2j\zeta\omega_n\omega}$$

and on rationalizing this yields

$$G(j\omega) = \frac{\omega_n^2((\omega_n^2 - \omega^2) - 2j\zeta\omega_n\omega)}{(\omega_n^2 - \omega^2)^2 + 4\zeta^2\omega_n^2\omega^2}.$$

Thus the system gain is

$$|G(j\omega)| = \sqrt{\left(\frac{(\omega_n^2 - \omega^2)^2 + 4\zeta^2\omega_n^2\omega^2}{((\omega_n^2 - \omega^2)^2 + 4\zeta^2\omega_n^2\omega^2)^2}\right)} \cdot \omega_n^2$$

$$= \frac{\omega_n^2}{\sqrt{((\omega_n^2 - \omega^2)^2 + (2\zeta\omega_n\omega)^2)}}$$

and the phase angle ϕ is

$$\underline{/G(j\omega)} = \tan^{-1}\left(\frac{-2\zeta\omega_n\omega}{\omega_n^2 - \omega^2}\right)$$

$$= -\tan^{-1}\left(\frac{2\zeta\omega_n\omega}{\omega_n^2 - \omega^2}\right).$$

From equation (3.45) the response to the sinusoidal input $\sin\omega t$ is

$$x(t) = \frac{\omega_n^2}{\sqrt{((\omega_n^2 - \omega^2)^2 + (2\zeta\omega_n\omega)^2)}} \cdot \sin(\omega t + \phi). \qquad (3.48)$$

As $\omega \to 0$, $|X(j\omega)| \to 1$, $\phi \to 0$

$\omega \to \infty$, $|X(j\omega)| \to 0$, $\phi \to -\pi$ rad, $-180°$.

The second order systems and those containing quadratic factors in general are frequently rewritten as

$$\frac{1}{1 + 2\zeta/\omega_n \cdot s + s^2/\omega_n^2}.$$

If the denominator factorizes then the factors immediately appear as $(1 + \tau_1 s)(1 + \tau_2 s)$ giving the time constants straight away.

Graphical representation of frequency response

The real significance of the frequency response and its graphical representation appears in the study of feedback control systems and it will be discussed further there. The representation of a systems dynamic behaviour in this way is still useful however when control is not primarily concerned, as conditions such as resonance may be predicted and the effect of damping to reduce resonance peaks may be observed.

The Polar plot

The polar plot is a plot of the amplitude of $G(j\omega)$ and the phase angle ϕ as the frequency is taken over its range zero to infinity. This

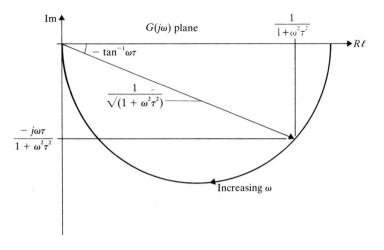

Fig. 3.26 Polar plot of first order system

information is shown for any point on the plot by a vector to that point from the origin. It may be plotted either from $|G(j\omega)|$ and ϕ or directly by plotting the real and imaginary parts of $G(j\omega)$ as ω varies, with the amplitude and phase angle subsequently being taken from the plot.

For the first order system we have

$$G(j\omega) = \frac{1}{1+\tau j\omega}$$

$$= \frac{1-\tau j\omega}{1+\tau^2\omega^2}$$

$$= \frac{1}{1+\tau^2\omega^2} - j\frac{\tau\omega}{1+\tau^2\omega^2}$$

$$= \frac{1}{\sqrt{(1+\tau^2\omega^2)}} \cdot \underline{/-\tan^{-1}\omega\tau}.$$

The polar plot is shown in Fig. 3.26. Note that ω values are shown around the curve.

The second order or 'quadratic' factor extends into a second quadrant of the $G(j\omega)$ plane. For

$$G(j\omega) = \frac{1}{1+2\zeta j\omega/\omega_n + (j\omega)^2/\omega_n^2}$$

the plot is shown in Fig. 3.27. Decreasing the damping ζ leads as expected to larger magnitudes and to changes in the phase angle. Note that the plot approaches the origin, $G(j\omega) = 0$, tangentially to the negative real axis, i.e. the phase lag at $\omega \to \infty$ is 180° (the phase angle

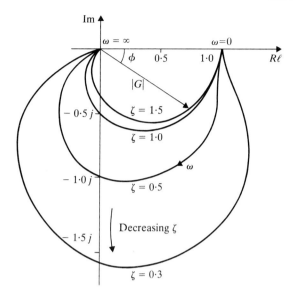

Fig. 3.27   Polar plot of quadratic factor ($\omega_n = 1$)

is $-180°$) as predicted by equation (3.48). Positive phase angles are measured counterclockwise about the origin.

The frequency response of a transportation lag (distance velocity and pure delay) is of considerable importance. For such a lag an input is delayed and appears at the output a time t_o later. For an input function $f(t)$ the output is $f(t-t_o)$, Fig. 3.28(a) and from the table of Laplace transforms we see that the transform of $f(t-t_o)$ is $e^{-st_o}\mathscr{L}[f(t)]$, i.e.

$$\mathscr{L}[f(t-t_o)] = e^{-st_o}\mathscr{L}[f(t)].$$

Thus the transfer function for the pure delay is $G(s) = e^{-st_o}$ and the frequency transfer function is

$$G(j\omega) = e^{-j\omega t_o}$$
$$= \cos \omega t_o - j \sin \omega t_o.$$

Therefore the magnitude $|G(j\omega)|$ is always unity and the phase angle is $\tan^{-1}(-\sin \omega t_0/\cos \omega t_0)$, i.e. $-\omega t_o$. $G(j\omega)$ may then be written

$$G(j\omega) = 1 \, . \, \underline{/-\omega t_o}.$$

The polar plot with ω increasing thus forms a set of overlapping unit circles of constant unity radius, Fig. 3.28(b). Note that at low frequencies the pure delay and simple lag plots are similar. The greater the pure delay t_o the greater is the phase lag at any particular frequency ω.

To construct the polar plots for more complex transfer functions may involve considerable effort. It will be seen that the construction of

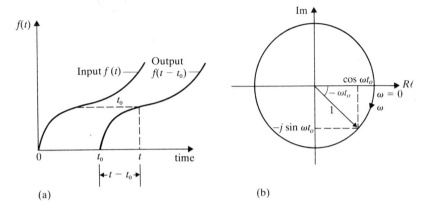

Fig. 3.28 Time response and polar plot for pure delay

the Bode plot utilizes the factorisation of complex transfer functions into the products of first and second terms, e.g. $G(j\omega) = G_1(j\omega) \cdot G_2(j\omega)$ etc. Then

$$|G(j\omega)| = |G_1(j\omega)| \cdot |G_2(j\omega)|$$

and

$$\underline{/G(j\omega)} = \underline{/G_1(j\omega)} + \underline{/G_2(j\omega)}. \tag{3.49}$$

Although polar graphical manipulation, Fig. 3.29(a), is possible it is just as, if not more, convenient to plot $G(j\omega)$ using the Bode method for more complex functions and then translate the results to the polar plot. As an example of a product consider

$$G(j\omega) = \frac{e^{-j\omega t_o}}{1 + \tau j\omega}.$$

For equation (3.49)

$$|G(j\omega)| = 1 \cdot \frac{1}{\sqrt{(1 + \tau^2 \omega^2)}}$$

$$\underline{/G(j\omega)} = -\omega t_o - \tan^{-1} \omega\tau.$$

These are shown in Fig. 3.29(b).

Bode plots

The Bode plot (corner plot or logarithmic plot) comprises a graph of the magnitude of $G(j\omega)$ against the frequency ω, plus a graph of the phase angle ϕ as a function of the frequency. The magnitude is plotted on a log scale and commonly expressed in decibels; the phase angle is plotted on a linear scale. The independent variable is on a log scale. Because the effects of most processes show up as attenuation of magnitude with the production of a phase lag the majority of both graphs fall below the abscissa. The Bode plot is of particular use when the frequency transfer function $G(j\omega)$ can be expressed in terms of the following simple factors; a constant K, a pole or zero at the origin $(j\omega)^{\mp 1}$, simple pole or zero $(1+j\omega\tau)^{\mp 1}$, and a quadratic factor $(1+2\zeta/\omega_n \cdot j\omega + (j\omega)^2/\omega_n^2)^{\mp 1}$. If the transfer function can be expressed in terms of the products of these factors then, since we are plotting to a logarithmic scale, the construction of the full magnitude plot is achieved by the process of addition. In addition the exact functions may be approximated by straight-line asymptotes. As it is easy to construct, the Bode plot may be used as the basis, as mentioned above, for the construction of the polar plot or of a magnitude vs phase shift plot (see Figs. 3.37 and 3.38).

The placing of the gain constant, poles, etc. in the complex plane helps in establishing the phase and magnitude values used below. Expressing the frequency transfer function in terms of its factors we may write

$$G(j\omega) = \frac{K(1+j\omega\tau_1)\ldots\ldots}{(j\omega)^n(1+j\omega\tau_2)(1+2\zeta/\omega_n \cdot j\omega + (j\omega)^2/\omega_n^2)\ldots}. \qquad (3.50)$$

The magnitude is the product of the magnitudes so that

$$|G(j\omega)| = \frac{|K|\cdot|1+j\omega\tau_1|\ldots\ldots}{|(j\omega)|^n\cdot|1+j\omega\tau_2|\cdot|(1+2\zeta/\omega_n \cdot j\omega + (j\omega)^2/\omega_n^2)|\ldots}. \qquad (3.51)$$

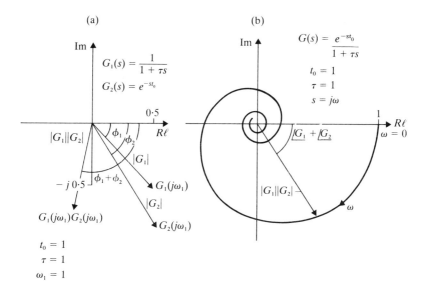

Fig. 3.29 Polar plot construction of the product of transfer functions

and taking logs

$$\ln |G(j\omega)| = \ln K + \ln |1 + j\omega\tau_1| + \ldots - n \ln |j\omega| - \ln |1 + j\omega\tau_2|$$
$$- \ln \ldots .$$

Defining the 'log magnitude' term more precisely as $\mathscr{L}m\, G(j\omega) = 20 \log_{10} |G(j\omega)|$ this equation may be written

$$\mathscr{L}m\, G(j\omega) = \mathscr{L}m\, K + \mathscr{L}m\,(1 + j\omega\tau) + \ldots - n\,\mathscr{L}m\,(j\omega)$$
$$- \mathscr{L}m\,(1 + j\omega\tau_2) \ldots . \quad (3.52)$$

With this notation the magnitude is expressed in decibels (db).

Using equation (3.49) the phase angle of $G(j\omega)$ is given by

$$\underline{/G(j\omega)} = \underline{/K} + \underline{/1 + j\omega\tau_1} + \ldots - n\underline{/j\omega} - \underline{/1 + j\omega\tau_2} \ldots \quad (3.53)$$

$$= 0 + \tan^{-1} \omega\tau_1 + \ldots - n \cdot \frac{\pi}{2} - \tan^{-1}\omega\tau_2 - \tan^{-1} \frac{2\zeta\omega/\omega_n}{1 - \omega^2/\omega_n^2}.$$

Equations (3.52) and (3.53) define the additions required to plot the Bode plot for $G(j\omega)$. Reference to Fig. 3.30 enables this division of a transfer function into its component parts to be seen in physical terms.

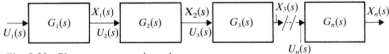

Fig. 3.30   Plant components in series

Consider the input to a plant component to be u_1, with output x_1. The transfer function relationship is

$$X_1(s) = G_1(s)U_1(s).$$

Similarly

$$X_2(s) = G_2(s)U_2(s)$$

and if the output from component one is the input to the second then

$$U_2(s) = X_1(s),$$

$$X_2(s) = G_2(s)G_1(s)U_1(s)$$

and

$$X_n(s) = G_n(s) \ldots G_2(s)G_1(s)U_1(s)$$
$$= G(s)U_1(s).$$

$G(s)$ has been established here from its components but equally if a given $G(s)$ is amenable to factorization then we may express it in terms of a product of its factors. The 'output' from each factor then becomes the 'input' to the next and so on, e.g.

$$G(s) = G_1(s)G_2(s).$$

If the first factor G_1 has a gain, or amplitude ratio which may be frequency dependent, of say $0 \cdot 5$, then the 'input' to G_2 is only one half of the initial input. If the gain of G_2 is also $0 \cdot 5$ then the output from it will be only one half of its input, i.e. one-quarter of the initial input. Thus we see that the gains in a physical sense are multiplied and for the frequency response,

$$|G(j\omega)| = |G_1(j\omega)| \cdot |G_2(j\omega)|$$

and so on,

$$|G(j\omega)| = |G_1(j\omega)| \cdot |G_2(j\omega)| \ldots |G_n(j\omega)|.$$

For the basic transfer function elements this leads to equation (3.51) above. Similarly the phase angle for the overall system transfer function $G(s)$ is contributed to from each factor or component. If the peak value of the output of factor G_1 lags ϕ_1° behind the input peak value, then the input to G_2 must be ϕ_1° behind the initial input also. If G_2 introduces a phase lag of ϕ_2° itself then its output must be $\phi_1^\circ + \phi_2^\circ$ behind the initial input, i.e.

$$\underline{/G(j\omega)} = -(\phi_1 + \phi_2)$$

and in general

$$\underline{/G(j\omega)} = \underline{/G_1(j\omega)} + \underline{/G_2(j\omega)} + \ldots \underline{/G_n(j\omega)} = \sum \phi_i.$$

Depending on the nature of each element of $G(j\omega)$ some of these factors will introduce phase lags (ϕ_i negative) and some phase leads (ϕ_i positive), equation (3.53) above.

If we look at each factor in turn and look especially at the values for high and low values of the input frequency ω it will become apparent how the Bode plots are built up. The individual factors are shown for a specific ω in the complex plane in Fig. 3.31.

(a) *Constant, K.*

$$\mathscr{L}\text{m } K = 20 \log_{10} K, \text{ db}$$

$$\underline{/K} = 0, \quad \text{i.e. no phase lag is introduced.}$$

Thus the $\mathscr{L}$m graph is independent of frequency and no phase angle is introduced by K itself.

(b) *Poles or zero at the origin,* $(j\omega)^{\mp 1}$.

These terms arise from pure derivative or integral factors in the model and transfer function.

$$\mathscr{L}\text{m } (j\omega)^{\mp n} = 20 \log |(j\omega)^{\mp n}|$$

$$= \mp 20n \log \omega, \text{ db}$$

$$\underline{/(j\omega)^{\mp n}} = \mp n \cdot \pi/2 \text{ rad, } \mp n \ 90^\circ.$$

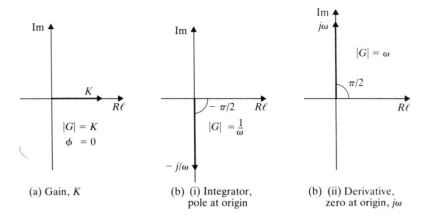

(a) Gain, K

(b) (i) Integrator, pole at origin

(b) (ii) Derivative, zero at origin, $j\omega$

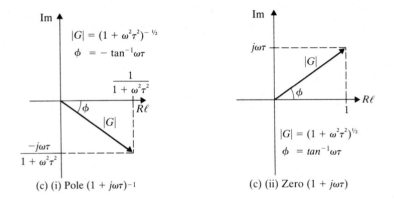

(c) (i) Pole $(1 + j\omega\tau)^{-1}$

(c) (ii) Zero $(1 + j\omega\tau)$

Fig. 3.31   Simple elements frequency transfer functions in the complex plane (for a single ω)

A log-log plot of magnitude is a straight line of slope $\pm n$. Each pure derivative term introduces a constant phase advance of $\pi/2$ radians and each integral term a phase lag of $\pi/2$ radians.

(c) *Simple pole or zero*, $(1+j\omega\tau)^{\mp 1}$.

For the simple zero

$$\mathscr{L}\mathrm{m}\,(1+j\omega\tau) = 20 \log |(1+j\omega\tau)|$$
$$= 20 \log \sqrt{(1+\omega^2\tau^2)}$$
$$\underline{/(1+j\omega\tau)} = \tan^{-1} \omega\tau.$$

The log-log plot is no longer a straight line but it approaches a straight line form asymptotically as $\omega \to$ zero or $\omega \to$ infinity.

For $\omega\tau \ll 1$, $\quad \mathscr{L}\mathrm{m}\,(1+j\omega\tau) \to 20\log 1 = 0$

$$\underline{/1+j\omega\tau} \to \tan^{-1} 0 = 0.$$

For $\omega\tau \gg 1$, $\quad \mathscr{L}\mathrm{m}\,(1+j\omega\tau) \to 20\log \omega\tau$

$$\underline{/1+j\omega\tau} \to \tan^{-1} \infty = \pi/2 \text{ rad, } 90°.$$

These two conditions form the low and high frequency magnitude asymptotes respectively and these asymptotes cross at a 'corner frequency' when $20\log \omega\tau$ is zero, i.e. ω equals $1/\tau$.

For the simple pole

$$\mathscr{L}\mathrm{m}\,(1+j\omega\tau)^{-1} = -\mathscr{L}\mathrm{m}\,(1+j\omega\tau)$$
$$= -20\log \sqrt{(1+\omega^2\tau^2)}$$

and

$$\underline{/(1+j\omega\tau)^{-1}} = -\underline{/1+j\omega\tau}$$
$$= -\tan^{-1} \omega\tau$$

so that for both the log magnitude and phase curves the only change is in the sign. The low frequency asymptote is the same as for the zero, and the high frequency asymptote is $-\mathscr{L}\mathrm{m}\,(1+j\omega\tau) \to -20\log \omega\tau$ and $-\underline{/1+j\omega\tau} \to -\tan^{-1}\infty$ which is $-\pi/2$ radians, i.e. a phase lag as distinct from the phase advance of the zero. The factors covered so far are shown in Fig. 3.32.

Addition of various plots is simply by numerical addition and plotting or by picking off and adding 'graphically' with dividers. The use of $\omega\tau$ as the ordinate enables all the above frequency transfer functions to be placed on to one set of co-ordinates. The use of this figure for specific values of ω and τ is shown below for a transfer function which factorizes into simple components. The transfer function for a plant with a pure integral and both lag (poles) and lead (zeros) term is

$$G(s) = \frac{K(1+s\tau_1)}{s(1+s\tau_2)}.$$

If the time constants are $\tau_1 = 2$ sec, $\tau_2 = 3$ sec, and the gain factor K is 2, the frequency transfer function is

$$G(j\omega) = \frac{2(1+2j\omega)}{j\omega(1+3j\omega)}.$$

What is the gain and phase angle for the plant at a forcing frequency of 2 rad . sec^{-1} (about 0·3 cycles sec^{-1})?

Using equation (3.52)

$$\mathscr{L}\mathrm{m}\,|G(j\omega)| = \mathscr{L}\mathrm{m}\,K + \mathscr{L}\mathrm{m}\,(1+2j\omega) - \mathscr{L}\mathrm{m}\,(j\omega) - \mathscr{L}\mathrm{m}\,(1+3j\omega)$$
$$= 20\log 2 + 20\log \sqrt{(1+4\omega^2)} - 20\log \omega$$
$$- 20\log \sqrt{(1+9\omega^2)}.$$

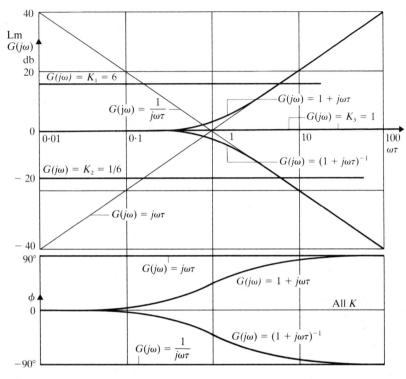

Fig. 3.32 Bode plots for simple factors

For $\omega = 2$ this value may be taken straight from Fig. 3.32 using values along the ordinate of 2 for the integral term, $4(=2\omega)$ for the zero and $6(=3\omega)$ for the pole,

$$\mathscr{L}m\,|G(j\omega)|_{\omega=2} = 6\cdot02 + 12\cdot30 + 6\cdot02 - 15\cdot68$$
$$= -3\cdot38\,\text{db.}$$

The phase angle is given by equation (3.53),

$$\underline{/G(j\omega)} = \underline{/K} + \underline{/1+2j\omega} - \underline{/j\omega} - \underline{/1+3j\omega}$$

and again using the respective values along the ordinate

$$\underline{/G(j\omega)} = 0 + 76 - 90 - 80 = -96^\circ\ (=1\cdot7\ \text{rad.})$$

Thus the gain of the frequency transfer function and of its physical system for a forcing frequency of ω of 2 rad. sec^{-1} is 0·68 and the output lags the input by 96°.

(d) *Quadratic factors*

If the damping coefficient, ζ, is greater than unity the quadratic factor may be factorized to two first order factors and each treated as above.

For values of ζ less than unity the quadratic factor is plotted without factoring. If the real and imaginary parts are grouped together,

$$\mathscr{L}\mathrm{m}\left\{1+\frac{2\zeta j\omega}{\omega_n}+\frac{(j\omega)^2}{\omega_n^2}\right\}^{\pm 1} = \mathscr{L}\mathrm{m}\left\{\frac{\omega_n^2-\omega^2}{\omega_n^2}+j\frac{2\zeta\omega}{\omega_n}\right\}^{\pm 1}$$

$$= \pm 20\log\left\{\left(1-\left(\frac{\omega}{\omega_n}\right)^2\right)^2+\left(\frac{2\zeta\omega}{\omega_n}\right)^2\right\}^{1/2}$$

$$= \pm 10\log\left\{\left(1-\left(\frac{\omega}{\omega_n}\right)^2\right)^2+\left(\frac{2\zeta\omega}{\omega_n}\right)^2\right\}$$

and the phase angle (or 'argument') is given by

$$\arg\left\{1+\frac{2\zeta j\omega}{\omega_n}+\frac{(j\omega)^2}{\omega_n^2}\right\}^{\pm 1} = \pm\tan^{-1}\left\{\frac{2\zeta\omega/\omega_n}{1-(\omega/\omega_n)^2}\right\}.$$

The value of ζ has a significant effect in the shape of the curves at intermediate frequencies but at high and low frequencies we still have asymptotic behaviour.

For $\omega/\omega_n \ll 1$

$$\mathscr{L}\mathrm{m}\left\{1+\frac{2\zeta j\omega}{\omega_n}+\frac{(j\omega)^2}{\omega_n^2}\right\}^{\pm 1} \rightarrow \pm 10\log 1 = 0$$

$$\arg\left\{1+\frac{2\zeta j\omega}{\omega_n}+\frac{(j\omega)^2}{\omega_n^2}\right\}^{\pm 1} \rightarrow \pm\tan^{-1}0 = 0$$

For $\omega/\omega_n \gg 1$

$$\mathscr{L}\mathrm{m}\left\{1+\frac{2\zeta j\omega}{\omega_n}+\frac{(j\omega)^2}{\omega_n^2}\right\}^{\pm 1} \rightarrow \pm 20\log(\omega/\omega_n)^2 = \pm 40\log(\omega/\omega_n)$$

$$\arg\left\{1+\frac{2\zeta j\omega}{\omega_n}+\frac{(j\omega)^2}{\omega_n^2}\right\}^{\pm 1} \rightarrow \tan^{-1}(-0) = \pm\pi \text{ rad}, \pm 180°.$$

As for the first order factors so here also the magnitude asymptotes intercept, in this case when $\log(\omega/\omega_n)$ is zero, i.e. ω is ω_n at the corner frequency. At this frequency the phase angle is $\pm\tan^{-1}(\infty)$, i.e. $\pm\pi/2$ rad. Because the quadratic factor is more common in the denominator the results for this case are shown in Fig. 3.33. The influence of ζ is shown and its effect may be compared with that shown in Fig. 3.22, the effect on transient response.

The frequency at which the magnitude peaks is the resonance frequency of the system, ω_r. Differentiation of magnitude with respect to frequency yields on equating the differential to zero,

$$\omega_r = \omega_n\sqrt{(1-2\zeta^2)}. \tag{3.54}$$

This resonance occurs only for $0<\zeta<0\cdot707$ (i.e. $1/\sqrt{2}$). Although there is no resonant frequency for higher values of ζ, overshoot and

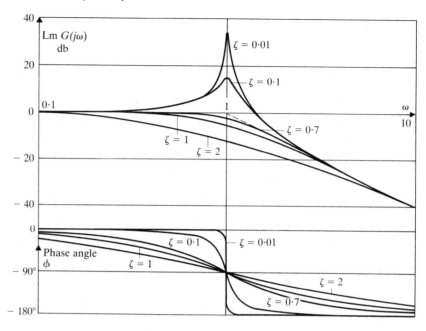

Fig. 3.33   Bode plot for $G(j\omega) = \{1 + 2\zeta j\omega/\omega_n + (j\omega)^2/\omega_n^2\}^{-1}$

oscillations occur in the transient response for all $\zeta < 1$. At the resonant frequency ω_r the magnitude of the quadratic factor is M_r where

$$M_r = \left| \left\{ 1 + \frac{2\zeta j\omega_r}{\omega_n} + \frac{(j\omega_r)^2}{\omega_n^2} \right\}^{-1} \right|$$

$$= \frac{1}{2\zeta\sqrt{(1-\zeta^2)}}. \tag{3.55}$$

Equation (3.55) is obtained by direct substitution from equation (3.54) in to the magnitude expression. For $\zeta > 1/\sqrt{2}$, there is no resonance and the maximum magnitude is unity as ω tends to zero. With $\zeta < 1/\sqrt{2}$ there is a resonance frequency and the maximum resonance magnitude tends to infinity as damping tends to zero. If the undamped system is continuously excited at its natural frequency the oscillations grow towards a 'steady state' value of infinite magnitude, or in practice until failure occurs, damping becomes a factor or some hard constraint or other deviation from linear behaviour is reached.

The resonance peak M_r and the resonant frequency ω_r are uniquely related to the damping ratio and the undamped natural frequency, ω_n of a system. However, only in the second order case as discussed here is it possible to obtain a direct relationship for these resonant characteristics in terms of the system parameters, ω_n and ζ. These relationships, equations (3.54) and (3.55), may be used in conjunction with those obtained for the damped natural frequency, $\omega_d = \omega_n \sqrt{(1-\zeta^2)}$, and the peak in the unit input transient curve, $1+M_p$ (where M_p the peak overshoot is given by equation (3.31)) to study the correlation between the frequency and time responses for a second order system.

Tabulating these again, we have

$$\omega_d = \omega_n \sqrt{(1-\zeta^2)}$$
$$\omega_r = \omega_n \sqrt{(1-2\zeta^2)}$$

Transient peak $= 1 + M_p$

$$= 1 + \exp\left(\frac{-\zeta\pi}{\sqrt{(1-\zeta^2)}}\right)$$

Resonance peak $= M_r$

$$= \frac{1}{2\zeta\sqrt{(1-\zeta^2)}}$$

and we see

(i) for systems with little damping, ω_d and ω_r are approximately the same;

(ii) for a given ζ, the larger the value of ω_r the larger is ω_n and the faster is the transient response of the system on account of the higher value of ω_d;

(iii) the smaller the value of ζ the larger are M_p and M_r. Thus the larger the value of M_r, the larger is M_p and for the higher values of ζ there is close correspondence between the transient peak value, $1+M_p$, and the resonance peak M_r, e.g. $\zeta = 0\cdot6$, $M_r = 1\cdot04$, transient peak $= 1\cdot09$.

Thus it is possible to obtain an approximate time response by knowing only M_r and ω_r of the frequency response. For higher order systems an effective ζ and ω_r may be used to get approximate relationships between frequency response and time response. For a system of any order, the larger ω_r, the faster is the response of the system and the transient peak approximates to the resonance peak M_r if the effective damping coefficient is in the range $0\cdot4 < \zeta < 0\cdot707$.

(e) *Pure delay*

The frequency transfer function of the pure delay may be expanded as

$$e^{-j\omega t_o} = \cos \omega t_o - j \sin \omega t_o$$

so that

$$\mathscr{L}m\, e^{-j\omega t_o} = 20 \log |\cos \omega t_o - j \sin \omega t_o|$$
$$= 20 \log 1 = 0$$

$$\underline{/e^{-j\omega t_o}} = \tan^{-1}\left(-\frac{\sin \omega t_o}{\cos \omega t_o}\right)$$
$$= -\omega t_o \text{ rad.}$$

The Bode plot of the pure delay has a magnitude constant value of zero db (unit gain) and a continuously increasing phase lag without limit as the frequency increases. Any system containing a pure delay as a factor of its overall transfer function will have this property of having possibly very large phase lags which are an important feature of stability studies in feedback systems.

Analogous to Fig. 3.31 for the simple elements the complex plane representation for the frequency transfer function of the quadratic factors and the pure delay is given in Fig. 3.34.

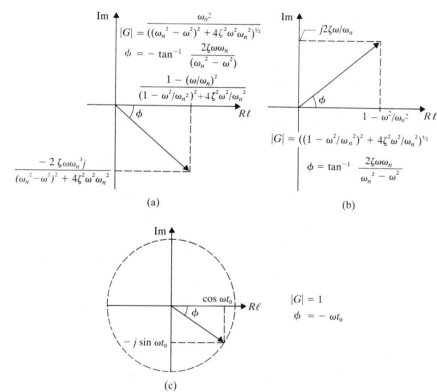

(a)

(b)

(c)

Fig. 3.34 Frequency transfer function of quadratic and pure delay elements in the complex plane (for a single ω)

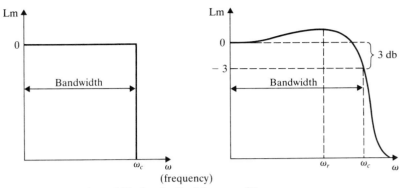

Fig. 3.35 Comparison of ideal and ralised low pass filters

In addition to specific specifications relative to feedback systems, such as gain and phase margins, which can be shown on the Bode plot and which are covered later, the performance of a system can be assessed in terms of its bandwidth and cut-off rate. It may be required that a system responds to inputs within a given operating frequency range but rejects high frequency noise, unwanted disturbances, that enter the system as well. Then a higher gain is required in the operating range of frequencies with a sudden drop off for higher frequencies. An ideal and a typical 'low pass' characteristic are shown in Fig. 3.35.

The bandwidth is the frequency ω_c at which the log magnitude has fallen by 3 db from its value at $\omega = 0$, i.e. the log magnitude graph drops from zero to -3 db, the gain dropping from unity to $0 \cdot 707$. Alternative choices from 3 db may be specified. The cut-off rate indicates the characteristics of the system in distinguishing the signal from the higher frequency noise. It is given by the slope of the log magnitude plot near the cut off (bandwidth) frequency.

The magnitude vs phase shift plot

This method of representation contains no more fundamental information than the Bode plot but is useful in the determination of the stability of closed loop (feedback) systems. The simple additive mode of construction used in the Bode plot cannot be used but, as with the construction of the polar plot, it may be quicker to start with the Bode plot construction. In Fig. 3.36, ω (or ω/ω_n) is the parameter varying along the curve and there is one curve for each value of ζ, the damping.

This method and the Bode and polar plots will be considered in greater detail when input–output feedback systems are discussed in Chapters 4 and 5.

Example When subject to a forcing input u a system output is displaced by an amount x such that the transfer function relating input

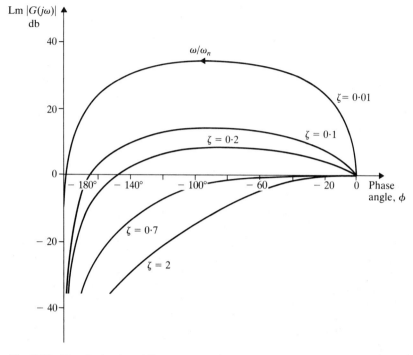

Fig. 3.36   Magnitude-phaseshift curve for $G(j\omega)=\{1+2\zeta j\omega/\omega_n+(j\omega/\omega_n)^2\}^{-1}$

and output is given by

$$G(s)=\frac{X(s)}{U(s)}$$

$$=\frac{s+3}{(s^2+4s+5)}.$$

Determine the expressions for the magnitude and phase angle of the frequency response of x and sketch the Bode plot and hence the Nyquist plot for the system.

It is convenient to rearrange $G(s)$, on substituting $s=j\omega$ to obtain the frequency response, as

$$G(j\omega)=\frac{3}{5}\cdot\frac{1+j\omega/3}{1+\frac{4}{5}j\omega+\frac{1}{5}(j\omega)^2}$$

In this way the steady state gain ($\omega=0$) is made explicit, as $K=\frac{3}{5}$, and subsequent plotting is simplified. This $G(j\omega)$ is comparatively simple and may easily be expressed in its gain and phase terms. However, it is still most easily expressed as the product of zero order, first order and second order functions, i.e.

$$G(j\omega) = KG_1(j\omega)G_2(j\omega)$$

where

$$K = \tfrac{3}{5},$$

$$G_1(j\omega) = 1 + \frac{j\omega}{3},$$

and

$$G_2(j\omega) = \frac{1}{1 + \tfrac{4}{5}j\omega + \tfrac{1}{5}(j\omega)^2}.$$

It is advisable to rationalize $G_2(j\omega)$ to

$$G_2(j\omega) = \frac{(1 - \omega^2/5) - \tfrac{4}{5}j\omega}{(1 - \omega^2/5)^2 + \tfrac{16}{25}\omega^2}$$

and the magnitude $|G(j\omega)|$ is then

$$|G(j\omega)| = |K| \cdot |G_1(j\omega)| \cdot |G_2(j\omega)|$$

$$= \frac{3}{5} \cdot \sqrt{\left(1 + \frac{\omega^2}{9}\right)} \cdot \frac{1}{\sqrt{((1 - \omega^2/5)^2 + \tfrac{16}{25}\omega^2)}}.$$

The log magnitude is

$$\mathscr{L}\mathrm{m}\, G(j\omega) = 20 \log |G(j\omega)|$$

$$= 20 \log 0.6 + 20 \log \sqrt{\left(1 + \frac{\omega^2}{9}\right)}$$

$$- 20 \log \sqrt{\left(\left(1 - \frac{\omega^2}{5}\right)^2 + \frac{16}{25}\omega^2\right)}$$

$$= 20 \log 0.6 + 10 \log \left(1 + \frac{\omega^2}{9}\right)$$

$$- 10 \log \left(\left(1 - \frac{\omega^2}{5}\right)^2 + \frac{16}{25}\omega^2\right).$$

The phase angle ϕ is the sum of the phase angle introduced by each component function,

$$\phi = \underline{/G(j\omega)} = \underline{/K} + \underline{/G_1(j\omega)} + \underline{/G_2(j\omega)}$$

$$= 0 + \tan^{-1}\frac{\omega}{3} + \tan^{-1}\frac{-4\omega}{(5 - \omega^2)}$$

$$= \tan^{-1}\frac{\omega}{3} - \tan^{-1}\frac{4\omega}{(5 - \omega^2)}.$$

The alternative is to multiply out the terms in the numerator of $G(j\omega)$ and after grouping these into a real part and an imaginary part determine the magnitude and phase angle as set out in this chapter.

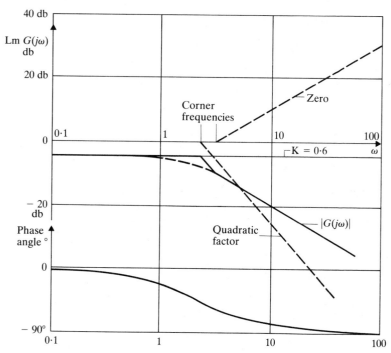

Fig. 3.37 Bode plot construction and sketch for

$$G(j\omega) = \frac{0.6(1 + j\omega/3)}{1 + \frac{4}{5} \cdot j\omega + (j\omega)^2/5}$$

This obviously involves more effort and cannot utilize the corner frequencies and asymptotes. The phase angle ϕ may be expressed in one term by use of the trigonometric sum of the terms $\tan^{-1} A$ and $\tan^{-1} B$ and this would then be that expression obtained by treating $G(j\omega)$ as a whole.

The most direct graphical representation is by the Bode plot and the construction is illustrated in Fig. 3.37. Note that the damping factor of the quadratic term is less than one. If it had been greater than unity then the quadratic term could have been factored into two first order terms.

The log magnitude asymptotes may be drawn for each term. For the simple zero as ω tends to infinity the log magnitude tends to $20 \log(\omega/3)$. For the quadratic term the log magnitude tends to $-20 \log(\omega^2/5)$, i.e. $-40 \log(\omega/\sqrt{5})$. The asymptotic phase angles are $90°$ and $-180°$ respectively. At low frequencies both first and second order magnitudes tend to zero. The gain term K is independent of frequency. The natural frequency ω_n of the quadratic factor is $\sqrt{5}$ and the damping coefficient ζ is given by $2\zeta/\omega_n = \frac{4}{5}$, so that ζ is $2/\sqrt{5}$ ($= 0.894$) and there is no resonant frequency.

We are now in a position to sketch the Bode plots for the system by combining the plots of the component factors. Each part of the plot is drawn separately and the complete response obtained by direct addition/substraction.

Although the magnitude and gain values may be evaluated directly for the polar plot, having plotted the Bode plot we may take corresponding values directly from this to give Fig. 3.38.

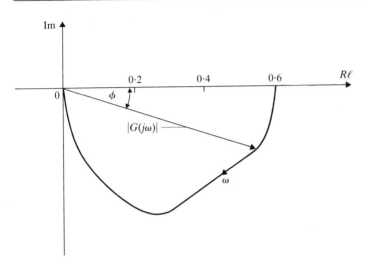

Fig. 3.38   Polar plot using gain and phase values taken from Fig. 3.37

3.4 Multi-input multi-output systems

So far in this chapter on system dynamics we have considered single input, single output systems. The analysis and synthesis for these cases have been established for some time compared with the fruitful applications of the theory of 'multivariable' systems or 'modern' control systems. In the remainder of this chapter the general solution of the multi-input multi-output linear system model equations will be considered, in particular their time domain solution. In the chapter on Modelling the ideas of state variable and general linear system equations were established and we can expand on these now. The more restrictive representation is by way of the transfer function matrix introduced in section 3.1. This retains the input–output concept as distinct from the state space representation which models the system in terms of its state variables and expresses the dynamics in first order differential equation form. The single input single output and multivariable and state variable treatments are complementary but the ability to handle and compute systems with many inputs, variables and outputs has switched much effort to the latter case. Since the majority of the application of system dynamics involves some attempt to 'control' the system, in whatever sense, the development of multivariable systems can be partially discussed to advantage later in terms of control.

We have seen that by suitable algebra and manipulation all linear time invariant lumped parameter (and some distributed parameter) systems can be solved, ideally, by the input–output Laplace transform approach. What is gained then by the use of multivariable theory? The restriction on time-varying and nonlinear systems is lifted and the number of variables which can be considered is considerably increased. The adaptive nature of control and the full specification and determination of an improved or an optimal performance can be pursued with more vigour. These points, and the concepts of controllability and observability which are basic to multivariable control systems will be detailed more in Chapter 7.

Transfer function matrix

In section 3.1 the idea of the block diagram was extended to cover the multivariable case using the transfer function matrix (or transfer matrix). The operation of this matrix on an input vector gives us the s-domain response, i.e. transform of the output. The direct manipulation of the vector-matrix algebra in feedback systems of equations will be described in Chapter 7. In this chapter we are concerned still with general system dynamics without detailed consideration of feedback, i.e. of the structure between the input and final output.

We have seen that the linear time invariant system may be represented (equation 2.69) by the equation

$$\dot{\mathbf{x}} = \mathbf{A}\mathbf{x} + \mathbf{B}\mathbf{u} \tag{3.56}$$

where $\mathbf{x}$ is the state vector, $\mathbf{A}$ the system matrix, $\mathbf{B}$ the distribution matrix and $\mathbf{u}$ the input (or control) vector. The output $\mathbf{y}$ is related to the state variables by the output matrix $\mathbf{C}$, equation (2.70), and we make this more general by including a further matrix $\mathbf{D}$ reflecting the direct influence, if any, of the input vector on the output vector $\mathbf{y}$, i.e.

$$\mathbf{y} = \mathbf{C}\mathbf{x} + \mathbf{D}\mathbf{u}. \tag{3.57}$$

Taking Laplace transforms throughout and with the stipulation of zero initial condition results in

$$s\mathbf{X}(s) = \mathbf{A}\mathbf{X}(s) + \mathbf{B}\mathbf{U}(s) \tag{3.58}$$

$$\mathbf{Y}(s) = \mathbf{C}\mathbf{X}(s) + \mathbf{D}\mathbf{U}(s) \tag{3.59}$$

since $\mathbf{A}$, $\mathbf{B}$, $\mathbf{C}$, $\mathbf{D}$ are constant matrices. $\mathbf{X}(s)$ means the vector of the transforms of the elements in the vector $\mathbf{x}(t)$. It is required to express the output $\mathbf{Y}(s)$ in terms of the input $\mathbf{U}(s)$, i.e. to determine $\mathbf{G}(s)$ in

$$\mathbf{Y}(s) = \mathbf{G}(s)\mathbf{U}(s). \tag{3.60}$$

$\mathbf{G}(s)$ is by definition the transfer function matrix. Rearranging equation (3.58) gives

$$(s\mathbf{I} - \mathbf{A})\mathbf{X}(s) = \mathbf{B}\mathbf{U}(s)$$

which on premultiplying by $(s\mathbf{I} - \mathbf{A})^{-1}$ and substitution into equation (3.57) gives

$$\mathbf{Y}(s) = \mathbf{C}(s\mathbf{I} - \mathbf{A})^{-1}\mathbf{B}\mathbf{U}(s) + \mathbf{D}\mathbf{U}(s),$$

i.e. $\mathbf{G}(s) = \mathbf{C}(s\mathbf{I} - \mathbf{A})^{-1}\mathbf{B} + \mathbf{D} \tag{3.61}$

in equation (3.60). In terms of the adjoint matrix

$$\mathbf{G}(s) = \mathbf{C}\frac{\text{adj}\,(s\mathbf{I} - \mathbf{A})}{|s\mathbf{I} - \mathbf{A}|}\mathbf{B} + \mathbf{D}.$$

If $\mathbf{D} = \mathbf{0}$ and $\mathbf{C}$ is the identity matrix, i.e. $\mathbf{y} = \mathbf{x}$

$$\mathbf{G}(s) = (s\mathbf{I} - \mathbf{A})^{-1}\mathbf{B}. \tag{3.62}$$

It can be seen that although the basic transfer function relationship is retained the elements of the transfer matrix become more complex as the order of the system is increased. These elements $G_{ij}(s)$ are the scalar transfer function relating each output y_i to each input u_j and each will require inversion. Its relationship with the state variable description will become apparent in the following section.

Example Establish the transfer function matrix between the state $\mathbf{x}$ and the input u of the second order system whose state equations are

$$\dot{x}_1 = -x_1 - 2x_2 + u$$
$$\dot{x}_2 = x_1 - 3x_2.$$

Hence determine the system response to a unit impulse and a unit step input.

The overall transfer function matrix $\mathbf{G}(s)$ is given by

$$\mathbf{X}(s) = \mathbf{G}(s)\mathbf{U}(s).$$

Writing the pair of equations as $\dot{\mathbf{x}} = \mathbf{A}\mathbf{x} + \mathbf{B}u$ and taking transforms yields

$$s\mathbf{X}(s) = \mathbf{A}\mathbf{X}(s) + \mathbf{B}\mathbf{U}(s)$$

so that

$$\mathbf{X}(s) = (s\mathbf{I} - \mathbf{A})^{-1}\mathbf{B}\mathbf{U}(s).$$

For the given equations

$$(s\mathbf{I} - \mathbf{A})^{-1} = \frac{1}{\begin{vmatrix} s+1 & 2 \\ -1 & s+3 \end{vmatrix}} \cdot \begin{bmatrix} s+3 & -2 \\ 1 & s+1 \end{bmatrix}$$

$$= \begin{bmatrix} \dfrac{s+3}{s^2+4s+5} & \dfrac{-2}{s^2+4s+5} \\ \dfrac{1}{s^2+4s+5} & \dfrac{s+1}{s^2+4s+5} \end{bmatrix}$$

and with $\mathbf{B} = \begin{bmatrix} 1 \\ 0 \end{bmatrix}$

$$\mathbf{G}(s) = (s\mathbf{I} - \mathbf{A})^{-1}\mathbf{B}$$

$$= \begin{bmatrix} \dfrac{s+3}{s^2+4s+5} \\ \dfrac{1}{s^2+4s+5} \end{bmatrix}$$

i.e. in this case $\mathbf{G}(s)$ is a single column matrix, a vector, and

$$\begin{bmatrix} X_1(s) \\ X_2(s) \end{bmatrix} = \begin{bmatrix} \dfrac{s+3}{s^2+4s+5} \\ \dfrac{1}{s^2+4s+5} \end{bmatrix} U(s).$$

The block diagram representations are as Fig. 3.39.
Note the order of the blocks $\mathbf{B}$ and $(s\mathbf{I} - \mathbf{A})^{-1}$ and compare with their positions in the matrix $\mathbf{G}(s)$.

Before proceeding to evaluate the solution from the input-state relationship we can already tell quite a lot about the system response to a bounded input. Each term of $\mathbf{G}(s)$ contains the denominator (s^2+4s+5) and the factors of this, or the roots of the equation $s^2+4s+5=0$, determine the nature of the response. The denominator's factors are $(s+2+j)(s+2-j)$, i.e. the roots are $s = -2 \pm j$. We shall see that these are the values of the system eigenvalues given by the equation $|s\mathbf{I} - \mathbf{A}| = 0$, equation 3.76. The negative real part indicates a solution which decays with time, i.e. positively damped and stable, and the presence of the imaginary part of the complex root indicates an oscillatory response.

Now take the impulse and step inputs in turn.
For the impulse input u, $U(s) = 1$,

$$\mathbf{X}(s) = \mathbf{G}(s)$$

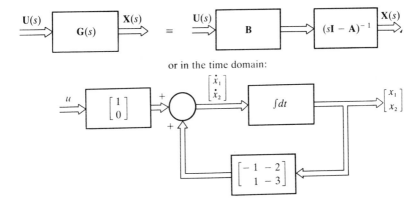

Fig. 3.39 Block diagrams for dynamic system

and

$$X_1(s) = \frac{s+3}{s^2+4s+5}$$

$$X_2(s) = \frac{1}{s^2+4s+5}$$

to give

$$x_1(t) = [-2 \cdot 24 e^{-2t} \sin{(t-0 \cdot 464)}] + [3e^{-2t} \sin{t}]$$
$$= e^{-2t}[3 \sin{t} - 2 \cdot 24 \sin{(t-0 \cdot 464)}], \quad (\text{i.e. } e^{-2t}(\sin{t} + \cos{t}))$$

and

$$x_2(t) = e^{-2t} \sin{t}$$

(using Table 1.1). Substitution back into the state equations confirms the solution and we see that a damped oscillatory behaviour is indeed found.

For the step input, $U(s) = \dfrac{1}{s}$

and

$$X_1(s) = \frac{1}{s^2+4s+5} + \frac{3}{s(s^2+4s+5)}$$

with

$$X_2(s) = \frac{1}{s(s^2+4s+5)}.$$

Again using Table 1.1 we obtain

$$x_1(t) = [e^{-2t} \sin{t}] + \tfrac{3}{5}[1 - 2 \cdot 24 e^{-2t} \sin{(t+0 \cdot 464)}]$$
$$= 0 \cdot 6 + e^{-2t}[\sin{t} - 1 \cdot 34 \sin{(t+0 \cdot 464)}]$$

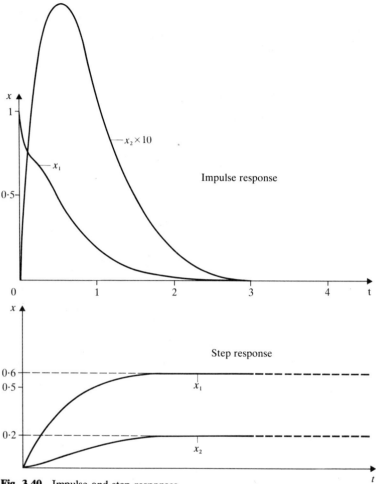

Fig. 3.40 Impulse and step responses

and

$$x_2(s) = 0 \cdot 2[1 - 2 \cdot 24e^{-2t} \sin(t + 0 \cdot 464)]$$
$$= 0 \cdot 2 - 0 \cdot 448e^{-2t} \sin(t + 0 \cdot 464).$$

Note that x_1 and x_2 as $t \to \infty$ have the values of 0·6 and 0·2 which may be obtained by putting $\dot{x}_1$, $\dot{x}_2$ to zero in the state equations with $u = 1$. We note also that as the impulse is the derivative of the step, so the linear system response to the impulse is the derivative of the response to the step. By evaluating the step response first the impulse response could have been found by differentiating.

The general form of the solution are sketched in Fig. 3.40. (The heavy damping masks the oscillations at higher times on this scale.)

The roots of the characteristic equation, or eigenvalues, $-2 \pm j1$ may now be looked at again. It is seen that the real part, $\sigma = -2$, gives the exponential damping term e^{-2t} and the imaginary part, $\omega = 1$, gives the frequency of the oscillation in the term $\sin(1 \cdot t + \phi)$.

State space representation

Transition matrix in the time domain

In Chapter 2 the state space model was developed,

$$\dot{x} = Ax + Bu$$

where x is an n-dimensional vector, u an $(m \leq n)$ dimensional vector, A an $n \times n$ matrix and B an $n \times m$ matrix. If A and B are not time invariant then, although the solution may be written in the form of an integral the actual integration requires numerical methods of solution. When the system is linear and the matrices A and B are time invariant, the solution may be expressed in a more precise form. The more general methods exist for the later case and most attention will be focussed on this, although one advantage of the state space description is that it can cope with the time-varying systems. We shall deal with the time invariant case in detail.

The problem is to obtain the solution $x(t)$ for the first order differential equation above. If there is a further relationship, $y = Cx$, then $y(t)$ follows simply once $x(t)$ has been established if C is a constant matrix. The solution of the matrix equation may be based on an experience of the solution of first order linear scalar differential equations, i.e. those of the form

$$\dot{x} = ax + bu.$$

We know that this can be made up from a consideration first of the homogeneous equation $\dot{x} = ax$, giving a general solution, and then by considering the forcing term, bu also to give the particular solution. Although this approach leads one along familiar paths it is not the only method of solution. By expressing the motion of a system as the sum of simple motions a 'modal' or 'spectral' form of dynamic behaviour is obtained (e.g. MacFarlane 1970) and solutions are effected directly in this way. This expression of motion as the sum of components will come out in the later sections of the treatment followed in this text.

Solution of the linear homogeneous equation $\dot{x} = ax$ may be effected by selection of a probable solution and then seeing if this fits the original equation. Basic differential equation literature gives us the solution

$$x = e^{at}x(0)$$

where $x(0)$ is the initial condition at $t = 0$. By differentiation it is seen immediately that this satisfies the equation $\dot{x} = ax$. Remembering that

$$e^{at} = 1 + at + \frac{(at)^2}{2!} + \frac{(at)^3}{3!} + \ldots$$

it is seen that we have assumed a solution of the general form

$$x(t) = a_0 + a_1 t + a_2 t^2 + a_3 t^3 + \ldots .$$

Now look at the matix equation

$$\dot{\mathbf{x}} = \mathbf{A}\mathbf{x} \tag{3.63}$$

and assume a solution (with $\mathbf{A}$ time invariant) of the form

$$\mathbf{x}(t) = \mathbf{a}_0 + \mathbf{a}_1 t + \mathbf{a}_2 t^2 + \mathbf{a}_3 t^3 + \ldots \tag{3.64}$$

where the $\mathbf{a}_i$ are n-dimensional constant vectors. Substitution of this 'solution' in equation (3.63) gives

$$\mathbf{a}_1 + 2\mathbf{a}_2 t + 3\mathbf{a}_3 t^2 + \ldots k\mathbf{a}_k t^{k-1} + \ldots = \mathbf{A}(\mathbf{a}_0 + \mathbf{a}_1 t + \ldots \mathbf{a}_k t^k + \ldots)$$

and if this is to hold for all t

$$\mathbf{a}_1 = \mathbf{A}\mathbf{a}_0$$
$$2\mathbf{a}_2 = \mathbf{A}\mathbf{a}_1$$
$$\cdot$$
$$\cdot$$
$$\cdot$$
$$k\mathbf{a}_k = \mathbf{A}\mathbf{a}_{k-1}.$$
$$\cdot$$
$$\cdot$$
$$\cdot$$

Rearranging the equations we see that

$$\mathbf{a}_1 = \mathbf{A}\mathbf{a}_0$$
$$\mathbf{a}_2 = \tfrac{1}{2} \cdot \mathbf{A}^2 \mathbf{a}_0$$
$$\mathbf{a}_3 = \frac{1}{3.2.1} \cdot \mathbf{A}^3 \mathbf{a}_0$$
$$\cdot$$
$$\cdot$$
$$\cdot$$
$$\mathbf{a}_k = \frac{1}{k!} \cdot \mathbf{A}^k \mathbf{a}_0.$$

The initial condition $\mathbf{x} = \mathbf{x}(0)$ at $t = 0$ gives, from equation (3.64) $\mathbf{x}(0) = \mathbf{a}_0$. Therefore equation (3.64) may be rewritten as

$$\mathbf{x}(t) = \left\{ \mathbf{I} + \mathbf{A}t + \frac{\mathbf{A}^2}{2!} t^2 + \ldots \frac{\mathbf{A}^k t^k}{k!} + \ldots \right\} \mathbf{x}(0)$$

By analogy with the scalar expansion for e^{at} the series in the brackets { } is called the 'matrix exponential' and is written $e^{\mathbf{A}t}$, and is an $n \times n$ matrix. The solution of the homogeneous equation is then

$$\mathbf{x}(t) = e^{\mathbf{A}t}\mathbf{x}(0). \tag{3.65}$$

This important matrix exponential is also known in this context as the 'transition' or 'fundamental' matrix and is denoted usually by $\boldsymbol{\Phi}$ or $\boldsymbol{\Omega}$. It is an $n \times n$ matrix but it reflects many properties of the scalar exponential;

e.g.

$$\frac{de^{\mathbf{A}t}}{dt} = \mathbf{A}e^{\mathbf{A}t}$$

$$= e^{\mathbf{A}t}\mathbf{A}$$

$$e^{\mathbf{A}(t_1+t_2)} = e^{\mathbf{A}t_1}e^{\mathbf{A}t_2}$$

$$= e^{\mathbf{A}t_2}e^{\mathbf{A}t_1}$$

i.e. $e^{\mathbf{A}t_1}$ and $e^{\mathbf{A}t_2}$ are two matrices which commute. If $t_1 = -t_2$ then the product is the identity matrix, $\mathbf{I}$, but the two general exponential matrices $e^{\mathbf{A}}$ and $e^{\mathbf{B}}$ only commute if $\mathbf{A}$ and $\mathbf{B}$ commute, i.e. if $\mathbf{AB}$ is equal to $\mathbf{BA}$.

On account of the exponential nature of the transition matrix, $e^{\mathbf{A}t}$ or $\boldsymbol{\Phi}(t)$, the following properties may also be deduced.

(i) $\quad \boldsymbol{\Phi}(0) = e^{\mathbf{A}0}$

$\qquad\qquad = \mathbf{I}$

(ii) $\quad \boldsymbol{\Phi}(-t) = e^{-\mathbf{A}t}$

$\qquad\qquad = (e^{\mathbf{A}t})^{-1}$

$\qquad\qquad = \boldsymbol{\Phi}^{-1}(t)$

(iii) $\boldsymbol{\Phi}(t_1 + t_2) = e^{\mathbf{A}(t_1+t_2)}$

$\qquad\qquad = \boldsymbol{\Phi}(t_1)\boldsymbol{\Phi}(t_2)$

$\qquad\qquad = \boldsymbol{\Phi}(t_2)\boldsymbol{\Phi}(t_1)$

(iv) $\boldsymbol{\Phi}(t_2 - t_0) = \boldsymbol{\Phi}(t_1 - t_0)\boldsymbol{\Phi}(t_2 - t_1)$.

Solution of the nonhomogeneous scalar case consists of its general solution determined above and its particular integral obtained by use of the multiplying factor e^{-at}. The solution is

$$x(t) = e^{at}x(0) + e^{at}\int_0^t e^{-a\tau}bu(\tau)\, d\tau.$$

May this principle be applied to obtain the solution of the vector-matrix equation $\dot{\mathbf{x}} = \mathbf{A}\mathbf{x} + \mathbf{B}\mathbf{u}$? Once again alternative procedural details are available but the most direct method is to invoke the linear scalar method.

Rearrange the equation so that

$$\dot{x} - Ax = Bu.$$

Premultiplying by the factor e^{-At} yields the pure derivative of a product

$$e^{-At}[\dot{x} - Ax] = e^{-At}Bu$$

$$= \frac{d}{dt}\{e^{-At}x\}$$

giving on integration from time $t = 0$ to time t

$$e^{-At}x(t) - x(0) = \int_0^t e^{-A\tau}Bu\, d\tau$$

where $x(0)$ is the initial condition at $t = 0$. Thus

$$x(t) = e^{At}x(0) + e^{At}\int_0^t e^{-A\tau}Bu(\tau)\, d\tau$$

$$= e^{At}x(0) + \int_0^t e^{A(t-\tau)}Bu(\tau)\, d\tau \tag{3.66}$$

or in terms of the transition matrix $\Phi(t)$

$$x(t) = \Phi(t)x(0) + \int_0^t \Phi(t-\tau)Bu(\tau)\, d\tau. \tag{3.67}$$

The use of the exponential notation may help initially in the use of these equations but the general transition matrix Φ has the advantage that it includes the time-variant case in its notation. Then the elements of A and B are functions of time themselves and the exponential expansion cannot be used. The Neumann series is one possible way of expanding and evaluating the time-variant case. The transition matrix now depends explicitly on the instant of starting the computation, t_0 as well as on the time that has expired from time $t = t_0$. Now $A = A(t)$, $B = B(t)$ and

$$x(t) = \Phi(t, t_0)x(t_0) + \int_{t_0}^t \Phi(t, t_0)B(\tau)u(\tau)\, d\tau$$

and $\Phi(t, t_0)$ may be expanded in the Neumann series (e.g. Friendly 1972)

$$\Phi(t, t_0) = I + \int_{t_0}^t A(\tau)\, d\tau + \int_{t_0}^t A(\tau)\left[\int_0^{\tau_1} A(\tau_2)\, d\tau_2\right] d\tau_1 + \dots .$$

Use of Laplace transform with the transition matrix

In section 3.3 the Laplace transform was used to give for the multi-input multi-output systems the equivalent of the scalar transfer function. This required the same conditions as transfer function formation including that of zero initial conditions. The Laplace transform as was seen in Chapter 1 can still be used if initial conditions are not zero but the transfer function as a ratio of transforms can then not be used. However, take the full transform of the time-invariant state equation $\dot{\mathbf{x}} = \mathbf{A}\mathbf{x} + \mathbf{B}\mathbf{u}$;

$$s\mathbf{X}(s) - \mathbf{x}(0) = \mathbf{A}\mathbf{X}(s) + \mathbf{B}\mathbf{U}(s).$$

Solving for $\mathbf{X}(s)$ yields

$$\mathbf{X}(s) = (s\mathbf{I} - \mathbf{A})^{-1}\mathbf{x}(0) + (s\mathbf{I} - \mathbf{A})^{-1}\mathbf{B}\mathbf{U}(s)$$

and

$$\mathbf{x}(t) = \mathscr{L}^{-1}[(s\mathbf{I} - \mathbf{A})^{-1}\mathbf{x}(0)] + \mathscr{L}^{-1}[(s\mathbf{I} - \mathbf{A})^{-1}\mathbf{B}\mathbf{U}(s)]. \tag{3.68}$$

This must be the same as the solution equations (3.66) and (3.67) and similarly a correspondence exists for the homogeneous case alone so that, since $\mathbf{x}(0)$ is a constant factor through the transformation

$$e^{\mathbf{A}t} = \mathscr{L}^{-1}[(s\mathbf{I} - \mathbf{A})^{-1}]. \tag{3.69}$$

Using this result in the convolution integral transform for the second half of equation (3.68), section 1.4 yields

$$\mathscr{L}^{-1}[(s\mathbf{I} - \mathbf{A})^{-1}\mathbf{B}\mathbf{U}(s)] = \int_0^t e^{\mathbf{A}(t-\tau)}\mathbf{B}\mathbf{u}(\tau)\, d\tau \tag{3.70}$$

which agrees with equation (3.66) also. Of these results equation (3.69) is particularly important as it provides an expression of significance for $e^{\mathbf{A}t}$ which may be obtained in explicit terms (in specific cases) rather than as a series matrix expansion.

For zero initial conditions but with a forcing input $\mathbf{u}$ equation (3.68) becomes

$$\mathbf{x}(t) = \mathscr{L}^{-1}[(s\mathbf{I} - \mathbf{A})^{-1}\mathbf{B}\mathbf{U}(s)]$$

or

$$\mathbf{X}(s) = (s\mathbf{I} - \mathbf{A})^{-1}\mathbf{B}\mathbf{U}(s). \tag{3.71}$$

Under these conditions the dynamics may be expressed by the transfer function matrix $\mathbf{G}(s)$. Where the output $\mathbf{y}$ is the state $\mathbf{x}$

$$\mathbf{X}(s) = \mathbf{G}(s)\mathbf{U}(s)$$

and so from equation (3.71),

$$\mathbf{G}(s) = (s\mathbf{I} - \mathbf{A})^{-1}\mathbf{B}. \tag{3.72}$$

If $\mathbf{y} = \mathbf{C}\mathbf{x}$, then $\mathbf{Y}(s) = \mathbf{C}\mathbf{X}(s)$ and the overall transfer function between the output $\mathbf{y}$ and the input vector $\mathbf{u}$ is

$$\mathbf{G}(s) = \mathbf{C}(s\mathbf{I} - \mathbf{A})^{-1}\mathbf{B}. \tag{3.73}$$

Referring back to the explicit treatment of the transfer function matrix it is seen that equations (3.72) and (3.73) are (3.62) and (3.61) respectively (with $\mathbf{D} = \mathbf{0}$).

These relationships between the transition matrix and convolution of input and output, and the Laplace transformation are critical in enabling easy evaluation of the transition matrices.

Example   A batch centrifuge bowl is mounted with its axis vertical (Fig. 3.41). When fully charged, and the removal of excess fluid is complete, the total moment of inertia J of the bowl and 'cake' is 100 kg m^2. The centrifuge is allowed to come to rest under natural damping forces and in 2 minutes the speed has dropped from 600 rpm to 30 rpm. A small brake is now applied by the centrifuge operator bringing it quickly to rest.

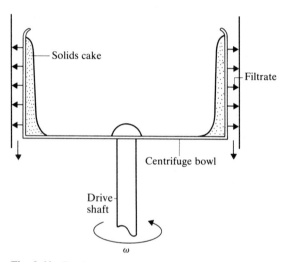

Fig. 3.41   Batch centrifuge

If the damping forces may be assumed to be proportional to the bowl speed, establish the state equations representing the dynamics of the centrifuge during the slowing down period. How many revolutions does the bowl make during this period?

What additional constant retarding torque would be necessary to reduce the time to slow down the centrifuge to 30 rpm in only one minute?

Let the damping torque be $K\omega$ where ω is the angular speed of the bowl. The equation of motion for the centrifuge under the initial conditions is thus

$$J\dot{\omega} = -K\omega. \tag{3.74}$$

As we are interested in the total revolutions also let us write this in terms of the angle of rotation θ, after positive drive is stopped

$$J\ddot{\theta} + K\theta = 0$$

Let the state variable x_1, be the position θ, and x_2 be the speed ω. Then

$$\dot{x}_1 = x_2$$

and

$$\dot{x}_2 = -K/J \cdot x_2$$

are the state equations. It remains only to find K from the data. The easiest way is to solve equation (3.74) for x_2 only,

$$x_2 = e^{-K/J \cdot t} x_2(0).$$

Standardizing on units:

At $t = 0$, $\quad x_2(0) = 10 \times 2\pi$ rad s^{-1}

$\quad\quad t = 120$ sec, $\quad x_2(t) = 0 \cdot 5 \times 2\pi$ rad s^{-1}.

Substituting these values into x_2 gives

$$\ln 20 = K \cdot 120/100$$

$$K = 2 \cdot 5 \text{ Nms}$$

so that the state equations are

$$\dot{x}_1 = x_2$$
$$\dot{x}_2 = -0 \cdot 025 x_2.$$

In the period of 2 minutes, the change in x_1 is

$$x_1 = \int_0^{120} x_2 \, dt$$

$$= x_2(0) \left[\frac{-e^{-0 \cdot 025t}}{0 \cdot 025} \right]_{t=0}^{t=120}$$

$$= 2388 \text{ rad}$$

$$= 380 \text{ rev.}$$

If there is an additional constant retarding torque T the equation of motion of the centrifuge bowl is

$$J\dot{\omega} + K\omega + T = 0$$

and the state equations are

$$\dot{x}_1 = x_2$$
$$\dot{x}_2 = -K/J \cdot x_2 - T/J$$

or

$$\begin{bmatrix} \dot{x}_1 \\ \dot{x}_2 \end{bmatrix} = \begin{bmatrix} 0 & 1 \\ 0 & -K/J \end{bmatrix} \begin{bmatrix} x_1 \\ x_2 \end{bmatrix} + \begin{bmatrix} 0 \\ -1/J \end{bmatrix} T$$

where T is the system 'input'. As it is constant it may be considered as a step input into the system.

For the system $\dot{\mathbf{x}} = \mathbf{Ax} + \mathbf{Bu}$ the full response, equations (3.69) and (3.70) is

$$\mathbf{x} = \mathscr{L}^{-1}[(s\mathbf{I} - \mathbf{A})^{-1}] \cdot \mathbf{x}(0) + \mathscr{L}^{-1}[(s\mathbf{I} - \mathbf{A})^{-1}\mathbf{B}U(s)]$$

so

$$\mathbf{x} = \mathscr{L}^{-1}\left[\begin{bmatrix} s & -1 \\ 0 & s+K/J \end{bmatrix}^{-1}\right]\mathbf{x}(0) + \mathscr{L}^{-1}\left[\begin{bmatrix} s & -1 \\ 0 & s+K/J \end{bmatrix}^{-1}\begin{bmatrix} 0 \\ -1/J \end{bmatrix}\frac{T}{s}\right]$$

$$\mathbf{x}(0) = [0, 20\pi]^T, \; J = 100, \; K = 2\cdot 5$$

so

$$\mathbf{x} = \mathscr{L}^{-1}\left[\begin{array}{cc} \dfrac{1}{s} & \dfrac{1}{s(s+0\cdot025)} \\ 0 & \dfrac{1}{s+0\cdot025} \end{array}\right]\mathbf{x}(0) + \mathscr{L}^{-1}\left[\begin{array}{c} \dfrac{-1}{s^2(s+0\cdot025)} \\ \dfrac{-1}{s(s+0\cdot025)} \end{array}\right]\dfrac{T}{100}$$

$$x_1 = \theta = \mathscr{L}^{-1}\left[\frac{20\pi}{s(s+0\cdot025)} - \frac{T}{100s^2(s+0\cdot025)}\right]$$

$$= (800\pi + 16T)(1 - e^{-0\cdot025t}) - 0\cdot4Tt$$

$$x_2 = \dot{\theta} = \omega = \mathscr{L}^{-1}\left[\frac{20\pi}{s+0\cdot025} - \frac{T}{100s(s+0.025)}\right]$$

$$= (20\pi + 0\cdot4T)e^{-0\cdot025t} - 0\cdot4T. \tag{3.75}$$

For $\dot{\theta}$ to be π at $t = 60$ sec, substitution into equation (3.75) yields

$$T = 35 \text{ Nm}$$

and x_1 (at 60 sec) = 1547 rad so that only 246 revolutions have been completed before the speed drops to the required 30 rpm.

Eigenvalues

The eigenvalues or characteristic roots of a matrix $\mathbf{A}$ are the roots of the equation

$$|\lambda\mathbf{I} - \mathbf{A}| = 0. \tag{3.76}$$

This equation is also known as the characteristic equation, and its relationship with the characteristic equation of the input–output model will become apparent when the single input–single output control case is studied.

Only for a square matrix **A** can eigenvalues be determined (i.e. exist) and expansion of equation (3.76) gives an equation in powers of λ up to the order of the matrix **A**. For an $n \times n$ matrix there is an nth order equation in λ and this equation has n roots or eigenvalues. For example, if **A** is the matrix $\begin{bmatrix} a_{11} & a_{12} \\ a_{21} & a_{22} \end{bmatrix}$ equation (3.76) is

$$\left| \lambda \begin{bmatrix} 1 & 0 \\ 0 & 1 \end{bmatrix} - \begin{bmatrix} a_{11} & a_{12} \\ a_{21} & a_{22} \end{bmatrix} \right| = \left| \begin{bmatrix} \lambda & 0 \\ 0 & \lambda \end{bmatrix} - \begin{bmatrix} a_{11} & a_{12} \\ a_{21} & a_{22} \end{bmatrix} \right|$$

$$= \left| \begin{matrix} \lambda - a_{11} & -a_{12} \\ -a_{21} & \lambda - a_{22} \end{matrix} \right|$$

$$= (\lambda - a_{11})(\lambda - a_{22}) - a_{12}a_{21}$$

$$= \lambda^2 - (a_{11} + a_{22})\lambda - a_{12}a_{21} + a_{11}a_{22} = 0.$$

The eigenvalues are the roots of this characteristic equation, in this case

$$\lambda_1 = [a_{11} + a_{22} + \sqrt{\{(a_{11} + a_{22})^2 + 4(a_{12}a_{21} - a_{11}a_{22})\}}]/2$$
$$\lambda_2 = [a_{11} + a_{22} - \sqrt{\{(a_{11} + a_{22})^2 + 4(a_{12}a_{21} - a_{11}a_{22})\}}]/2$$

and the equation may be written as

$$|\lambda \mathbf{I} - \mathbf{A}| = (\lambda - \lambda_1)(\lambda - \lambda_2)$$
$$= 0. \tag{3.77}$$

It will be assumed at present that all eigenvalues are distinct, i.e. two or more eigenvalues do not have the same value. Eigenvalues may be complex and very frequently are and they then occur in conjugate complex pairs; $\lambda_1, \lambda_2 = \sigma \pm j\omega$.

The physical significance of the eigenvalues in a dynamic system model is made clear when we look again at the dynamics of the time invariant system equation (3.66), and at the equation for the exponential matrix, equation (3.69). Rewriting equation (3.69), the transition matrix $e^{\mathbf{A}t}$ is given by

$$e^{\mathbf{A}t} = \mathscr{L}^{-1} \left[\frac{adj(s\mathbf{I} - \mathbf{A})}{|s\mathbf{I} - \mathbf{A}|} \right]$$

$$= \mathscr{L}^{-1} \begin{bmatrix} \alpha_{11} & \cdots & \alpha_{1n} \\ \cdot & & \\ \cdot & & \\ \cdot & & \\ \alpha_{n1} & \cdots & \alpha_{nn} \end{bmatrix}. \tag{3.78}$$

The determinant $|s\mathbf{I}-\mathbf{A}|$ appears as the denominator in every term α_{ij} to be inverted. With the change in notation of s for λ this determinant $|s\mathbf{I}-\mathbf{A}|$ is just that which we have considered in the determination of the eigenvalues. It can therefore be factorized, like equation (3.77) into the terms

$$|s\mathbf{I}-\mathbf{A}| = (s-\lambda_1)(s-\lambda_2)\ldots(s-\lambda_n)$$

where $\lambda_1, \lambda_2, \ldots \lambda_n$ are the eigenvalues of the $n \times n$ matrix $\mathbf{A}$. Each element of the matrix in equation (3.78) may then be expressed as partial fractions so that

$$\alpha_{ij} = \frac{a_{ij}}{|s\mathbf{I}-\mathbf{A}|}$$

$$= \frac{a_{ij}}{(s-\lambda_1)(s-\lambda_2)\ldots(s-\lambda_n)}$$

$$= \frac{a_1}{s-\lambda_1} + \frac{a_2}{s-\lambda_2} + \ldots \frac{a_n}{s-\lambda_n}. \tag{3.79}$$

Each a_{ij} the elements of the adjoint matrix will be a function of s but the coefficients a_1, a_2, will be constants. This is because the adjoint matrix will have as its elements terms which are at least one order less, in terms of their highest powers of s, than n since it is formed from the co-factors of the $n \times n$ matrix $(s\mathbf{I}-\mathbf{A})$ (see Chapter 1).

Inversion of a transform matrix, equation (3.78), is by inversion of each element so that each element of the matrix $e^{\mathbf{A}t}$ will have the form:

$$\mathscr{L}^{-1}\left[\frac{a_1}{s-\lambda_1} + \frac{a_2}{s-\lambda_2} + \ldots \frac{a_n}{s-\lambda_n}\right] = a_1 e^{\lambda_1 t} + a_2 e^{\lambda_2 t} + \ldots a_n e^{\lambda_n t} \tag{3.80}$$

The inversion of equation (3.70) involves the same operation, so that the response to external disturbances as well as the movement of the system from an initial condition in the absence of an input vector $\mathbf{u}$ depends on the eigenvalues of the system matrix. The addition of the factor $\mathbf{U}(s)$ does not affect the significance of the eigenvalues or their place in the time domain solution.

If there are repeated roots in the characteristic equation additional terms in the elements of $e^{\mathbf{A}t}$ are obtained of the form $te^{\lambda_i t}, \ldots,$ $t^{k-1}e^{\lambda_i t}/(k-1)!$ where the k is the multiplicity of the root (see Table 1.1). Any complex roots $\lambda_i = \sigma \pm j\omega$ occur in conjugate pairs and on inversion the complex components of opposite sign cancel each other out but an oscillatory term is left, $e^{\sigma t} \sin(\omega t + \phi)$ where ϕ is a phase angle.

Example For the system equation $\dot{\mathbf{x}} = \mathbf{Ax} + \mathbf{Bu}$ let $\mathbf{u}$ be a constant unit input and the full equation be

$$\begin{bmatrix} \dot{x}_1 \\ \dot{x}_2 \end{bmatrix} = \begin{bmatrix} -2 & 1 \\ 1 & -2 \end{bmatrix} \begin{bmatrix} x_1 \\ x_2 \end{bmatrix} + \begin{bmatrix} 1 \\ 2 \end{bmatrix} [1].$$

Determine the state response to the step input u. The eigenvalues of $\mathbf{A}$ are given by $|\lambda \mathbf{I} - \mathbf{A}| = 0$,

i.e.

$$\begin{vmatrix} \lambda + 2 & -1 \\ -1 & \lambda + 2 \end{vmatrix} = 0$$

so that

$$(\lambda + 2)(\lambda + 2) - 1 = 0$$

i.e.

$$\lambda^2 + 4\lambda + 3 = (\lambda + 1)(\lambda + 3)$$
$$= 0.$$

The roots, or eigenvalues, are

$$\lambda_1 = -1$$
$$\lambda_2 = -3.$$

Now the matrix inverse of $(s\mathbf{I} - \mathbf{A})$ is

$$\frac{1}{\begin{vmatrix} s+2 & -1 \\ -1 & s+2 \end{vmatrix}} \cdot \begin{bmatrix} s+2 & 1 \\ 1 & s+2 \end{bmatrix}$$

so that for equation (3.69)

$$e^{\mathbf{A}t} = \mathscr{L}^{-1} \begin{bmatrix} \dfrac{s+2}{(s+1)(s+3)} & \dfrac{1}{(s+1)(s+3)} \\ \dfrac{1}{(s+1)(s+3)} & \dfrac{s+2}{(s+1)(s+3)} \end{bmatrix}$$

$$= \begin{bmatrix} \frac{1}{2}(e^{-t} + e^{-3t}) & \frac{1}{2}(e^{-t} - e^{-3t}) \\ \frac{1}{2}(e^{-t} - e^{-3t}) & \frac{1}{2}(e^{-t} + e^{-3t}) \end{bmatrix}.$$

Thus the unforced free response of the system is given by

$$\mathbf{x} = \begin{bmatrix} \frac{1}{2}(e^{-t} + e^{-3t}) & \frac{1}{2}(e^{-t} - e^{-3t}) \\ \frac{1}{2}(e^{-t} - e^{-3t}) & \frac{1}{2}(e^{-t} + e^{-3t}) \end{bmatrix} \mathbf{x}(0).$$

For a unit step input $U(s) = 1/s$, and the second term of the response equation (3.67) is given by equation (3.70),

$$\mathscr{L}^{-1}[(s\mathbf{I}-\mathbf{A})^{-1}\mathbf{B}U(s)] = \mathscr{L}^{-1}\left[\left[\begin{array}{cc} \dfrac{s+2}{(s+1)(s+3)} & \dfrac{1}{(s+1)(s+3)} \\[2ex] \dfrac{1}{(s+1)(s+3)} & \dfrac{s+2}{(s+1)(s+2)} \end{array}\right]\begin{bmatrix}1 \\ 2\end{bmatrix}\dfrac{1}{s}\right]$$

$$= \mathscr{L}^{-1}\begin{bmatrix} \dfrac{s+4}{s(s+1)(s+3)} \\[2ex] \dfrac{2s+5}{s(s+1)(s+3)} \end{bmatrix}$$

$$= \begin{bmatrix} \tfrac{4}{3} - \tfrac{3}{2}e^{-t} + \tfrac{1}{6}e^{-3t} \\[1ex] \tfrac{5}{3} - \tfrac{3}{2}e^{-t} - \tfrac{1}{6}e^{-3t} \end{bmatrix}$$

so that the full response for a system starting from the non-equilibrium position $\mathbf{x}(0)$ with a unit step input response superimposed on its free motion is

$$\mathbf{x} = \begin{bmatrix} \tfrac{1}{2}(e^{-t}+e^{-3t}) & \tfrac{1}{2}(e^{-t}-e^{-3t}) \\[1ex] \tfrac{1}{2}(e^{-t}-e^{-3t}) & \tfrac{1}{2}(e^{-t}+e^{-3t}) \end{bmatrix}\mathbf{x}(0) + \begin{bmatrix} \tfrac{4}{3}-\tfrac{3}{2}e^{-t}+\tfrac{1}{6}e^{-3t} \\[1ex] \tfrac{5}{3}-\tfrac{3}{2}e^{-t}-\tfrac{1}{6}e^{-3t} \end{bmatrix}.$$

Note that the exponent coefficients which appear in both parts of the equation are the eigenvalues of $\mathbf{A}$ so that just the determination of the eigenvalues indicates the rate at which the system approaches a steady value and indeed if a steady value is approached at all! For an eigenvalue with a positive real part the values of $\mathbf{x}$ will continue to grow without limit and we have an unstable system. If the eigenvalues have negative real parts then the system is stable in that it tends to a

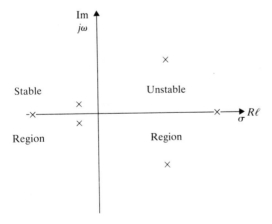

Fig. 3.42 Eigenvalues, $\sigma \pm j\omega$, in the complex plane. (Note that complex values occur in conjugate pairs)

steady value. In addition if some of the eigenvalues are complex then the complex exponentials are evaluated as an oscillatory response. These oscillations decrease or grow in amplitude according to whether the real part of the eigenvalue is negative or positive respectively. This criterion for stability may be shown by positioning the eigenvalues in the complex plane (Fig. 3.42). The region to the right of the imaginary axis is a prohibited region for a 'stable' response from the system. If the eigenvalues fall on the imaginary axis then this may be classed as unstable or marginally unstable. The criterion of stability is very important in control systems and this will be expanded in Chapters 4 to 7.

The state variables chosen for a system are not unique in that one set may be replaced by another made up of linear combinations of the first set. Such a 'linear transformation' in which the new state variables $\mathbf{z}$ are related to the original set $\mathbf{x}$ is given by

$$\mathbf{x} = \mathbf{Tz} \tag{3.81}$$

where $\mathbf{T}$ is the transformation matrix. If $\mathbf{z}$ and $\mathbf{x}$ are both n^{th} order, $\mathbf{T}$ is $n \times n$ and for the ith element

$$x_i = t_{i1}z_1 + t_{i2}z_2 + \ldots t_{in}z_n$$

i.e. a linear combination of the z_i. Substituting this transformation in $\dot{\mathbf{x}} = \mathbf{Ax} + \mathbf{Bu}$ gives the new system equation

$$\mathbf{T\dot{z}} = \mathbf{ATz} + \mathbf{Bu}$$

i.e.

$$\dot{\mathbf{z}} = \mathbf{T}^{-1}\mathbf{ATz} + \mathbf{T}^{-1}\mathbf{Bu}. \tag{3.82}$$

Since we have used a different state vector which is a linear transformation of the previous one but the system itself is still the same we might expect that the dynamics of the new state variables have the same characteristics as the old state variables, i.e. the same eigenvalues. That this is so may be simply shown (e.g. Ogata, 1970). These eigenvalues are determined by $|\lambda\mathbf{I} - \mathbf{A}| = 0$ or in the new equation (3.82) by

$$|\lambda\mathbf{I} - \mathbf{T}^{-1}\mathbf{AT}| = 0.$$

Using the relation $\mathbf{T}^{-1}\mathbf{T} = \mathbf{I}$

$$\begin{aligned}
|\lambda\mathbf{I} - \mathbf{T}^{-1}\mathbf{AT}| &= |\lambda\mathbf{T}^{-1}\mathbf{T} - \mathbf{T}^{-1}\mathbf{AT}| \\
&= |\mathbf{T}^{-1}(\lambda\mathbf{T} - \mathbf{AT})| \\
&= |\mathbf{T}^{-1}(\lambda\mathbf{I} - \mathbf{A})\mathbf{T}| \\
&= |\mathbf{T}^{-1}| \, |\lambda\mathbf{I} - \mathbf{A}| \, |\mathbf{T}| = 0. \tag{3.83}
\end{aligned}$$

Thus, provided $\mathbf{T}$ is not singular, i.e. $|\mathbf{T}| \neq 0$, the eigenvalues of equation (3.83) are still given by $|\lambda\mathbf{I} - \mathbf{A}| = 0$, i.e. the eigenvalues of the system are invariant under the linear transformation.

Eigenvectors

For the homogeneous system, equation (3.80) and the example which followed that showed that each element of the transition matrix $e^{\mathbf{A}t}$ was comprised of the components $a_1 e^{\lambda_1 t}, \ldots, a_n e^{\lambda_n t}$. With a change in notation of the coefficients a_j to a subsequently more convenient and informative form each state variable x_i can be expressed as

$$x_1(t) = e_{11} e^{\lambda_1 t} + \ldots e_{1i} e^{\lambda_i t} + \ldots e_{1n} e^{\lambda_n t}$$

.
.
.

$$x_i(t) = e_{i1} e^{\lambda_1 t} + \ldots e_{ii} e^{\lambda_i t} + \ldots e_{in} e^{\lambda_n t}$$

.
.
.

$$x_n(t) = e_{n1} e^{\lambda_1 t} + \ldots e_{ni} e^{\lambda_i t} + \ldots e_{nn} e^{\lambda_n t}.$$

Each $x_i(t)$ contains, for systems without multiple eigenvalues, up to n terms in $e^{\lambda_1 t}, \ldots, e^{\lambda_n t}$. However, some of the e_{ij} may be zero.

Let $\mathbf{e}^{\lambda t}$ be defined as the vector $[e^{\lambda_1 t} \ldots e^{\lambda_i t} \ldots e^{\lambda_n t}]^T$ and $\mathbf{E}$ be defined as the matrix of coefficients e_{ij}. That is

$$\mathbf{E} = \begin{bmatrix} e_{11} & e_{12} & \cdots & e_{1n} \\ \cdot & & & \cdot \\ \cdot & & e_{ij} & \cdot \\ \cdot & & & \cdot \\ e_{n1} & & \cdots & e_{nn} \end{bmatrix}. \tag{3.84}$$

Then we may write this set of equations for the individual $x_i(t)$ by the shorter notation of the composite vector-matrix equation

$$\mathbf{x} = \mathbf{E}\mathbf{e}^{\lambda t}. \tag{3.85}$$

The pre-exponent coefficients in the $n \times n$ matrix $\mathbf{E}$ are determined by only n initial conditions of $\mathbf{x}(0)$. For the n^2 elements with n initial conditions there are n degrees of freedom and it is possible to express all solutions $\mathbf{x}$ in terms of n linearly independent groups. The groups chosen are the n groups which are the columns of matrix $\mathbf{E}$ and these are known as the eigenvectors, $\mathbf{e}$, i.e. write

$$\mathbf{E} = [\mathbf{e}_1 \mathbf{e}_2 \ldots \mathbf{e}_n] \tag{3.86}$$

where each eigenvector

$$\mathbf{e}_k = \begin{bmatrix} e_{1k} \\ \cdot \\ \cdot \\ \cdot \\ e_{ik} \\ \cdot \\ \cdot \\ \cdot \\ e_{nk} \end{bmatrix}$$

in the notation of equation (3.84).

Substituting equation (3.86) into (3.85) the state vector $\mathbf{x}$ may be written as

$$\mathbf{x} = [\mathbf{e}_1 \mathbf{e}_2 \dots \mathbf{e}_n] \begin{bmatrix} e^{\lambda_1 t} \\ \cdot \\ \cdot \\ \cdot \\ e^{\lambda_n t} \end{bmatrix}$$

$$= \mathbf{e}_1 e^{\lambda_1 t} + \mathbf{e}_2 e^{\lambda_2 t} + \dots \mathbf{e}_n e^{\lambda_n t}$$

$$= \sum_{i=1}^{n} \mathbf{e}_i e^{\lambda_i t}. \tag{3.87}$$

Because of its linearity each term of the full solution (3.87) satisfies the original system equation $\dot{\mathbf{x}} = \mathbf{A}\mathbf{x}$. Thus taking each term of equation (3.87) in turn and substituting it into the original (unforced) system equation yields

$$\lambda_i \mathbf{e}_i e^{\lambda_i t} = \mathbf{A} \mathbf{e}_i e^{\lambda_i t}$$

for each $i = 1, \dots n$. The term $e^{\lambda_i t}$ is a scalar so this equation may be reduced to

$$\lambda_i \mathbf{e}_i = \mathbf{A} \mathbf{e}_i$$

or

$$(\lambda_i \mathbf{I} - \mathbf{A})\mathbf{e}_i = \mathbf{0}. \tag{3.88}$$

This equation highlights the algebraic relationship between the eigenvector $\mathbf{e}_i$ and the specific eigenvalue λ_i. Each term in equation (3.87) is formed with one specific eigenvalue and its eigenvector and the term $\mathbf{e}_i e^{\lambda_i t}$ is the ith natural mode associated with the ith eigenvalue. Equation (3.87) is the modal, or spectral, form of the system dynamics and shows that the full dynamics may be represented as the sum of its modes, which are simpler forms of response.

This result may be expressed in the general terms that if λ_i is a solution of the characteristic equation $|\lambda \mathbf{I} - \mathbf{A}| = 0$, i.e. λ_i is an eigenvalue of $\mathbf{A}$, then corresponding to each distinct λ_i there exists a solution to the equation $\lambda_i \mathbf{x} = \mathbf{A}\mathbf{x}$, say $\mathbf{x}_i$ which is an eigenvector of $\mathbf{A}$.

This particular solution is denoted by e_i. This is also referred to as the right eigenvector of A. (The left eigenvector of A corresponding to λ_i is the corresponding eigenvector of A^T.) It can be seen from equation (3.88) that any e_i can be multiplied by a scalar and still satisfy the equation. Thus this relationship gives the ratios between the elements of e_i only. The full specifications of the eigenvectors requires the use of the boundary conditions. The eigenvector 'direction' may be specified in terms of unit eigenvectors. A scalar factor by which the unit eigenvectors is multiplied is then determined by the initial conditions. Let us illustrate this using the system of the previous example.

Example   For the homogeneous part of the previous example the system equation is

$$\begin{bmatrix} \dot{x}_1 \\ \dot{x}_2 \end{bmatrix} = \begin{bmatrix} -2 & 1 \\ 1 & -2 \end{bmatrix} \begin{bmatrix} x_1 \\ x_2 \end{bmatrix}$$

and the eigenvalues of A are $\lambda_1 = -1$, $\lambda_2 = -3$. Using equation (3.88) and each eigenvalue in turn

$$-1 . e_1 = A e_1$$

and

$$-3 e_2 = A e_2$$

where

$$e_1 = \begin{bmatrix} e_{11} \\ e_{21} \end{bmatrix} \quad \text{and} \quad e_2 = \begin{bmatrix} e_{12} \\ e_{22} \end{bmatrix}.$$

Expanding the first eigenvector equation (for $\lambda_1 = -1$) gives us

$$-\begin{bmatrix} e_{11} \\ e_{21} \end{bmatrix} = \begin{bmatrix} -2 & 1 \\ 1 & -2 \end{bmatrix} \begin{bmatrix} e_{11} \\ e_{21} \end{bmatrix}$$

or

$$-e_{11} = -2e_{11} + e_{21}$$
$$-e_{21} = e_{11} - 2e_{21}.$$

Both of these relationships give $e_{11} = e_{21}$, so that only the ratio between e_{11} and e_{21} is available and in this case we can replace e_{21} by e_{11}, i.e.

$$e_1 = \begin{bmatrix} e_{11} \\ e_{11} \end{bmatrix}.$$

The second eigenvalue, $\lambda_2 = -3$, yields

$$-3e_{12} = -2e_{12} + e_{22}$$

or

$$e_{22} = -e_{12}$$

i.e.

$$\mathbf{e}_{22} = \begin{bmatrix} e_{12} \\ -e_{12} \end{bmatrix}.$$

The two eigenvectors $\mathbf{e}_1$ and $\mathbf{e}_2$ may be expressed in unit eigenvectors $\bar{\mathbf{e}}_i$ and scalars,

$$\mathbf{e}_1 = \alpha_1 \begin{bmatrix} 1/\sqrt{2} \\ 1/\sqrt{2} \end{bmatrix}$$

$$= \alpha_1 \bar{\mathbf{e}}_1$$

and

$$\mathbf{e}_2 = \alpha_2 \begin{bmatrix} 1/\sqrt{2} \\ -1/\sqrt{2} \end{bmatrix}$$

$$= \alpha_2 \bar{\mathbf{e}}_2$$

i.e. the sum of the squares of all elements in a unit eigenvector is unity. This is called normalizing the eigenvectors. This two dimensional case is shown in Fig. 3.43. the scalars α_i, that is the magnitudes of the eigenvalues, are still unspecified by the information so far available. The solution of the dynamic equation is then, using equation (3.87)

$$\mathbf{x} = \alpha_1 \begin{bmatrix} 1/\sqrt{2} \\ 1/\sqrt{2} \end{bmatrix} e^{-t} + \alpha_2 \begin{bmatrix} 1/\sqrt{2} \\ -1/\sqrt{2} \end{bmatrix} e^{-3t},$$

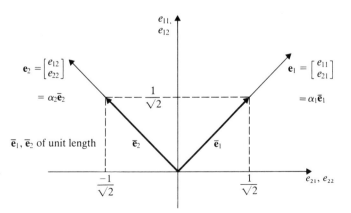

Fig. 3.43 Eigenvectors and unit eigenvectors

and α_1 and α_2 are determined by the initial conditions, say $\mathbf{x}(0) =$ $[1 \quad 2]^T$. Substituting $t = 0$ and $\mathbf{x} = \mathbf{x}(0)$ yields

$$\begin{bmatrix} 1 \\ 2 \end{bmatrix} = \alpha_1 \begin{bmatrix} 1/\sqrt{2} \\ 1/\sqrt{2} \end{bmatrix} + \alpha_2 \begin{bmatrix} 1/\sqrt{2} \\ -1/\sqrt{2} \end{bmatrix}$$

giving

$$1 = \alpha_1/\sqrt{2} + \alpha_2/\sqrt{2}$$

and

$$2 = \alpha_1/\sqrt{2} - \alpha_2/\sqrt{2}.$$

Hence

$$\alpha_1 = 3\sqrt{2}/2$$

$$\alpha_2 = -\sqrt{2}/2$$

and

$$\mathbf{x} = \frac{3}{\sqrt{2}} \begin{bmatrix} 1/\sqrt{2} \\ 1/\sqrt{2} \end{bmatrix} e^{-t} - \frac{1}{\sqrt{2}} \begin{bmatrix} 1/\sqrt{2} \\ -1/\sqrt{2} \end{bmatrix} e^{-3t}$$

or

$$\begin{bmatrix} x_1 \\ x_2 \end{bmatrix} = \begin{bmatrix} \frac{3}{2} \\ \frac{3}{2} \end{bmatrix} e^{-t} + \begin{bmatrix} -\frac{1}{2} \\ \frac{1}{2} \end{bmatrix} e^{-3t}$$

$$= \mathbf{e}_1 e^{\lambda_1 t} + \mathbf{e}_2 e^{\lambda_2 t}.$$

Example   A linear unforced time invariant system is described by three state variables and the equations

$$\dot{x}_1 = -2x_1$$
$$\dot{x}_2 = x_1 + x_2$$
$$\dot{x}_3 = -x_1 + x_2 - x_3$$

and starts from the initial position $\mathbf{x}(0)$. What is the transition matrix $e^{\mathbf{A}t}$ and is the system stable or unstable? The state equation is

$$\begin{bmatrix} \dot{x}_1 \\ \dot{x}_2 \\ \dot{x}_3 \end{bmatrix} = \begin{bmatrix} -2 & 0 & 0 \\ 1 & 1 & 0 \\ -1 & 1 & -1 \end{bmatrix} \begin{bmatrix} x_1 \\ x_2 \\ x_3 \end{bmatrix}.$$

The eigenvalues are obtained from $|\lambda\mathbf{I}-\mathbf{A}|=0$,

$$|\lambda\mathbf{I}-\mathbf{A}| = \begin{vmatrix} \lambda+2 & 0 & 0 \\ -1 & \lambda-1 & 0 \\ 1 & -1 & \lambda+1 \end{vmatrix}$$

$$= (\lambda+2)(\lambda-1)(\lambda+1)$$

i.e.

$\lambda_1 = -2$

$\lambda_2 = 1$

$\lambda_3 = -1$.

The presence of the real positive eigenvalues shows, as can be seen by inspection of the equations in this case, an 'unstable' system. The values of at least some of the state variables increasing continuously with time and without oscillation as all eigenvalues are real.

From the λ_i we may proceed to the transition matrix, $e^{\mathbf{A}t}$:

$$e^{\mathbf{A}t} = \mathscr{L}^{-1}[(s\mathbf{I}-\mathbf{A})^{-1}]$$

$$= \mathscr{L}^{-1}\left[\frac{adj[s\mathbf{I}-\mathbf{A}]}{|s\mathbf{I}-\mathbf{A}|}\right]$$

$$= \mathscr{L}^{-1}\left[\frac{1}{(s+2)(s-1)(s+1)}\right.$$

$$\times \left. \begin{bmatrix} (s-1)(s+1) & 0 & 0 \\ (s+1) & (s+2)(s+1) & 0 \\ (2-s) & (s+2) & (s+2)(s-1) \end{bmatrix} \right]$$

$$= \mathscr{L}^{-1}\begin{bmatrix} \dfrac{1}{s+2} & 0 & 0 \\ \dfrac{1}{(s+2)(s-1)} & \dfrac{1}{s-1} & 0 \\ \dfrac{2-s}{(s+2)(s-1)(s+1)} & \dfrac{1}{(s-1)(s+1)} & \dfrac{1}{s+1} \end{bmatrix}$$

$$= \begin{bmatrix} e^{-2t} & 0 & 0 \\ \frac{1}{3}(e^t-e^{-2t}) & e^t & 0 \\ \frac{4}{3}e^{-2t}+\frac{1}{6}e^t-\frac{3}{2}e^{-t} & \frac{1}{2}(e^t-e^{-t}) & e^{-t} \end{bmatrix}$$

and

$$\mathbf{x}(t) = e^{\mathbf{A}t}\mathbf{x}(0).$$

Note that instability does not mean that all state variables continue to increase. In this case x_1 tends to a finite value (zero) as time increases, but x_2 and x_3 continue to increase because of the positive exponential in time. This is possible in the given equations because the dynamics of the x_1 variable are independent of both x_2 and x_3.

Example   As a practical example of the use of eigenvalues reconsider the centrifuge in the example of Fig. 3.41, the dynamic equations of which may be solved via the eigenvectors. The state equations are, with $K = 2.5$ Nms,

$$\dot{x}_1 = x_2$$
$$\dot{x}_2 = -0.025x_2.$$

The system matrix **A** is $\begin{bmatrix} 0 & 1 \\ 0 & -0.025 \end{bmatrix}$ with eigenvalues 0, -0.025.
Taking each eigenvalue in turn leads to the eigenvectors. In this case the unit eigenvectors will not be used but each element is expressed in proportion to the first element in the vector, which is set to one. A scalar is then used with each vector and this is determined by the initial conditions.

For $\lambda_1 = 0$:

$$0 \cdot \begin{bmatrix} e_{11} \\ e_{21} \end{bmatrix} = \begin{bmatrix} 0 & 1 \\ 0 & -0.025 \end{bmatrix} \begin{bmatrix} e_{11} \\ e_{21} \end{bmatrix}$$

giving $e_{21} = 0$ and an indeterminate e_{11} so

$$\mathbf{e}_1 = \begin{bmatrix} 1 \\ 0 \end{bmatrix}.$$

For $\lambda_2 = -0.025$:

$$-0.025 \begin{bmatrix} e_{12} \\ e_{22} \end{bmatrix} = \begin{bmatrix} 0 & 1 \\ 0 & -0.025 \end{bmatrix} \begin{bmatrix} e_{12} \\ e_{22} \end{bmatrix}$$

giving $-0.025e_{12} = e_{22}$ and thus

$$\mathbf{e}_2 = \begin{bmatrix} 1 \\ -0.025 \end{bmatrix}.$$

The state is thus given by

$$\mathbf{x} = \alpha \mathbf{e}_1 e^{\lambda_1 t} + \beta \mathbf{e}_2 e^{\lambda_2 t}$$

$$= \alpha \begin{bmatrix} 1 \\ 0 \end{bmatrix} + \beta \begin{bmatrix} 1 \\ -0.025 \end{bmatrix} e^{-0.025t}. \tag{3.89}$$

The initial conditions at $t = 0$ are $x_1 = 0$, $x_2 = 20\pi$ so

$$\alpha + \beta = 0$$
$$-0 \cdot 025\beta = 20\pi$$

and $\alpha = -\beta = 800\pi$ to give from (3.89) the equations

$$\theta = x_1 = 800\pi(1 - e^{-0 \cdot 025t})$$
$$\omega = x_2 = 20\pi e^{-0 \cdot 025t}.$$

Note that although we have $\lambda = 0$ for one eigenvalue the same procedure may be used to determine the eigenvectors.

Diagonalization of a matrix

Above it was seen that under linear transformation the eigenvalues of the system remained unchanged. How can we make best use of this transformation by a suitable choice of the matrix $\mathbf{T}$?

If $\mathbf{T}$ is put equal to $\mathbf{E}$, which is the matrix whose columns are the eigenvectors of $\mathbf{A}$, then equation (3.82) becomes

$$\dot{\mathbf{z}} = \mathbf{E}^{-1}\mathbf{A}\mathbf{E}\mathbf{z} + \mathbf{E}^{-1}\mathbf{B}\mathbf{u}. \tag{3.90}$$

Defining a diagonal matrix $\mathbf{\Lambda}$ whose elements are the eigenvalues λ_i,

$$\mathbf{\Lambda} = \begin{bmatrix} \lambda_1 & & 0 \\ & \cdot & \\ & & \cdot \\ 0 & & \lambda_n \end{bmatrix}$$

then

$$\mathbf{E}\mathbf{\Lambda} = \begin{bmatrix} \lambda_1 e_{11} & \lambda_2 e_{12} \dots & \lambda_n e_{1n} \\ \lambda_1 e_{21} & \lambda_2 e_{22} \dots & \lambda_n e_{2n} \\ \cdot & & \\ \cdot & & \\ \cdot & & \\ \lambda_1 e_{n1} & \lambda_2 e_{n2} \dots & \lambda_n e_{nn} \end{bmatrix}.$$

Rewrite equation (3.88) for all i,

$$\lambda_1 \mathbf{e}_1 = \mathbf{A}\mathbf{e}_1$$
$$\cdot$$
$$\cdot$$
$$\cdot$$
$$\lambda_n \mathbf{e}_n = \mathbf{A}\mathbf{e}_n$$

and these terms may be rewritten as the columns of the two matrices

$$[\lambda_1 \mathbf{e}_1 \quad \lambda_2 \mathbf{e}_2 \dots \quad \lambda_n \mathbf{e}_n] = [\mathbf{A}\mathbf{e}_1 \quad \mathbf{A}\mathbf{e}_2 \dots \quad \mathbf{A}\mathbf{e}_n]. \tag{3.91}$$

The left-hand side of equation (3.91) is

$$[\mathbf{e}_1 \quad \mathbf{e}_2 \ldots \mathbf{e}_n] \begin{bmatrix} \lambda_1 & & & \\ & \lambda_2 & & \\ & & \ddots & \\ & & & \lambda_n \end{bmatrix}$$

$$= \mathbf{E}\mathbf{\Lambda}.$$

The right-hand side of equation (3.91) is

$$\mathbf{A}[\mathbf{e}_1 \quad \mathbf{e}_2 \ldots \mathbf{e}_n]$$

$$= \mathbf{A}\mathbf{E}.$$

Thus

$$\mathbf{E}\mathbf{\Lambda} = \mathbf{A}\mathbf{E}$$

or

$$\mathbf{E}^{-1}\mathbf{A}\mathbf{E} = \mathbf{\Lambda} \tag{3.92}$$

so that the transformation in equation (3.90) can be written

$$\dot{\mathbf{z}} = \mathbf{\Lambda}\mathbf{z} + \mathbf{E}^{-1}\mathbf{B}\mathbf{u} \tag{3.93}$$

by using the matrix $\mathbf{E}$ of eigenvectors for the transformation matrix $\mathbf{T}$.

The reason for this choice of transformation is now apparent. Because $\mathbf{\Lambda}$ is diagonal, for each new state variable z_i the dynamics of the system variable, $\dot{z}_i$, are expressed solely in terms of z_i and the ith eigenvalue plus the contribution from the forcing vector $\mathbf{u}$, i.e.

$$\dot{z}_i = \lambda_i z_i + \text{contribution from } \mathbf{E}^{-1}\mathbf{B}\mathbf{u}.$$

This transformation is the 'canonical transformation' into canonical co-ordinates $\mathbf{z}$.

If there are multiple eigenvalues then a diagonal form is not possible and a Jordan canonical form containing off-diagonal elements is formed (Birkhoff and MacLane 1941). This transformation will be returned to when considering observability and controllability of a dynamic system.

3.5 Summary

The equations of dynamic systems may themselves be represented in block diagram and signal flow graphical form and both forms are used extensively for input–output representation of system behaviour. These aids may be used also for the early stages of analysis by state space methods but their significance and use is not so direct. The input–output representation has held sway for many years, particularly in process systems and the Laplace transformation plays a dominant part in the literature of systems handled in this way. Input–output responses are generally observed by forcing the system with some discontinuity like a pulse or step change, although in practice it is necessary to use a combination of sharp ramps. Alternatively continuous sinusoidal signals may be used to give the powerful frequency response methods of analysis.

Systems of higher orders or with many variables are represented more conveniently by the state space methods and by the solution of equations in the time domain. Although the general equations may be formed for nonlinear and time-varying systems, linearization is usual and more complete solutions are obtainable in an analytical and closed form for the linear time-invariant systems. These will form the basis of the bulk of the control studies which follow. (Although continuous time studies have been considered here, computational methods frequently require a discrete time representation. System equations of this form will be considered separately.)

The system dynamics discussed so far are general in that they are independent of the structure of the system. In the following chapters on control the structure of the system will become of importance, especially in the presence of feedback and there will be further consideration of stability. However, most of the analytical requirements have been covered in Chapters 1–3 and Chapters 4 onwards on control will assume a knowledge of these chapters.

Part 2

Control – A-Linear continuous input–output control systems

Chapter 4
Single input–single output feedback control: time domain

4.1 Feedback

In Chapter 3 we looked at the representation of general continuous time system dynamics both in terms of the input–output relation and using state variable notation. This treatment was general in the sense that the structure of the system was not in itself important in our considerations. In feedback control systems an important concept is that the system input is influenced in some way by the system response to that or some other input, i.e. information is fed back from the output to an input and that input is thereby modified. This 'feedback loop' forms an important structural part of the overall system, e.g. its presence can make unstable systems stable, or cause instability in basicly stable systems if incorrectly used and designed. It also reduces the effects of both external and internal system disturbances, e.g. variations in voltages, flowrates. The feedback loop was introduced in sections 1.1–1.3 when it was stated that most of our attention would be focussed on the regulator problem where we are concerned with maintaining, in the presence of disturbances, a desired steady output. The difference between the desired output and the actual output is the error so that the purpose of control is to minimize this in some way.

In this chapter we shall look at the single input–single output feedback theory (the 'classical' approach) and utilize and develop the general theory of Chapter 3 for such systems. Despite the rapid growth of the state space representation and the increased computing power available a study on the basis of this chapter still has considerable benefit and much detailed presentation of this work is available (e.g. D'Azzo and Houpis 1960).

By single input–single output systems we mean that large number of cases where an attempt is made to influence one of the system outputs by changing one input. The dynamic relationship between the input and output will be as a first, second or higher order differential equation. Thus although we consider only the input–output relation, i.e. the dependence of one variable on the input, the system if expressed in the state variable form would normally contain more than one state variable. These other variables are not explicity determined, or their dynamics studied, by the classical approach which is largely based on experimental methods.

The feedback or closed loop control, when compared with open loop behaviour of a system, is less dependent on the system component characteristic, less sensitive to load disturbances (which it may be specifically required to deal with) and enables the system in many cases to be considerably modified without a subsequent major change in the controller. Open loop controllers generally require much more information about the system's internal structure or behaviour.

The general output–feedback control system structure is shown in Fig. 4.1. In this chapter we are concerned with the relation between a

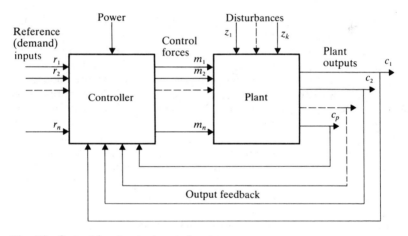

Fig. 4.1　General input–output control system

single reference input and a single output. With this condition Fig. 4.1 may be reduced but drawn to show all the basic elements for the simple feedback control system, e.g. Fig. 4.2.

We know from Chapter 3 that additional manipulation of the blocks and the information they contain and the use of the Laplace transform enable us to reduce the general representation to that of Fig. 4.3, where the feedback element in them may frequently be adequately represented as unity. For normal operation we have 'negative feedback', the error being the difference between the reference and the output (Fig. 4.3).

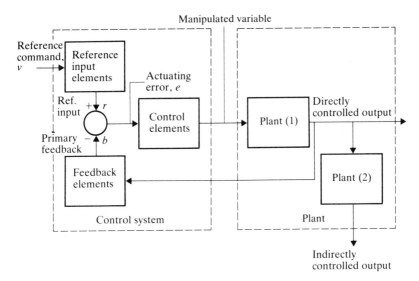

Fig. 4.2 Basic elements in feedback control – the various elements operate on and change the form of the signals shown

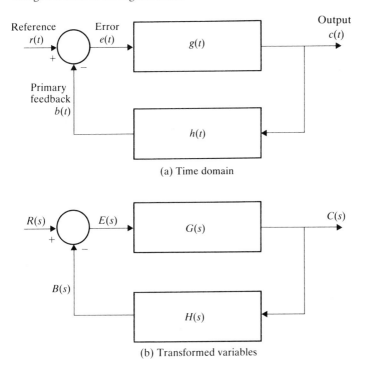

Fig. 4.3 General block diagram for study of feedback systems

4.2 Closed loop responses

The closed loop transfer function

The Laplace transform notation will be used principally and the closed loop transfer function is established for the condensed representation of Fig. 4.3.

$R(s)$ is the Laplace transform of the reference variable
$C(s)$ is the Laplace transform of the controlled output
$E(s)$ is the Laplace transform of the system error signal
$B(s)$ is the Laplace transform of the primary feedback variable
$G(s)$ is the forward path transfer function
$H(s)$ is the feedback path transfer function

We have seen that in a linear system the effects of additional inputs may be handled by the principle of superposition so that by concentrating on the form of Fig. 4.3 we are not eliminating any generalities for linear systems. The transfer functions $G(s)$ and $H(s)$ will usually be a product of a gain term K, which contains the dimensional and steady state magnitude aspects and a non-dimensional transfer function expressing the dynamic nature of the system.

Using the basic transfer function relations and Fig. 4.3 gives,

$$C(s) = G(s)E(s) \tag{4.1}$$

$$E(s) = R(s) - B(s)$$
$$= R(s) - H(s)C(s) \tag{4.2}$$

Combining these two equations to eliminate $E(s)$ we obtain the closed loop transfer function

$$\frac{C(s)}{R(s)} = \frac{G(s)}{1 + G(s)H(s)}. \tag{4.3}$$

With unity feedback, $H(s) = 1$ the error signal is

$$E(s) = R(s) - C(s)$$

and the closed loop transfer function reduces to

$$\frac{C(s)}{R(s)} = \frac{G(s)}{1 + G(s)}. \tag{4.4}$$

Because of the block diagram algebra available to us it is always possible and frequently most convenient to base our arguments around equation (4.4) rather than (4.3).

The 'characteristic equation' of the general feedback system is obtained from the denominator of the closed loop transfer function and is

$$1 + G(s)H(s) = 0. \tag{4.5}$$

It will be shown that this determines the stability and response of the closed loop system and some indication of this has already been given in section 3.3.

In addition to the closed loop transfer function it is usual to refer to the forward transfer function as the ratio

$$\frac{C(s)}{E(s)} = G(s)$$

and the open loop transfer function as the ratio

$$\frac{B(s)}{E(s)} = G(s)H(s).$$

For unity feedback the open loop and forward transfer function become the same function, $G(s)$.

From equation (4.3) it is seen that the order of the system (highest power of s in the transfer function) depends on the product $G(s)H(s)$ or solely on $G(s)$ for unity feedback systems. The physical nature of processes is such that the order of the denominator of the transfer function is usually equal to or greater than the order of the numerator. The order of the system is unchanged by the introduction of unity feedback but increased powers of s in $H(s)$ add to the order of the system and change consequently the nature of the system considerably.

The transient and frequency responses of first and second order systems with unity feedback will therefore follow the general first and second order system results given in sections 3.2 and 3.3. The general form remains unchanged but the natural frequency and damping are affected. Consider, for example, the system of Fig. 4.4, with and without feedback.

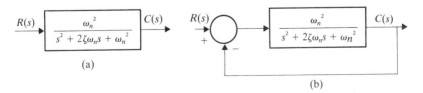

Fig. 4.4 Second order system (a) without and (b) with feedback

Without feedback

$$\frac{C(s)}{R(s)} = G(s)$$

$$= \frac{\omega_n^2}{s^2 + 2\zeta\omega_n s + \omega_n^2}$$

and with feedback

$$\frac{C(s)}{R(s)} = \frac{G(s)}{1 + G(s)}$$

$$= \frac{\omega_n^2}{s^2 + 2\zeta\omega_n s + 2\omega_n^2}$$

$$= \frac{\omega_n^2}{s^2 + 2\zeta'\omega_n' s + \omega_n'^2}.$$

Note that the steady state gain of this sytem ($t \to \infty$) has been halved, the new natural frequency ω_n' is $\sqrt{2}\omega_n$ and the new damping coefficient is given by $\zeta'\omega_n'$ equal to $\zeta\omega_n$, i.e. ζ' is $\zeta/\sqrt{2}$. Thus the dynamics have been speeded up, as shown by the new natural frequency and the damping decreased. However, the general form of the equation for $C(s)/R(s)$ is unchanged by the addition of unity feedback. Hence we can describe unity feedback systems in terms of the transient and frequency responses of Chapter 3. Similarly the time domain specifications, rise time, overshoot etc. are defined as for the general system and so are the frequency response characteristics. However, the interpretation of these specifications and values have added significance for the feedback system and the difference between the input and output, i.e. the error, is now a definite measure of feedback system performance.

Steady state errors

In general the object of a feedback control whether of the regulator or servo type, is to remove any difference between the desired output as given by the reference input and the actual output, i.e. to remove the error. Thus we can consider performance in terms of this error. In particular, whether or not an error exists will depend on the system type and the form of input to the system.

Consider the unity feedback system (Fig. 4.5). The transfer function $G(s)$ may be expressed as the ratio of two polynomials in s (equation 3.34)

$$G(s) = \frac{C(s)}{E(s)}$$

$$= \frac{k_n(1 + a_1 s + \dots a_v s^v)}{s^n(1 + b_1 s + \dots b_u s^u)}. \tag{4.6}$$

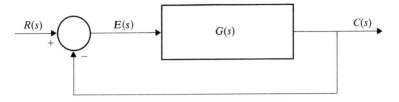

Fig. 4.5 Unity feedback control system

k_n is the overall gain of $G(s)$, a_i, b_j are constants and the value of n determines the system 'type'; zero for type 0 system, unity for type 1 system etc. As mentioned above, due to the energy storage or dissipative nature of physical systems, the powers of s in the denominator are higher than in the numerator and poles at the origin, i.e. factors of s^{-1}, are more common than zeros at the origin, factors of s.

Rearranging equation (4.6) gives

$$
\begin{aligned}
C(s)s^n &= \frac{k_n(1+a_1s+\ldots a_v s^v)}{(1+b_1s+\ldots b_u s^u)} E(s) \\
&= \mathcal{L}\left[\frac{d^n c(t)}{dt^n}\right].
\end{aligned}
\tag{4.7}
$$

Using this relationship we can determine the value of the nth derivative of the output for a given actuating signal error $e(t)$ and in particular we can find the steady state value of this derivative by using the final value theorem

$$
\lim_{t\to\infty} f(t) = \lim_{s\to 0} sF(s).
$$

Using the final value theorem with equation (4.7)

$$
\begin{aligned}
\left.\frac{d^n c(t)}{dt^n}\right|_{\text{s.s}} &= \lim_{s\to 0} sC(s)s^n \\
&= \lim_{s\to 0}\left[\frac{sk_n(1+a_1s+\ldots a_v s^v)}{(1+b_1s+\ldots b_u s^u)} \cdot E(s)\right].
\end{aligned}
$$

If the actuating signal is a constant E so that $E(s) = E/s$ then

$$
\begin{aligned}
\left.\frac{d^n c(t)}{dt^n}\right|_{\text{s.s.}} &= \lim_{s\to 0}\left[\frac{sk_n(1+a_1s+\ldots a_v s^v)}{(1+b_1s+\ldots b_u s^u)} \cdot \frac{E}{s}\right] \\
&= k_n E.
\end{aligned}
\tag{4.8}
$$

Bearing in mind that the steady state output of a closed loop linear system with feedback will be of the same form as the input although different in magnitude and phase (i.e. will lag the input) we can see how the steady state error is related to the system type, i.e. to the value of n, and to the actuating signal.

System type 0, $n = 0$

If n is zero the nth derivative is the steady state itself so that for a constant actuating signal E

$$\left. \frac{d^0 c(t)}{dt^0} \right|_{\text{s.s.}} = c(t)_{\text{s.s}}$$

$$= k_0 E. \tag{4.9}$$

Also

$$e(t)_{\text{s.s}} = r(t)_{\text{s.s}} - c(t)_{\text{s.s}}$$

$$= E. \tag{4.10}$$

Thus to produce a constant output $c(t)_{\text{s.s}}$ there must be a constant signal E, equation (4.9). If $c(t)$ and $e(t)$ are constant then from equation (4.10) this is only so if the reference input $r(t)$ is also constant, say R. Thus a constant value input R produces a steady state constant value output C but in the type 0 system there must exist a constant error E between them, given by C/k_0.

System type 1, $n = 1$

Equation (4.8) now yields

$$\left. \frac{dc(t)}{dt} \right|_{\text{s.s}} = k_1 E \tag{4.11}$$

for a constant actuating signal E. But if the reference input is constant, the steady state output will become constant, i.e. $dc/dt = 0$, so that to satisfy equation (4.11) the steady state error E must be zero. For a type 1 system the steady state output will follow a constant input with zero steady state error. Now integration of equation (4.11) shows

$$e(t)_{\text{s.s}} = k_1 Et + C_1$$

where C_1 is a constant of integration. At the steady state with a ramp output of this form

$$e(t)_{\text{s.s}} = r(t)_{\text{s.s}} - c(t)_{\text{s.s}}$$

$$= r(t)_{\text{s.s}} - k_1 Et - C_1$$

$$= E$$

so that $r(t)_{\text{s.s}}$ must also be a ramp, $k_1 Et + (C_1 + E)$. For a type 1 system a ramp input produces a ramp output but there is a constant error, E between them.

System type 2, n = 2

This analysis can be extended indefinitely on the above pattern but we shall stop at the type 2 system for which equation (4.8) shows

$$\left.\frac{d^2c(t)}{dt^2}\right|_{s.s} = k_2 E. \tag{4.12}$$

Integration twice yields, remembering our stipulation of constant actuating signal E,

$$c(t)_{s.s} = k_2 E\frac{t^2}{2} + C_2 t + C_0$$

and

$$e(t)_{s.s} = r(t)_{s.s} - k_2 E\frac{t^2}{2} - C_2 t - C_0$$
$$= E$$

with C_2 and C_0 constants of integration. With either a constant (step) or a ramp input, the output has the same form as the input and in both cases $d^2c(t)/dt^2$ is zero, i.e. E is zero. Thus for a type 2 system both step and ramp inputs will be followed in the steady state by the same form of output and with zero steady state error. If the input is parabolic, $r(t) = R_2 t^2/2$, substitution in the last equation shows that $R_2 = k_2 E$ and C_2 is zero. Thus in a type 2 system this parabolic input produces a steady state output of the same form but there is a constant error E between them.

Where a steady state error does exist it is reduced by increasing the gain, k_0, k_1, or k_2 as the case may be. The initial form of response depends on the individual system but the step and ramp responses at the longer times are shown in Fig. 4.6.

Static error coefficients

The conclusions on steady state errors arrived at above may be obtained from a more general consideration of static error coefficients. These give a measure of a system's steady state accuracy for a given desired output that is constant or only slowly varying. The following definitions are independent of system type (for the special cases substitution is made in the general case) and are defined for specific forms of the input, $r(t)$.

Static step (position) error coefficient, K_p, is defined by

$$K_p = \frac{\text{steady state output}}{\text{steady state actuating signal}}$$

$$= \frac{c(t)_{s.s}}{e(t)_{s.s}}$$

TYPE '0'

TYPE '1'

TYPE '2'

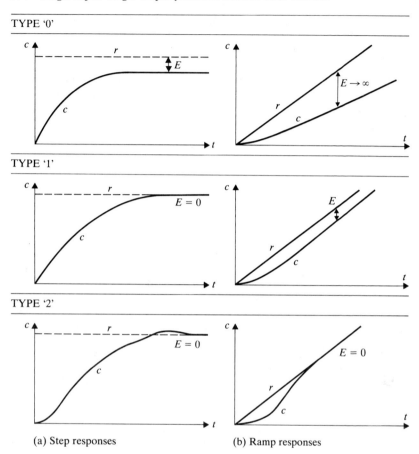

(a) Step responses   (b) Ramp responses

Fig. 4.6 'Steady state' step and ramp responses for stable type 0, 1, 2 systems (actual transient depends on $G(s)$)

and applies only for a step input $r(t) = Ru(t)$. With a step input the steady state output must also be constant and the error will also be constant, E. Using the final value theorem again

$$c(t)_{s.s} = \lim_{s \to 0} sG(s)E(s)$$

$$= \lim_{s \to 0} sG(s)\frac{E}{s}$$

and K_p becomes

$$K_p = \frac{1}{E} \cdot \lim_{s \to 0} G(s)E$$

$$= \lim_{s \to 0} G(s).$$

For a type 0 system K_p will be finite as there is no factor of s^{-1} in $G(s)$, but for type 1 systems and above K_p is infinite. This means that except for type 0 systems the steady state output follows the step input with zero error. This is as deduced above.

Static ramp (velocity) error coefficient K_v is defined as

$$K_v = \frac{\text{steady state derivative of output}}{\text{steady state actuating signal}}$$

$$= \frac{[dc(t)/dt]_{\text{s.s}}}{e(t)_{\text{s.s}}}$$

and applies only for a ramp input, $r(t) = Rt$.

Now

$$\left.\frac{dc(t)}{dt}\right|_{\text{s.s}} = \lim_{s \to 0} s(sG(s)E(s))$$

and with a constant actuating error E, $E(s) = E/s$ and

$$\left.\frac{dc(t)}{dt}\right|_{\text{s.s}} = \lim_{s \to 0} sG(s)E$$

and

$$K_v = \lim_{s \to 0} sG(s).$$

Thus for a type 0 system K_v will be zero, it will be finite for a type 1 system and infinite for type 2 systems and above.

Static parabolic (acceleration) error coefficient K_a is defined as

$$K_a = \frac{\text{steady state second derivative of output}}{\text{steady state actuating signal}}$$

$$= \frac{[d^2c(t)/dt^2]_{\text{s.s}}}{e(t)_{\text{s.s}}}$$

and applies only for a parabolic input $r(t) = Rt^2$. Using the final value theorem again with the second derivative

$$\left.\frac{d^2c(t)}{dt^2}\right|_{\text{s.s}} = \lim_{s \to 0} s(s^2G(s)E(s))$$

and with a constant actuating signal E

$$\left.\frac{d^2c(t)}{dt^2}\right|_{\text{s.s}} = \lim_{s \to 0} s^2G(s)E$$

so that

$$K_a = \lim_{s \to 0} s^2G(s).$$

Now for type 0 and type 1 systems K_a must be zero, it is finite for type 1 and infinite for higher type systems. This is seen by consideration of the factors of s in $G(s)$ for each type of system.

The state error coefficients are summarized in Tables 4.1 and 4.2. They are static or steady state coefficients since they apply after any initial transients have died out and indicate the ability of the various systems to eliminate steady state errors. Although higher type systems have greater ability to reduce the steady state errors which result from the increasing order form in the input, the addition of integral terms to a system can be significant also in reducing system stability.

Table 4.1 Static error coefficients

Error coefficient	Definition	Value of coefficient	Form of input signal
K_p	$[c(t)/e(t)]_{s.s}$	$\lim_{s \to 0} G(s)$	$R \cdot u(t)$, step
K_v	$\left[\dfrac{(dc(t)/dt)}{e(t)} \right]_{s.s}$	$\lim_{s \to 0} sG(s)$	$R \cdot t$, ramp
K_a	$\left[\dfrac{(d^2c(t)/dt^2)}{e(t)} \right]_{s.s}$	$\lim_{s \to 0} s^2G(s)$	$R \cdot t^2$, parabolic

Table 4.2 Relationship between system type and error coefficients

System type	Open loop gain constant	Step error coefficient, K_p	Ramp error Coefficient, K_v	Parabolic error coefficient, K_a
0	k_0	k_0	0	0
1	k_1	∞	k_1	0
2	k_2	∞	∞	k_2

The open loop gain constants k_0, k_1, k_2 are given by equation (4.6)

Linear control actions

The most basic control action is proportional action. The actuating signal m applied to the plant is directly proportional to the error signal e so that, referring to Fig. 4.7.

$$m(t) = k_p e(t)$$

or

$$M(s) = k_p E(s).$$

The transfer function of this control action, $M(s)/E(s)$, is thus the proportional constant k_p or 'proportional gain' of the controller. Note that the control action is a function of the error signal. Although proportional action goes a long way to satisfying a natural intuition as to what a controller should be it is usually insufficient on its own. Both

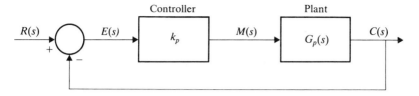

Fig. 4.7 Proportional control action

the presence of the steady state error which arises and the dynamics in terms of speed of response and overshoot may be unacceptable or incompatible for a stipulated overall plant performance and additional compensatory control actions are introduced. The three basic additional modes of control are integral, derivative and velocity feedback and these are used normally in conjunction with the proportional action.

For a type 0 system it was seen above that there is a continual difference between a constant reference input and the output. This 'offset' is a characteristic of proportional control systems. The actual output may be referred to as the control point. For the system of Fig. 4.7 the closed loop transfer function is

$$\frac{C(s)}{R(s)} = \frac{k_p G_p(s)}{1 + k_p G_p(s)} \tag{4.13}$$

and

$$E(s) = \frac{R(s)}{1 + k_p G_p(s)}.$$

For a unit step input, $R(s) = 1/s$ and the steady state error is by way of the final value theorem

$$e(t)_{s.s} = \lim_{s \to 0} sE(s)$$

$$= \lim_{s \to 0} \left[s \cdot \frac{1}{s(1 + k_p G_p(s))} \right].$$

If $G_p(s)$ is a type 0 transfer function, $k_0 \dfrac{(1 + a_1 s + \dots a_v s^v)}{(1 + b_1 s + \dots b_u s^u)}$ then

$$e(t)_{s.s} = \frac{1}{1 + k_p k_0}.$$

This steady state offset is reduced by increasing the controller gain k_p since k_0 may be a fixed parameter of the plant. The gain k_p is limited, however, since it affects also the dynamics of the system, equation (4.13), and the range may be limited, in particular normally at the upper level, by stability considerations. To eliminate this offset when the reference input is constant, i.e. under constant desired output conditions, integral action may be used in place of, or in combination with, proportional action.

If integral action is added to proportional action then the block diagram takes on the form of Fig. 4.8 although of course the controller action may be reduced to a single block with the function $k_p + k_i/s$.

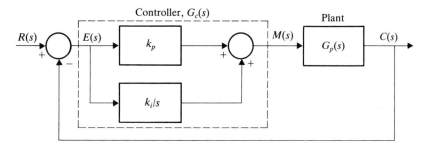

Fig. 4.8 Proportional plus integral control action

The closed loop transfer function is now

$$\frac{C(s)}{R(s)} = \frac{G_c(s)G_p(s)}{1 + G_c(s)G_p(s)} \qquad (4.14)$$

where the controller transfer function $M(s)/E(s)$ is

$$G_c(s) = k_p + \frac{k_i}{s}$$

$$= \frac{k_p s + k_i}{s}.$$

For a unit step input, $R(s) = 1/s$ and

$$e(t)_{s.s} = \lim_{s \to 0} \left[s \cdot \frac{1}{s} \cdot \frac{1}{1 + G_c(s)G_p(s)} \right]$$

$$= \lim_{s \to 0} \frac{s}{s + (k_p s + k_i)G_p(s)}$$

$$= 0.$$

i.e. the offset is eliminated. If integral action is used alone this is still true, i.e. if k_p is zero, and the closed loop transfer function is

$$\frac{C(s)}{R(s)} = \frac{k_i G_p(s)/s}{1 + k_i G_p(s)/s}. \qquad (4.15)$$

Note that with integral action the order of the system equation and of the characteristic equation $1 + G_c(s)G_p(s) = 0$ is increased. Whereas first and second order negative feedback systems are nearly always stable, third and higher order systems are less likely to be, so that the introduction of integral action also increases the possibility of instability. This may need to be counteracted by a reduced value of proportional gain. This will be illustrated when we consider empirical values of controller settings.

The effect of integral action is to give a controller output signal $M(s)$ which changes at a rate proportional to the error, so that for the integral action component,

$$\frac{dm(t)}{dt} = k_i e(t)$$

or

$$M(s) = \frac{k_i E(s)}{s}$$

or

$$m(t) = \int_0^t k_i e(t)\, dt.$$

For a step error signal, E, the controller action starts from zero and continues to increase because of the integral action but the component due to the proportion action remains constant. At some time these are equal so that

$$k_p E = k_i \int_0^t E\, dt$$

$$= k_i E t.$$

The time at which the two contributions are equal is thus

$$t = \frac{k_p}{k_i}$$

$$= T_i.$$

This T_i is the integral action time. Integral action is also known as 'reset' because of its capability of eliminating the offset in the ouput.

The second mode used with proportional control is derivative action. The error signal is now acted upon to produce an output $k_p + k_d \cdot de/dt$, Fig. 4.9. This action may be considered as anticipatory, noting the instantaneous trend of the error and applying correcting action ahead of the error. It acts so as to reduce overshooting in an oscillatory response and thus reduces the time to come within chosen limits near to the desired steady state value.

The closed loop transfer function is

$$\frac{C(s)}{R(s)} = \frac{(k_p + k_d s)G_p(s)}{1 + (k_p + k_d s)G_p(s)}. \tag{4.16}$$

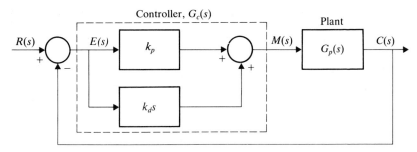

Fig. 4.9 Proportional plus derivative control action

As with integral action the closed loop transfer function is modified and the characteristic equation will have changed roots and hence the dynamics are changed. Unlike integral action derivative action cannot be used alone since it produces a control action only when the error is changing and takes no action to act against steady errors whatever their magnitude. In the presence of rapidly changing errors ('noise') then the output from the derivative control will change rapidly and may tend to swing from one extreme of its range to the other.

Similarly to that for integral action we may defined a derivative action time, T_d. This is the time required, in the presence of an error signal changing at a constant rate, for the proportional action component to become equal to the derivative action component. Thus at some time, t

$$k_p e = k_d \frac{de}{dt}.$$

If de/dt is constant, and $e(0)$ is zero, then $e = \left(\frac{de}{dt}\right) . t$ so that

$$k_p t = k_d$$

and

$$t = \frac{k_d}{k_p}$$
$$= T_d.$$

All three actions may be included in one controller and conventional process controllers normally are provided with the three actions. Then

$$m = k_p e + k_i \int e \, dt + k_d \frac{de}{dt}$$

or

$$m = k_p \left(e + \frac{1}{T_i} \int e \, dt + T_d \frac{de}{dt}\right). \tag{4.17}$$

The controller transfer function is

$$G_c(s) = \frac{M(s)}{E(s)}$$

$$= k_p\left(1 + \frac{1}{T_i s} + T_d s\right). \tag{4.18}$$

This proportional plus integral plus derivative action is usually referred to as P.I.D. control.

An effect similar to that of derivative action is obtainable by rate feedback in conjunction with direct feedback. Figure 4.10 shows rate feedback with proportional action. Because of the inner loop the system appears slightly more involved but the closed loop transfer function is shown by algebra or block diagram reduction to be

$$\frac{C(s)}{R(s)} = \frac{k_p G_p(s)}{1 + (k_p + s k_t) G_p(s)}. \tag{4.19}$$

The effect is to reduce overshoot as with derivative action but the derivative action also gives a reduction in rise time on account of the additional zero in the transfer function. This zero is absent with rate feedback.

These combinations of controller actions and their effects on overall system dynamics will become apparent in the following sections. Equation (4.14) shows that they appear in the closed loop transfer functions and may affect poles and zeros. In addition to these fundamental forms of control action the controller may contain other functions designed to specifically change the characteristics of the dynamic system. These 'phase lag' and 'phase lead' compensators are in a sense more general combinations of the basic actions considered and are discussed in section 4.6. The general form of the controller transfer function is then

$$\frac{M(s)}{E(s)} = \frac{1 + \alpha s}{1 + \beta s}$$

where the relative values of α and β govern the use of the controller.

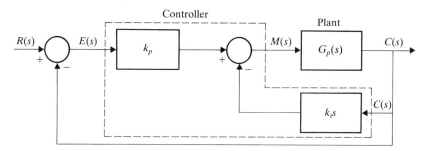

Fig. 4.10 Control with rate (velocity, tachometer) feedback

Generation by physical means of the desired control action is not always straightforward although good reproduction of the desired form is now generally possible. In particular derivative action may be difficult to generate precisely.

Transient responses

In section 3.3 and above we saw that the response of the system is governed by the factors of its transfer function and in particular by the factors of its denominator, its poles. For the closed loop system the denominator has the form $1 + G(s)H(s)$ and equating this to zero gives the characteristic equation

$$1 + G(s)H(s) = 0. \tag{4.20}$$

The roots of s in this equation are the closed loop poles of the system.

Section 3.2 dealt with the transient response of first and second order systems and since the overall transfer function was being used this is applicable to systems both with and without feedback. The main features, rise time etc., are still relevant to the feedback systems. We can extend the transient response work, however, to illustrate the effect of the fundamental control modes described above and this is best done using a second order system.

Let

$$G_p(s) = \frac{\omega_n^2}{s^2 + 2\zeta\omega_n s + \omega_n^2}$$

be the plant transfer function. Let us consider in turn the effect of (i) unity feedback with proportional control, (ii) proportional plus integral control, and (iii) proportional plus derivative action. (Note that the chosen $G_p(s)$ could itself represent a unity feedback system with a forward transfer function $\omega_n^2/s(s + 2\zeta\omega_n)$.)

Let us also chose by way of example $\zeta = 0·5$, i.e. underdamped, and $\omega_n = 1$ so that without feedback

$$\frac{C(s)}{R(s)} = G_p(s)$$

$$= \frac{1}{s^2 + s + 1}.$$

For a unit step input $R(s) = 1/s$ and we know from equation (3.25) that the time domain solution for the output is then

$$c(t) = 1 - \frac{e^{-t/2}}{\sqrt{0·75}} \sin(\sqrt{0·75} \cdot t + \tan^{-1} 2\sqrt{0·75}). \tag{4.21}$$

(i) With unity feedback and proportional gain k_p.

$$\frac{C(s)}{R(s)} = \frac{k_p G_p(s)}{1 + k_p G_p(s)}$$

$$= \frac{k_p}{(k_p + 1) + s + s^2}$$

$$= \frac{k_p}{1 + k_p} \cdot \frac{1 + k_p}{s^2 + s + (1 + k_p)}.$$

We can see that the second order form is unchanged but that the natural frequency ω_n has changed from unity to $\sqrt{(1 + k_p)}$, the steady state gain has changed from unity to $k_p/(1 + k_p)$ and the damping coefficient has changed from 0.5 to $0.5/\sqrt{(1 + k_p)}$, so that the step response is

$$c(t) = \frac{k_p}{1 + k_p} \left\{ 1 - \frac{e^{-t/2}}{\sqrt{(1 - 0.25/(1 + k_p))}} \right.$$

$$\left. \times \sin\left[\sqrt{(0.75 + k_p)} \cdot t + \tan^{-1} 2\sqrt{(0.75 + k_p)}\right] \right\}. \quad (4.22)$$

Thus we see that proportional control has reduced the steady state gain to $k_p/(1 + k_p)$, i.e. there will be a steady state offset, $c(t) - r(t)$, which is obtained by equation (4.22) as $t \to \infty$; it has decreased the damping coefficient so that the response will be more oscillatory and it has increased the natural frequency, damped natural frequency, i.e. decreased the period of the oscillations. The phase angle has also been increased. However, these changes result in a more rapid initial response (Fig. 4.11). Although some of these changes may appear to be detrimental it must be remembered that feedback control has been used not because we know absolutely the relationship between input and output and any other system effects, but because it can cope with unexpected changes within the system itself, e.g. parameter changes and with unmeasured disturbances. If we wish to make further improvements then we need to consider additional control actions.

(ii) In the same way let us consider proportional plus integral action, expressing the compound action in terms of the proportional gain and integral action time so that $G_c(s) = k_p(1 + 1/T_i s)$ (Fig. 4.8). The closed loop transfer function is then

$$\frac{C(s)}{R(s)} = \frac{k_p(1 + T_i s)}{T_i s^3 + T_i s^2 + (k_p T_i + T_i)s + k_p}.$$

We see that the steady state gain will now be unity for a stable system, so that the integral action has removed the offset introduced with proportional action alone, but the system has been increased to third

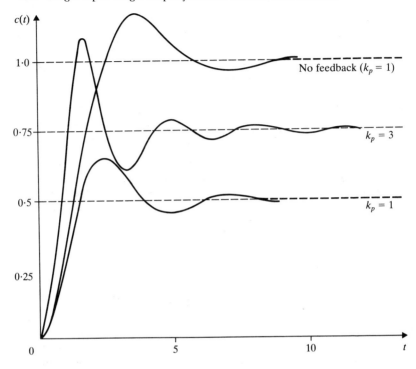

Fig. 4.11 Response of system of Fig. 4.7 with and without unity feedback and $G_p(s) = (s^2 + s + 1)^{-1}$

order. Depending on the coefficients of the powers of s in the denominator there may be roots of the characteristic equation with positive real parts so that the system may be unstable. The choice of the controller variables k_p, T_i, is now the major problem, even if the structure of the controller, e.g. $P + I$, has been decided on. Trial and error selection of the variables has been the general method in the absence of the definition of a specified performance criterion and also full use may be made of the root locus method described in section 4.3. The alternative is to specify a performance criterion and analytically determine the controller variables which 'optimize' the performance (Chapter 10). The integral action results in increased dynamic gain and phase lag in the system and the addition of the integral mode may require reduction of the proportional action of the controller in order to maintain stability. At low values of controller gain we expect stability but increasing gain takes us through a critical point into the unstable region of behaviour, the output tending to move continuously or in an oscillatory manner away from the desired operating point, i.e. the reference input. The critical value of controller gain depends on the parameters (time constants) of the system and will be influenced by those components which may carelessly be neglected, measurement elements, transmitters, control valves etc. It can be seen that the emphasis of our considerations has moved away from the response of arbitrary systems to that of feedback systems and in particular to the problem of stability. This will dominate the rest of this chapter.

(iii) Before moving on to stability as such consider the addition of the derivative mode to the controller. The closed loop transfer function is

$$\frac{C(s)}{R(s)} = \frac{k_p(1+T_d s)}{s^2+(1+k_p T_d)s+(1+k_p)}.$$

The second order form is retained so instability by increasing the order alone is not a factor. However, the steady state offset still exists with a steady state gain of $k_p/(1+k_p)$, the natural frequency is $\sqrt{(1+k_p)}$ and the damping coefficient in the denominator is $0\cdot5(1+k_p T_d)/\sqrt{(1+k_p)}$. Compare these factors with the forward function and with proportional control alone. The offset is the same as with proportional action alone and the natural frequency is unchanged also. The damping is increased however by the factor $(1+k_p T_d)$ so that the oscillations (if any) in the response are of lower magnitude. The damped frequency $\omega_d(=\sqrt{(1-\zeta^2)}\,.\,\omega_n)$ is, for our chosen case $\sqrt{0\cdot75}$ for the forward transfer function, $(\sqrt{0\cdot75+k_p})$ for the proportional action system and for the proportional plus derivative action system it is $\sqrt{(0\cdot75+k_p-0\cdot25(2k_p T_d+k_p^2 T_d^2))}$. Thus derivative action decreases the frequency of oscillations, i.e. increases the period. Thus derivative action improves performance by limiting oscillation magnitude but the system also has longer time constants than if proportional action is used alone.

It can be seen that the combination of all three of these modes, i.e. P.I.D. control will give a controller exhibiting the features of all actions to some extent, the closed loop transfer function being, for the chosen $G_p(s)$

$$\frac{C(s)}{R(s)} = \frac{k_p(T_d s^2+s+1/T_i)}{s^3+(T_d k_p+1)s^2+(1+k_p)s+(1+k_p)}.$$

If proportional action plus rate (velocity) feedback is applied then from equation (4.19) with $G_p(s)=1/(s^2+s+1)$

$$\frac{C(s)}{R(s)} = \frac{k_p}{s^2+(1+k_t)s+(1+k_p)}.$$

There is still a steady state offset, the natural frequency is $\sqrt{(1+k_p)}$ but the damping coefficient ζ is increased to $0\cdot5(1+k_t)/\sqrt{(1+k_p)}$. The major difference compared with derivative action is the absence of the zero in the numerator. The damping effect is similar with the rate constant k_t playing a similar part to the derivative action coefficient

$k_d (= k_p T_d)$. The damped natural frequency is affected as in the derivative case.

Example   Derivative action, with a derivative action time of $0 \cdot 5$ units is added to the controller shown in Fig. 4.12. What will be the change in rise time and the qualitative effect on maximum overshoot and the rate of decay of oscillations following a unit step input to the system.

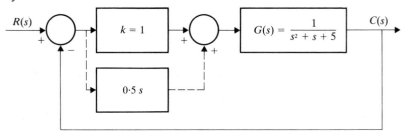

Fig. 4.12   Addition of derivative action to form proportional plus derivative controller

The derivative action is shown dotted in. Without the derivative action the closed loop transfer function is

$$\frac{C(s)}{R(s)} = \frac{1}{s^2 + s + 6}.$$

With the addition of derivative action

$$\frac{C(s)}{R(s)} = \frac{(1 + 0 \cdot 5 s) G(s)}{1 + (1 + 0 \cdot 5 s) G(s)}$$

$$= \frac{1 + 0 \cdot 5 s}{s^2 + 1 \cdot 5 s + 6}.$$

(*a*) *No derivative action.*   The closed loop transfer function has the standard form of equation (3.19), with only the gain different, so that in this case the rise time, overshoot and decay ratio may be determined directly from equations (3.29), (3.31), and (3.32).

Natural frequency $\omega_n = \sqrt{6} = 2 \cdot 45$

Damping coefficient $\zeta = 0 \cdot 204$

Damped natural frequency $\omega_d = \sqrt{(1 - \zeta^2)} \cdot \omega_n = 2 \cdot 40$

The rise time $t_r = \frac{1}{\omega_d} \tan^{-1} \left\{ \frac{-\sqrt{(1 - \zeta^2)}}{\zeta} \right\}$

$$= 0 \cdot 74.$$

Peak overshoot $M_p = \exp \left\{ \dfrac{-\zeta\pi}{\sqrt{(1-\zeta^2)}} \right\}.$

$$= 0\cdot52$$

Decay ratio $\gamma = \exp \left\{ \dfrac{-2\pi\zeta}{\sqrt{(1-\zeta^2)}} \right\}.$

$$= 0\cdot27.$$

(*b*) *With derivative action.* The transfer function is now a little more complex and it is not possible to make use of the above expressions. For a unit step input $R(s) = 1/s,$

$$C(s) = \frac{1+0\cdot5s}{s(s^2+1\cdot5s+6)}$$

and using Table 1.1

$$c(t) = \frac{1}{\omega_n^2} \left\{ 1 - \frac{e^{-\zeta\omega_n t}}{\sqrt{(1-\zeta^2)}} [\sin(\omega_n\sqrt{(1-\zeta^2)} \cdot t + \phi) \right.$$

$$\left. - 0\cdot5\omega_n \sin(\omega_n\sqrt{(1-\zeta^2)}t)] \right\}.$$

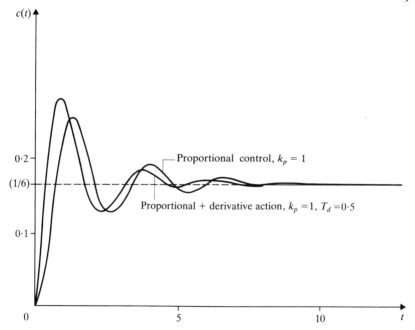

Fig. 4.13 Effect of derivative action addition

$c(t)$ as $t \to \infty$ is $1/\omega_n^2$ so that the rise time is the time at which $c(t)$ first reaches this value. Putting $c(t) = 1/\omega_n^2$ leads to

$$\sin(\omega_d t + \phi) - 0.5\omega_n \sin(\omega_d t) = 0$$

where $\phi = \tan^{-1}\{\sqrt{(1-\zeta^2)}/\zeta\}$. Substitution of the new values for ζ and ω_d leads to the trial and error solution for the rise time. With derivative action $\omega_n = 2.45$ as before, $\zeta = 0.306$. The damped natural frequency is $\omega_d = 2.34$. The rise time t_r is thus the first solution of

$$\sin(2.34t + 1.25) = 1.23 \sin 2.34t$$

and $t_r \simeq 0.35$.

The peak overshoot might be determined by differentiation of $c(t)$ with respect to time in the standard function maximization way. However, we can establish the qualitative effect by argument. The anticipatory form of the derivative action and the additional damping lead to a reduction of the oscillatory nature and with the reduced damped frequency the oscillations are naturally of longer period. The effect of the additional control mode is thus as in Fig. 4.13.

4.3 Stability of single input–single output systems

Stability and pole location

The response of a system may either reach a limited value with increasing time, continue to increase, or oscillate about a mean value with ever-increasing amplitude. The later two conditions constitute an unstable situation. It was seen in Chapter 3 that this condition is shown by the presence of an exponent with a positive real part in the expression for the output as a function of time. This in turn, when the behaviour of the system is expressed as a transfer function, is related to the roots of the characteristic equation of the transfer function. These roots may be complex and may be plotted in the imaginary plane. Consequently on the imaginary plane we may show regions of 'stability' and of 'instability'. A system can be defined as stable if the output response to a bounded input is finite at all times.

To recap, the roots of the characteristic equations are the transfer function poles; any value of s in the transfer function $G(s)$ which causes $G(s)$ to become infinite is a pole of $G(s)$, any value of s causing $G(s)$ to vanish, i.e. equal zero, is a zero of $G(s)$. From the above we see that if all the system poles contain negative real parts, i.e. are located in the left half of the s-plane (Fig. 4.14), the system is stable as the response decays with increasing time. Poles in the right-half of the plane lead to ever-increasing transient values of the form e^{at}, $\text{Rl } a > 0$, i.e. the system is unstable. The stability or instability of a system is not

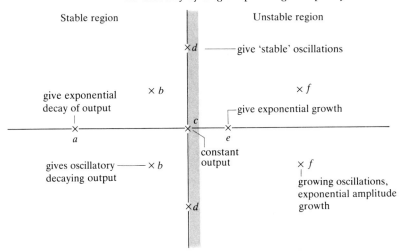

Fig. 4.14 Stability and pole locations in the s-plane and response of equivalent simple systems to impulse disturbances.

influenced by the input forcing function applied (provided the system behaviour remains linear). Positive real-part zeros affect the values and polarities of the partial fraction expansion coefficients of the transfer function. Open loop zeros in the right-half plane may lead to closed loop poles in that half of the plane and hence to closed loop instability. Note that complex poles (roots) always occur in conjugate pairs, $\sigma \pm j\omega$. The relationships between the physical nature of the response and the location of the poles are summarized in Table 4.3.

Table 4.3 The location of poles and the nature of the impulse response

Position in s-plane diagram	Transfer function partial fraction	Response form	Nature of response
a	$\dfrac{k}{s+\alpha}$	$ke^{-\alpha t}$	damped exponential
b, b	$\dfrac{k}{s+(\alpha+j\beta)}, \dfrac{k}{s+(\alpha-j\beta)}$	$ke^{-\alpha t}\sin(\beta t+\phi)$	exponentially damped sinusoid
c	$\dfrac{k}{s}$	k	constant (normally undesirable)
d, d	$\dfrac{k}{s+j\beta}, \dfrac{k}{s-j\beta}$	$k\sin(\beta t+\phi)$	constant amplitude sinusoid (undesirable)
e	$\dfrac{k}{s-\alpha}$	$ke^{\alpha t}$	increasing exponential (unstable)
f, f	$\dfrac{k}{s-(\alpha+j\beta)}, \dfrac{k}{s-(\alpha-j\beta)}$	$ke^{\alpha t}\sin(\beta t+\phi)$	exponentially increasing sinusoid (unstable)

Example   To illustrate some of the features of Table 4.3 consider the following example. A sliding block of mass 2 kg is restrained by a damper of viscous force constant of 40 $\mathrm{Nm^{-1}}$ s and a retaining spring with a spring constant k of 100 $\mathrm{Nm^{-1}}$. In addition a force P N is brought to bear to bring the block to a chosen position. This force is proportional to the error term $r - x$, where r is the chosen (reference) position and x is the actual displacement of the block from its natural equilibrium position. The equation of the motion for the block of mass m is

$$m\ddot{x} + \lambda\dot{x} + kx = P$$

or

$$2s^2 X(s) + 40s X(s) + 100 X(s) = P(s)$$

and

$$P(s) = k(R(s) - X(s)).$$

How does the nature of the response of the system vary as k is varied?

The block diagram is shown in Fig. 4.15. The closed loop transfer function is

$$\frac{X(s)}{R(s)} = \frac{kG(s)}{1 + kG(s)}$$

$$= \frac{k/2}{s^2 + 20s + (50 + k/2)}.$$

The poles of the transfer function are the roots of the characteristic equation,

$$s^2 + 20s + (50 + k/2) = 0$$

and are

$$s_1, s_2 = -10 \pm \tfrac{1}{2}\sqrt{(200 - 2k)}.$$

The transfer function may be expressed as the partial fractions

$$\frac{X(s)}{R(s)} = \frac{k}{2}\left\{\frac{\alpha}{s - s_1} + \frac{\beta}{s - s_2}\right\}.$$

corresponding to terms in the second column of Table 4.3.

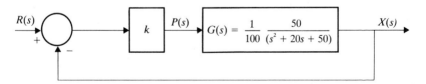

Fig. 4.15 Mechanical system block diagram

Table 4.4 Specific pole location and nature of impulse response

Value of k	Transfer function poles	Position in s-plane diagram	Nature of response
0	$-17\cdot1, -2\cdot9$	$\left.a_1, a_2\right\}$	Overdamped
28	$-16\cdot0, -4\cdot0$	$\left.a_1, a_2'\right\}$	non-oscillatory
100	$-10\cdot0, -10\cdot0$	a_2''	Critically damped, non-oscillatory
150	$-10\cdot0 \pm j5\cdot0$	$\left.b_1, b_2\right\}$	Underdamped,
172	$-10 \pm j6\cdot0$	$\left.b_1', b_2'\right\}$	oscillatory
-100	$-20, 0$	a_3, c	As e^{-20t} decreases tends to constant output
-142	$-21, 1$	a_4, e	Grows exponentially as e^t term dominates

By inspection the roots s_1, s_2 are real for k less than 100, are real and equal for k equal to 100, and are complex for k greater than 100. Only if k is negative, i.e. $P = k'(x - r)$, k' positive, which is an example of positive feedback, does the real part of the roots become positive leading to unstable behaviour. By substituting in various values of k we may construct a particular form of Table 4.3, Table 4.4 and Fig. 4.16.

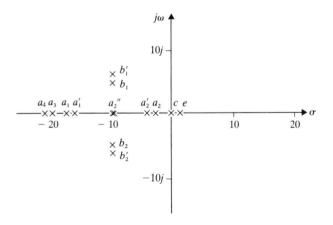

Fig. 4.16 Pole locations in the s-plane (Table 4.4)

We shall now discuss stability in terms of certain techniques and criteria; the Routh–Hurwitz criterion, root-locus, Nyquist criterion and the Bode plot representation of frequency response. This criteria of absolute stability will lead on to the relative stability of systems, including gain- and phase-margins and representation by Nichols' charts.

The Routh–Hurwitz criterion

This criterion (Routh 1877; Hurwitz 1895) is based on the relationship which holds between the coefficients of an nth order algebraic equation if all n roots are to have negative real parts. It supplies the necessary and sufficient conditions for stability, in terms of the coefficients, without the need of actually solving for the roots of the characteristic equations. This can save much time and effort in computation.

The closed loop characteristic equation in its general form

$$1 + G(s)H(s) = 0$$

may be expressed as a power series in s,

$$a_0 s^n + a_1 s^{n-1} + a_2 s^{n-2} + \ldots a_{n-1} s + a_n = 0. \tag{4.23}$$

For example, for $G(s) = 1/(s^2 + 1)$, $H(s) = 1/(s+2)$

$$1 + G(s)H(s) = 1 + \frac{1}{(s^2+1)(s+2)}$$

$$= 1 + \frac{1}{s^3 + 2s^2 + s + 2}$$

$$= 0$$

or

$$s^3 + 2s^2 + s + 3 = 0.$$

The conditions we are seeking may be expressed in terms of the Hurwitz determinants or more simply by the Routh array. Certain basic conditions are necessary also and enable us to detect purely by inspection whether the system is definitely unstable or whether stability is possible.

For stability, none of the n roots of equation (4.23) must have positive real parts and in order that this is so it is necessary (but not on its own sufficient) that

(a) All the coefficients a_i have the same sign, and
(b) none of the coefficients a_i is zero.

The third, necessary and sufficient, condition is

(c) that for all roots of the nth order polynomial to lie in the left half of the s-plane, the polynomial's Hurwitz determinants D_k $(k = 1, 2, \ldots, n)$ must all be positive.

The Hurwitz determinants are

$$D_1 = a_1$$

$$D_2 = \begin{vmatrix} a_1 & a_3 \\ a_0 & a_2 \end{vmatrix}$$

$$D_n = \begin{vmatrix} a_1 & a_3 & a_5 & \cdots & a_{2n-1} \\ a_0 & a_2 & a_4 & \cdots & a_{2n-2} \\ 0 & a_1 & a_3 & \cdots & a_{2n-3} \\ \vdots & & & & \vdots \\ 0 & 0 & 0 & & a_n \end{vmatrix}$$

where the coefficients, e.g. a_{2n-1}, with indices larger than n or with negative indices are replaced by zeros. The formal stability criterion may then be stated in terms of the characteristic equation roots:

'The necessary and sufficient condition that all roots of the polynomial in s, equation (4.23) lie in the left half of the s plane is that $a_0 > 0$, $D_1 > 0, \ldots, D_n > 0$ where $D_1, \ldots, D_n$ are the Hurwitz determinants defined as above.'

It would appear at first sight that the use of this criterion involves the evaluation of high order determinants. For example, for a fifth order equation this would involve the evaluation of second, third, fourth and fifth order determinants. This evaluation is avoided by using the Routh array which incorporates the same criterion regarding the sign of the real parts of the roots of the characteristic equation.

(a) The coefficients are first arranged in two rows

$a_0 \quad a_2 \quad a_4 \quad \cdots$

$a_1 \quad a_3 \quad a_5 \quad \cdots$

(b) Using these, and a first column s^n, $s^{n-1}, \ldots, s^1$, s^0, as a basis the Routh array is built up as below (here shown for a fifth order system). The pattern of multiplication and subtraction is repeated as we progress down the array.

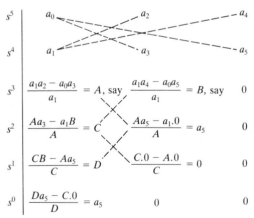

(c) The condition for all the roots of the polynomial to be in the left half of the s-plane is that all elements of the first column in the array are positive, i.e. a_0, a_1, A, C, D, $a_5 > 0$. If there are negative values in this column then the number of sign changes indicates the number of roots with positive real parts. This is because the first column in the array is related to the Hurwitz determinants by, in this example,

$$a_0 = a_0, \qquad a_1 = D_1, \qquad A = D_2/D_1, \qquad C = D_3/D_2$$
$$D = D_4/D_3 \quad \text{and} \quad a_5 = D_5/D_4.$$

The array provides a very rapid method to check on stability and to determine certain requirements for stability, but special cases may easily arise which cause minor deviations to be required. Also, to reduce the effort in evaluating the array, the numerical coefficients which one obtains in any row may be multiplied (or divided) by a positive number without changing the signs in the first column.

Rules for special cases

(*a*) *Zero value as the first element of a row.*   This difficulty may be overcome either by

- (i) substituting a small positive number ε for the zero and proceeding as usual,
- (ii) substituting $s = 1/x$ in the original equation and 'solving' for the roots of x with positive real parts, which will be the same number as roots of s with positive real parts, or
- (iii) by multiplying the original polynomial by $(s + a)$ where a is any real positive number and proceeding as usual. If the first chosen a leads to zeros also then a different value may be selected.

(*b*) *Row of zeros.*   When in the construction of the array a row appears which is all zeros this indicates that there are pairs of real roots with opposite signs, pairs of conjugate roots on the imaginary axis or two pairs of complex conjugate roots. The equation formed using the coefficients in the row just above the row of zeros is the 'auxiliary equation' and it is always of even powers only (see example). It indicates the number of root pairs equal in magnitude but opposite in sign. The values of these roots are then found by solving the auxiliary equation. To progress beyond the row of zeros the procedure is to: (i) form the auxiliary equation; (ii) replace the all-zero row by the coefficients obtained by differentiating the auxiliary equation; and (iii) complete the array and inspect the first column as usual.

As well as being used to check on the stability of a system this criterion may be used to determine the conditions of stability for a linear feedback control system. An undetermined parameter, say k, appearing in the characteristic equation will be carried through the array and will appear at one or more point is the first column. The condition for all the first column values to be positive then gives a number of inequalities $f(k) > 0$, which k must satisfy. Solution of these inequalities gives the value or range of k which will give a stable system.

Example   Using the Routh–Hurwitz criterion investigate the stability of the system for which the closed loop characteristic equations are

(i) $s^4 + s^3 + s^2 + s + K = 0$
(ii) $2s^3 + 2s^2 + s + 1 = 0$
(iii) $s^4 + s^3 + 5s^2 + Ks + K = 0$

None of these expressions is ruled out as unstable merely by inspection. Use the array on each in turn.

(i) $s^4 + s^3 + s^2 + s + K = 0$

$$\left.\begin{array}{c|ccc} s^4 & 1 & 1 & K \\ s^3 & 1 & 1 & \\ s^2 & 0 & K & \end{array}\right\}$$

To overcome the zero in the first column replace it by the small positive value ε.

$$\left.\begin{array}{c|ccc} s^2 & \varepsilon & K \\ s^1 & \dfrac{\varepsilon - K}{\varepsilon} & 0 \\ s^0 & K & \end{array}\right\}$$

Conditions for stability are

$$\frac{\varepsilon - K}{\varepsilon} > 0 \quad \text{as} \quad \varepsilon \to 0$$

$$K > 0.$$

The first of these requires $-K > 0$, $K < 0$. Thus the two conditions show that for any K the system is unstable as for any value of K one of the conditions will be violated. Thus there is always a root in the right-hand plane.

(ii) $2s^3 + 2s^2 + s + 1 = 0$

$$\begin{array}{c|cc} s^3 & 2 & 1 \\ s^2 & 2 & 1 \\ s^1 & 0 & 0 \end{array}$$

A complete row is now zero. Form the auxiliary equation

$$2s^2 + 1 = 0$$

and the roots of this $s = \pm j/\sqrt{2}$ show that there are two roots equal in magnitude but opposite in sign and that these are located on the imaginary axis. Taking the first derivative of the auxiliary eqution,

$$4s = 0$$

we replace the s^1 row by the new coefficients and continue the array

$$\begin{array}{c|cc} s^1 & 4 & 0 \\ s^0 & 1. & \end{array}$$

Thus the system is not unstable but will oscillate continuously due to the pure imaginary poles.

(iii) $s^4 + s^3 + 5s^3 + Ks + K = 0$

s^4	1	5	K
s^3	1	K	
s^2	$5-K$	K	
s^1	$\dfrac{(5-K)K-K}{5-K}$	0	
s^0	K		

The conditions for stability are thus

(a) $5-K>0$

(b) $\dfrac{(5-K)K-K}{5-K}>0$

(c) $K>0$

Thus from (c) $K>0$ and from (a) $K<5$. With the restrictions (b) gives $4K-K^2>0$, $4-K>0$, i.e. $K<4$. To satisfy all these conditions for stability

$$0<K<4$$

The root-locus

The root-locus method, introduced by Evans (Evans 1948, 1954), is not restricted to absolute stability determination but is aimed at predicting general closed loop performance without full solution of the system equations. It is an analytical graphical method for the choice of control or plant parameters for satisfactory plant behaviour. The root locus (or roots-loci) is a plot of the roots of the characteristic equation of the closed loop system as a function of some system variable. In most cases this variable is the system or controller open loop gain but the method may be used for other system parameters which are variable also. The method is dependent on the relationships between the poles of the closed loop transfer function, the open loop poles and zeros, i.e. of $G(s)H(s)$, the gain K and the roots of the characteristic equation. Also, as stated above, the correct location of the closed loop poles, i.e. of all roots, $s_1, \ldots, s_n$, of the characteristic equation in the complex plane is a necessary and sufficient condition to specify the absolute system stability. As the forward transfer function gain K will be the parameter on which we focus most attention it is convenient to express $G(s)$ as $KG_1(s)$ (Fig. 4.17). The characteristic equation is now written

$$1 + KG_1(s)H(s) = 0$$

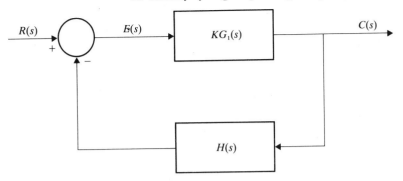

Fig. 4.17 Basic system for root locus study

and as K is varied the roots of this equation in s, s_i, vary also. It is the plots of these roots as K varies which gives the root locus. The terms 'closed loop poles' and 'characteristic equation roots' are interchangeable. Because there is more than one root (depending on the system order) for each K there will be more than one locus. The aggregate of the loci are known as the 'roots-loci'.

The notation is based on the poles and zeros of the open loop transfer function. In general we may write

$$G(s)H(s) = KG_1(s)H(s)$$
$$= \frac{K(s+a_1)(s+a_2)\ldots(s+a_v)}{s^n(s+b_1)(s+b_2)\ldots(s+b_u)} \quad (4.24)$$

where a_i and b_i may be real or complex.

Denoting the values of s which give rise to poles and zeros of the open loop transfer function by p_i and z_i respectively it is seen that these occur when

$$z_1 = -a_1; \quad z_2 = -a_2 \quad \text{etc.}$$
$$p_1 = -b_1; \quad p_2 = -b_2 \quad \text{etc.}$$

and equation (4.24) becomes (see equation 3.34)

$$G(s)H(s) = \frac{K(s-z_1)(s-z_2)\ldots(s-z_v)}{s^n(s-p_1)(s-p_2)\ldots(s-p_u)}. \quad (4.25)$$

Example To clarify this consider a unity feedback system with $H(s) = 1$ and with

$$G(s) = \frac{C(s)}{E(s)}$$
$$= \frac{K}{s(s+a)}.$$

The closed loop transfer function is

$$\frac{C(s)}{R(s)} = \frac{G(s)}{1 + G(s)H(s)}$$

$$= \frac{K}{s^2 + as + K}.$$

The roots of the characteristic equation, $s^2 + as + K = 0$, are $s_1, s_2 = -a/2 \pm \sqrt{(a^2/4 - K)}$. That is

$$\frac{C(s)}{R(s)} = \frac{K}{(s - s_1)(s - s_2)}.$$

Letting $a = 1$ then $s_1, s_2 = -0\cdot5 \pm \sqrt{(0\cdot25 - K)}$ and for each value of K from $-\infty$ to $+\infty$ the roots s_1, s_2 will have the values which are plotted in Fig. 4.18. Note the relationship to Fig. 4.16. The continuous heavy lines are the loci for positive K. The broken line shows the locus for $K < 0$ but where K is the gain of the system it is usual only to consider its practical, i.e. positive values, from zero to plus infinity. Figure 4.18 may be plotted easily, as the function is simple, by evaluating the roots s_1 and s_2 from the characteristic equation as K varies, but the roots loci are not normally constructed in this manner when used as an

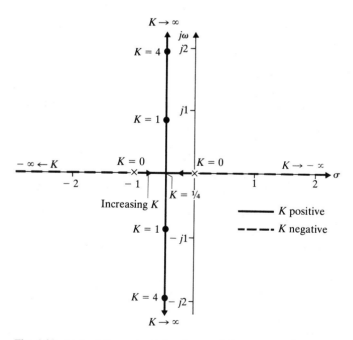

Fig. 4.18 Loci of the roots of the characteristic equation $s^2 + s + K = 0$, as a function of K

analytical aid. Instead they are built up from a knowledge of the open loop poles and zeros. When K is zero we see that the roots of the characteristic equation are also the poles of the open loop transfer function and thus the roots-loci 'branches' are said to start at the open loop poles. The arrows on the branches show the direction of increasing K. This will be one of the general rules for the construction of a root-locus covered below.

There are two basic conditions which are satisfied by any point on the root-locus. The characteristic equation may be rearranged as

$$G(s)H(s) = -1$$

and to satisfy the equation in a complex quantity s we can use the algebra of complex numbers and equate the magnitude (modulus) and phase angle (argument) of both sides, i.e.

$$|G(s)H(s)| = 1, \text{ the magnitude condition} \tag{4.26}$$

and, to give the negative sign of the real quantity -1,

$$\underline{/G(s)H(s)} = (k \cdot 360 + 180)°$$
$$= (2k + 1)\pi \text{ radians, the angle condition} \tag{4.27}$$

This is shown in Fig. 4.19.

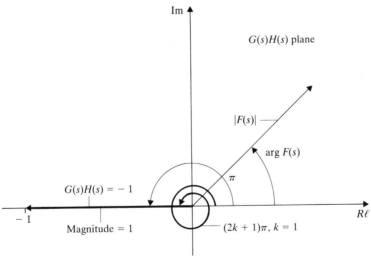

Fig. 4.19 Magnitude and angle condition for $G(s)H(s) = -1$, and for the general function $F(s)$

Thus the magnitude of $G(s)H(s)$ must always be unity and its phase angle must be an odd multiple of 180° (π radians) if a particular value of s is to be a root of $1 + G(s)H(s) = 0$. The root-locus plot provides a plot in the complex s plane of the variation of each of these roots, or poles of the closed loop transfer function, as the gain K is varied from zero to infinity, such that both these magnitude and angle conditions hold. Using equations (4.25) and (4.26) the magnitude condition may be expressed

$$|G(s)H(s)| = \frac{K \prod\limits_{i=1}^{v} |s - z_i|}{\prod\limits_{j=1}^{n+u} |s - p_j|} = 1 \tag{4.28}$$

and the angle condition

$$\underline{/G(s)H(s)} = \sum_{i=1}^{v} \underline{/s - z_i} - \sum_{j=1}^{n+u} \underline{/s - p_j}$$
$$= (2k + 1)\pi \tag{4.29}$$

where allowance has been included for roots at the origin. (Π indicates the product of terms.) The magnitude relation (4.28) may be expressed explicitly in K,

$$K = \frac{\prod\limits_{j=1}^{n+u} |s - p_j|}{\prod\limits_{i=1}^{v} |s - z_i|}. \tag{4.30}$$

The values of s satisfying both conditions are the roots of the characteristic equation for the given K. The number of roots is the order of the equation (4.30) in powers of s. Thus there are $n + u$ or v roots depending upon whether $n + u$ or v is the higher power. As K varies each root varies and hence a plot of each of their values gives a branch of the locus.

The construction of the roots-loci

The following rules open the way to the construction of roots-loci for even the most complicated transfer functions. They also enable rapid sketches of the full loci to be made very quickly so that an assessment of system behaviour and its dependence on K can be easily seen. Remember that from the loci the roots of the characteristic equation can be 'picked off' for values of K and these of course define the system dynamics, decay times, oscillary behaviour etc.

The rules are developed from the relations between the poles and zeros of the open loop transfer function and the roots of the characteristic equation (e.g. Dransfield and Haber 1973). Some of them may appear obvious but a systematic use is essential to the use of the root-locus method.

(i) *Starting point of the root loci* ($K = 0$). The root-loci start at the poles of $G(s)H(s)$. They are considered to start at the points at which the gain K is zero.

As K tends to zero the equation (4.30) reduces to

$$\prod |s - p_j| = 0$$

so that at this value of K the roots of equation (4.30) approach the values $s = p_j$, i.e. the poles of $G(s)H(s)$.

(ii) *End points of the root-loci* $(K \to \infty)$. From equation (4.30) as K tends to infinity the expression in s tends to infinity requiring that s tends to z_i. That is, the loci end at the zeros of $G(s)H(s)$. This is more readily seen by rearranging equation (4.30) as

$$\frac{1}{K} = \frac{\prod |s - z_i|}{\prod |s - p_j|}$$

which reduces as K increases to the limiting equation

$$\prod |s - z_i| = 0.$$

The roots of this equation are thus $s = z_i$, i.e. the zeros of $G(s)H(s)$.

(iii) *Root-loci on the real axis.* As a direct result of the angle condition the root-loci may be found on a given section of the real axis only if the total numbers of poles and zeros of $G(s)H(s)$ on the real axis to the right of the section is odd.

(iv) *The number of branches of the loci.* The number of branches of the root-loci is equal to the number of poles (or zeros) of the open loop transfer function. The characteristic equation for the open loop transfer function of equation (4.24) is of order $n + u$, giving $n + u$ roots which are functions of K and these trace $n + u$ curves as K varies over its range zero to infinity. To allow for the possibility of a higher order numerator than denominator, i.e. a higher number of zeros of $G(s)H(s)$ the simple rule is that if N is the number of separate loci, Z is the number of finite zeros of $G(s)H(s)$, and P is the number of finite poles of $G(s)H(s)$ then

$N = Z$ if $Z > P$

$N = P$ if $Z < P$.

This is consistent with rules (i) and (ii). If the finite poles and zeros are equal in number then the whole locus is generally possible on a drawing. Where the number of poles (zeros) exceeds the number of zeros (poles) the branches terminate (start) at infinity, i.e. off the drawing. There are then $P - Z$, or $Z - P$, branches which tend asymptotically to straight line sections of loci.

(v) *Symmetry of the roots-loci.* The roots-loci are symmetrical with respect to the real axis, since any complex roots appear in complex conjugate pairs always.

(vi) *Asymptotes of roots-loci.* For large values of the roots, $P-Z$ branches of the root-loci are asymptotic to straight lines with angles to the real axis given by $(2m+1)\pi/(P-Z)$ rads where $m=0$, 1, 2,..., $P-Z$. The asymptotes do not necessarily pass through the origin but because of symmetry they do intercept on the real axis. For a rapid indication of the approximate shape only of the loci the asymptotes may frequently be sketched through the origin.

(vii) *Intersection of the asymptotes.* The point of intersection of the asymptotes on the real axis is given

$$\sigma = \frac{\sum \text{Poles of } G(s)H(s) - \sum \text{Zeros of } G(s)H(s)}{P-Z}.$$

Since the imaginary parts of poles and zeros are in conjugate pairs we may reduce this to

$$\sigma = \frac{\sum\limits_{j=1}^{u} Rl(p_j) - \sum\limits_{i=1}^{v} Rl(z_i)}{P-Z}$$

where Rl stands for the real part of the following term.

(viii) *Intersection of the loci with the imaginary axis.* The intersection of the loci with the imaginary axis marks the stage at which the real parts of the roots change from negative to positive and hence the system changes from a stable to an unstable system. The real part of the root at the point of intersection is zero and the value of ω, in $s = \sigma \pm j\omega$, and the gain K at this point may be determined by the Routh–Hurwitz criterion or by direct algebraic considerations.

(ix) *Breakaway point on the real axis.* The points in the s-plane where multiple roots of the characteristic equation are found are called the 'breakaway points' of the root-locus diagram. At such a point two or more roots loci branches branch away or meet. Consistent with the symmetry of the plot these points must either lie on the real axis or exist in conjugate pairs. Where branches between two poles meet on the real axis the loci then branch away, i.e. with increasing K, and between two zeros on the real axis the loci move in to the real axis if loci exist on the axis according to rule (iii).

The loci approach and breakaway on the real axis at an angle of π/n radians apart, where n is the number of branches approaching (and also leaving) the point. The actual position of the breakaway point is given by the following rules.

(x) *Breakaway point due to poles and zeros on the real axis only.* If there are only real poles and zeros of $G(s)H(s)$ and $s = -q$ is assumed to be the breakaway point on the real axis, the value of q is determined by

$$\sum_{\substack{\text{Zeros to the} \\ \text{right of } -q}} \frac{1}{q-a_i} - \sum_{\substack{\text{Poles to the} \\ \text{right of } -q}} \frac{1}{q-b_i} =$$

$$\sum_{\substack{\text{Zeros to the} \\ \text{left of } -q}} \frac{1}{a_i-q} - \sum_{\substack{\text{Poles to the} \\ \text{left of } -q}} \frac{1}{b_i-q}$$

using the convention of equations (4.24) and (4.25). This rule is derived by applying the angle condition at points s_1 (Fig. 4.20) and taking the limit as ε tends to zero, i.e. as s_1 tends to $-q$, replacing the angles in the limiting area by their tangents.

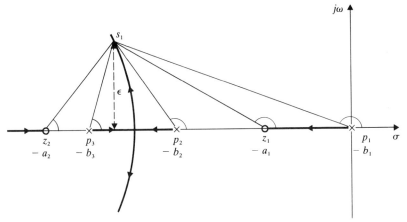

Fig. 4.20 Breakaway point on the real axis due to real poles and zeros of $G(s)H(s)$

(xi) *Breakaway point due to complex poles and zeros.* The contribution to the breakaway point on the real axis from the complex poles and zeros of $G(s)H(s)$ is determined from the equation

$$\sum_{\substack{\text{Complex zeros} \\ \text{to right of } -q}} \frac{2(q-\alpha_i)}{(q-\alpha_i)^2+\beta_i^2} - \sum_{\substack{\text{Complex poles} \\ \text{to right of } -q}} \frac{2(q-\alpha_i)}{(q-\alpha_i)^2+\beta_i^2} =$$

$$\sum_{\substack{\text{Complex zeros} \\ \text{to left of } -q}} \frac{2(\alpha_i-q)}{(\alpha_i-q)^2+\beta_i^2} - \sum_{\substack{\text{Complex poles} \\ \text{to left of } -q}} \frac{2(\alpha_i-q)}{(\alpha_i-q)^2+\beta_i^2}$$

where the complex roots are at $s = \alpha_i \pm j\beta_i$. The factor 2 is carried right through as the roots appear in conjugate pairs and the total expression when complex and real poles and zeros both exist is obtained by just adding this condition with the one above (x). Again this rule comes from the angle condition (Fig. 4.21).

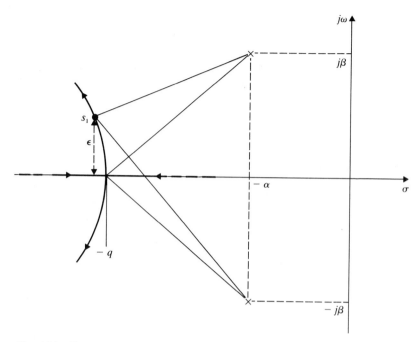

Fig. 4.21 Contribution to the breakaway point on the real axis due to complex zeros and poles of $G(s)H(s)$

(xii) *Breakaway point not on the real axis.* The above equations for locating the breakaway point can still be used by choosing a new origin and a new set of co-ordinates (Fig. 4.22). Initially the break-away point on the real axis is determined as above and a new real axis is chosen to be on the line joining the two breakaway points to be determined. In complex systems this may be difficult but due to the symmetry of the loci it is only necessary to determine one of the two points. The above rules are then applied to the new set of co-ordinates $(\sigma', j\omega')$, in the example shown the new origin is chosen at one of the poles. The breakaway point having been located along the new 'real' axis the new co-ordinates are then transferred back to the original set.

(xiii) *General analytical method for breakaway points.* The characteristic equation may be written explicitly in K, i.e.

$$K = f(s).$$

The breakaway points, both real and complex, of the root locus diagram are the roots of the equation which is obtained by taking the first derivative of K with respect to s and setting it equal to zero, i.e.

$$\frac{dK}{ds} = 0.$$

The algebraic roots of this equation in s are the points of breakaway on the roots-loci.

(xiv) *Angle of departure from a complex pole and zero.* The direction of the locus as it leaves the pole or zero is determined by adding up, by the angle condition, all the angles of all vectors from all

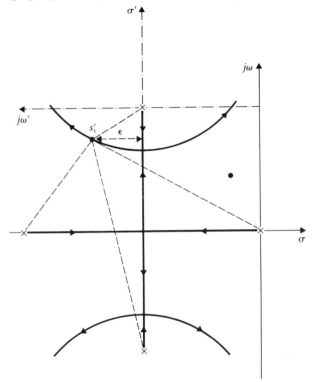

Fig. 4.22 New co-ordinate system σ', $j\omega'$ to evaluate complex breakaway points

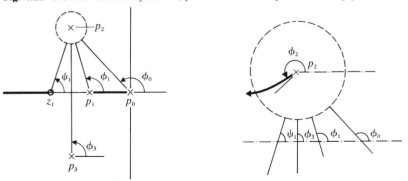

Fig. 4.23 Angular condition near a pole.

other poles and zeros to the pole or zero in question, and subtracting this sum from 180° to give the required direction. The angular conditions in the vicinity of a complex pole are shown in detail in Fig. 4.23. Applying the angle condition we have

$$\phi_0 + \phi_1 + \phi_2 + \phi_3 - \psi_1 = 180(2m + 1)°$$

i.e.

$$\phi_2 = 180(2m + 1)° - \phi_0 - \phi_1 - \phi_3 + \psi_1.$$

(xv) *Determination of roots on the roots-loci diagram.* The determination of all roots for a given K from a root locus diagram is illustrated in detail by the example at the end of the chapter. Initially the specification of system performance is used to determine the dominant roots (those closest to the imaginary axis and the slowest to decay in a transient). The remaining may then be found by:

(a) trial and error search for roots on each locus that satisfies the static loop gain found for the dominant roots or

(b) dividing the characteristic equation by the factors that are known and determining the remaining roots from the remainder resulting from this division.

(xvi) *Calculation of the gain K for a specific point on the loci.* Once the loci are constructed the value of K at any point, s_1, say, on the loci can be determined from the magnitude condition

$$K = |G(s_1)H(s_1)|^{-1}$$

evaluated either graphically or analytically. From a root-locus already constructed the graphical method is convenient. From Fig. 4.24

$$K_1 = |G(s_1)H(s_1)|^{-1}$$
$$= \frac{A \cdot B \cdot C}{D}$$

using the magnitude condition. A, B, C, D are the lengths of the vectors from the open loop poles and zeros to the point s_1. Sequential use of the above rules enables the root-locus to be drawn in detail. Selective use of the rules enables the root-locus to be sketched rapidly. The 'construction rules' are all based on the magnitude and angle conditions.

Some properties of the root-locus method

The root locus method as used in control studies is a graphical representation of the closed loop poles, and hence of the dynamic behaviour of the system, established on the basis of the open loop behaviour as manifest by its poles and zeros. The use of the method in control systems comes from its ability to show readily the effects of

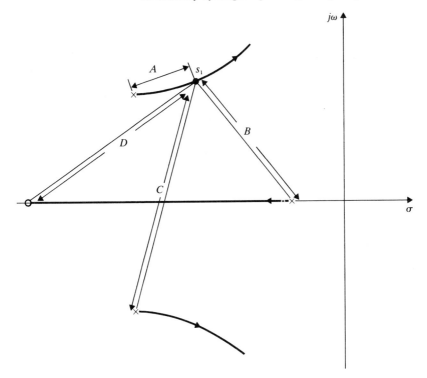

Fig. 4.24 Graphical evaluation of K for a point on the root locus

adding additional control modes to the system and the effects of the variation of other parameters as well as of the open loop gain K. This aids in the correct choice of control if the physical behaviour corresponding to the set of roots obtained from the loci is correctly interpreted.

The mere addition of an additional mode (pole or zero) effects a basic change in the root-locus appearance but the magnitudes of the parameters in this mode also have a dramatic effect on the root-locus plot appearance, i.e. on the system dynamics for a given gain. This effect can be described in terms of increasing stability or decreasing stability and also in terms of the speed of the system response.

We shall consider first the successive addition of poles and zeros and then the effect of the movement of the poles and zeros in terms of integral and derivative control action.

(a) *The effect of adding open loop poles.* One way in which a pole can be added will be shown later but the effect of the general addition of real or complex conjugate poles is to decrease the closed loop stability of the system. This is shown by the bending of the roots-loci more rapidly towards the imaginary axis so that they cut this axis at lower values of gain K. Even if the effect does not actually cause instability by intersection with the imaginary axis the movement of the locus, or some of its branches towards this axis indicates that there is less option on the change of system parameters before instability might become a factor to be dealt with.

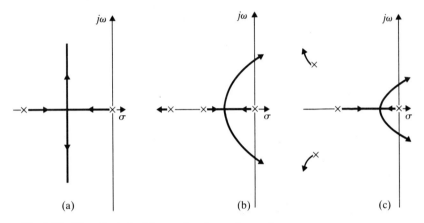

Fig. 4.25   The effect of adding real and complex conjugate open loop poles

(b) *The effect of additional zeros.*   Open loop zero addition in the left half plane increases the closed loop system tendency to stability. This is true for the addition of either real zeros or conjugate pairs. Figure 4.26 shows the effect on the root-locus plot.

If the forward transfer function $G(s)$ is comprised of a controller transfer function $G_c(s)$ and a plant transfer function the open loop transfer function $G(s)H(s)$ may be written

$$G_c(s)G_p(s)H(s).$$

As we have seen the steady state gain K may also be taken as a separate factor and it will usually be associated with the controller. If integral action is added to a controller then we saw (section 4.2) that $G_c(s)$ has the form

$$G_c(s) = k_p\left(1 + \frac{1}{T_i s}\right)$$

$$= k_p\left(\frac{T_i s + 1}{T_i s}\right)$$

so that integral action adds both a pole at the origin and a zero at $s = -1/T_i$. The effect of this is shown in Fig. 4.27 for four values of the integral action time T_i. As well as the absolute presence of poles and zeros having an effect on stability their position in the complex plane

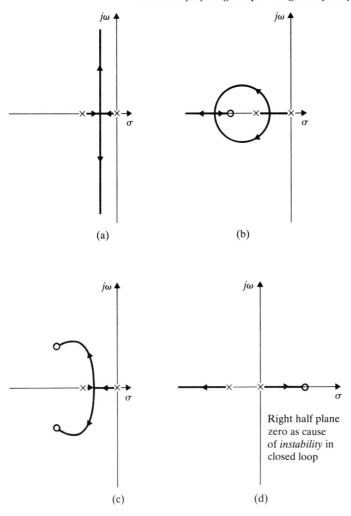

(a)

(b)

(c)

(d)

Right half plane
zero as cause
of *instability* in
closed loop

Fig. 4.26 The effect of adding real and complex open loop zeros

determines both their effect on response time and on critical values of
gain. Decreasing the real part of a pole, i.e. moving it closer to the
origin decreases stability. It will be remembered also that moving the
pole nearer to the origin in the case of a stable system, slows up the
response time. The effect of pole movement on the root locus is shown
in Fig. 4.28, where a pole is brought in from 'infinity' to the origin.
The pattern of the root locus diagram is changed considerably but
these changes can be easily observed using the guide rules for plotting
as given above. The locus shows that with the double pole at the origin
the system is unstable for all values of the gain K.

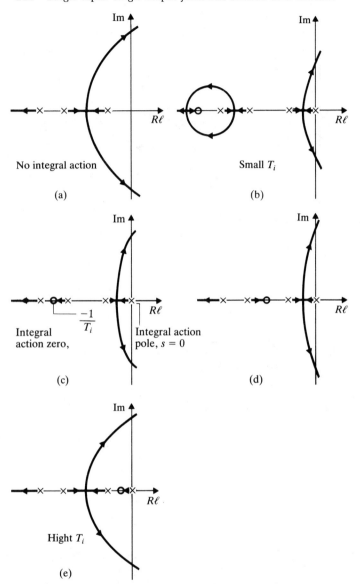

Fig. 4.27 Qualitative effect of integral action on roots loci configuration

The reverse is true for the 'movement' of an open loop zero. As the zero approaches the origin, from the left-hand plane, stability is increased until the reverse effect is achieved, the same initial system becomes fully stable for all values of gain (Fig. 4.29). Further right-hand plane movement of the zero reintroduces instability by roots on

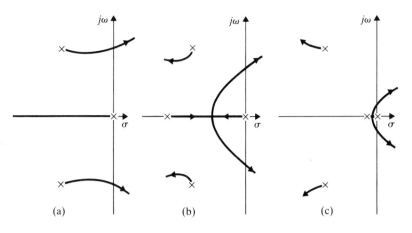

Fig. 4.28 Effect of pole movement

the real axis. When talking of the addition of poles and zeros we are usually talking in physical terms of the addition of controller actions (modes). Movement of poles and zeros infers a change in some variable or parameter, e.g. a time constant or gain, in the system.

Variable parameters other than gain K

A generalized root-locus ('root contour') technique may be used when parameters other than K are varied in a feedback system. The root contour is still the plot of the closed loop poles but K is held constant for each contour while, say, the open loop transfer function poles and zeros are varied. Repetition for alternative values of K gives a series of root contours.

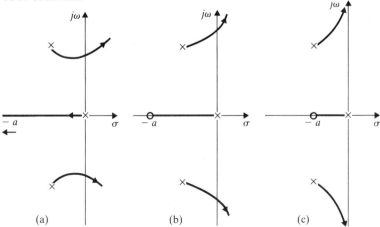

Fig. 4.29 Effect of zero movement $(s + a)$, $a > 0$

(a) *Variable poles of $G(s)H(s)$.* Consider when one pole in the open loop transfer function is variable, perhaps as a result of a controller action being variable or changes being possible in a plant parameter. Then this factor may be 'pulled out' from the transfer function so that

$$G(s)H(s) = \frac{W(s)}{1 + T_1 s}$$

where T_1 is the variable and $W(s)$ is the 'constant' part of the function and is independent of T_1.

The closed loop transfer function is then

$$\frac{C(s)}{R(s)} = \frac{G(s)}{1 + G(s)H(s)}$$

$$= \frac{G(s)(1 + T_1 s)}{1 + T_1 s + W(s)}. \tag{4.31}$$

This general change may be made without difficulty even if $(1 + T_1 s)^{-1}$ is also a factor of $G(s)$ alone. To reduce the closed loop form to a recognized form define the two new functions

$$G_1(s) = \frac{G(s)(1 + T_1 s)}{1 + W(s)}$$

$$H_1(s) = \frac{s T_1}{G(s)(1 + T_1 s)}$$

and then in equation (4.31)

$$\frac{C(s)}{R(s)} = \frac{G_1(s)}{1 + G_1(s)H_1(s)}.$$

This implies the equivalent system of Fig. 4.30 with open loop gain

$$G_1(s)H_1(s) = \frac{s T_1}{1 + W(s)}.$$

Thus the conventional root-locus may be drawn but instead of using K the branches of the contours now begin at T_1 equal to zero at the poles of $G_1(s)H_1(s)$ and terminate at T_1 tending to infinity at the zeros of $G_1(s)H_1(s)$. The rules of construction of the root-locus still apply and instability and the degree of stability for a given T_1 is given by the path of the loci and the approach to or passage into the right half plane.

(b) *Variable zero of $G(s)H(s)$.* An equivalent procedure may be followed when

$$G(s)H(s) = (1 + T_2 s) W(s).$$

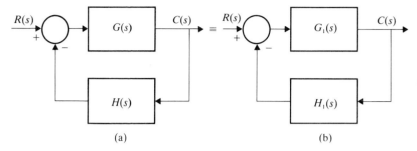

Fig. 4.30 Single loop feedback (a) and the equivalent system (b)

Substituting

$$G_2(s) = \frac{G(s)}{1 + W(s)}$$

and

$$H_2(s) = \frac{sT_2 W(s)}{G(s)}$$

gives once more the form

$$\frac{C(s)}{R(s)} = \frac{G_2(s)}{1 + G_2(s)H_2(s)}$$

and

$$G_2(s)H_2(s) = \frac{T_2 s W(s)}{1 + W(s)}.$$

Once again an equivalent system has been formed and in this case the contours start at T_2 equal to zero at the poles and end at T_2 tending to infinity at the zeros of $G_2(s)H_2(s)$.

(c) *Both poles and zeros variable.* Since only one parameter can be varied at a time to give a root locus or root contour plot it is necessary to have more than one plot, i.e. a family of contours to show both the variation of T_1 and T_2 (D'Azzo and Houpis 1960). If the gain K is also variable then the plotting becomes even more complex since only a restricted amount of information can be given by a single plot in a plane.

Conditional stability

So far stability has been thought of in the sense that a value of gain or an equivalent variable will be reached above which the system is unstable. In fact further increase in gain might restore stability to the system and this means that the closed loop poles return into the left half of the complex plane (Fig. 4.31). This conditional stability is also picked out by the Routh–Hurwitz method.

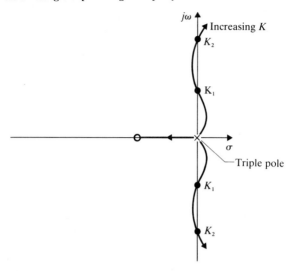

Fig. 4.31 Conditionally stable system (stable for $K_1 < K < K_2$)

The root locus method and Routh–Hurwitz criteria are based on the time domain. We shall return in Chapter 5 to the frequency domain to consider the Nyquist stability criterion and the use of the Bode plots for stability studies.

Example An excellent extensive example of the root-locus is given in D'Azzo and Houpis (D'Azzo and Houpis 1960). In keeping with the overall pattern of this text we shall consider a shorter example. Sketch the root locus for the system whose open loop transfer function is

$$G(s)H(s) = \frac{K}{(s+3)(s+1)(s^2+2s+2)}.$$

There are no open loop zeros and the open loop poles are at $s = -3$, $s = -1$, $s = -1 \pm j$.

We can now follow the 'rules' to arrive at the full roots-loci.

(i) The roots-loci start at the poles with $K = 0$.

(ii) There are no open loop zeros so the loci will continue to move away towards the outer regions of the plot as K increases, i.e. no branch will finish in the finite plane.

(iii) There will be a branch on the real axis to the left of the real pole at $s = -1$ and to the right of the pole at $s = -3$.

(iv) There will be four branches which

(v) will be arranged symmetrically about the real axis.

This is as far as we can go in the plotting without resorting to calculations. Determination of the asymptote angles will enable us to complete a rough sketch of the loci.

(vi) The angles of the asymptotes are given by

$$\frac{(2m+1)180°}{P-Z} = \frac{(2m+1)}{4}180°$$

$$= 45°, 135°, 225°, 315°.$$

As a first approximation we may now sketch the probable loci as in Fig. 4.32 (this alone indicates the unstable nature with increasing gain K). Note that as a first approximation the asymptotes have been sketched through the origin but their exact location is not yet known and the loci in the vicinity of the poles are not defined, nor are the crossing points on the imaginary axis or the breakaway point on the real axis. However, having this idea of how the plot will look in general is a guide during the calculations and serves as a check of some sort on our calculations. Proceed now to determining the precise positions of the main features of the loci. By rule (vii) establish the intersection of the asymptotes. This is at $\sigma = \{\sum Rl(p_j)\}/P$ as there are no zeros.

$$\sigma = (-3-1-1-1)/4 = -1{\cdot}5.$$

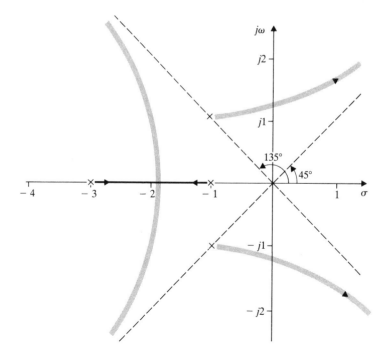

Fig. 4.32 Approximate first sketch of roots loci

(vii) Intersection with the imaginary axis. This may be determined from the characteristic equation of the closed loop and the Routh–Hurwitz criteria:

The characteristic equation is

$$(s+3)(s+1)(s^2+2s+2)+K=0$$

i.e.

$$s^4+6s^3+13s^2+14s+(6+K)=0.$$

The array is

s^4	1	13	$6+K$
s^3	6	14	
s^2	$\dfrac{78-14}{6}=\dfrac{64}{6}$	$6+K$	
s^1	$14-\dfrac{6(6+K)6}{64}$	0	
s^0	$6+K$		

When K reaches a value so that the roots of the characteristic equation cross the imaginary axis we have a limiting K for stability and the roots of the characteristic at this value of K are determined by the auxiliary equation in s^2. The limiting value of K is given by the first column of the array:

$$K>-6$$

and

$$170-9K>0$$

i.e.

$$K<170/9=18\cdot9.$$

Substituting the latter value of K in the auxiliary equation:

$$64s^2+36+170\times6/9=0$$

i.e.

$$s=\pm j1\cdot53.$$

(ix) The breakaway point on the real axis will cause the branches to leave the breakaway point at $180°/2=90°$. Rules (x) and (xi) summed together give the condition for finding the breakaway point. Since there are only poles

$$-\frac{1}{q-1}-\frac{2(q-1)}{(q-1)^2+1}=-\frac{1}{3-q}-\frac{2(1-q)}{(1-q)^2+1}.$$

Solution for q in this way requires the solution of a cubic in q. Use of the general analytical expression from the characteristic equation, rule (xiii) also results in a cubic

$$4s^3 + 18s^2 + 26s + 14 = 0.$$

This is in a more tractable form. Solution yields three values but only one falls in the required region at $-2\cdot4$ by trial and error solution.

(xiv) To determine the departure direction of the branches from the complex poles we need only consider one of the complex poles because of symmetry. Taking the upper pole,

$$\phi_{\text{departure}} = 180(2m+1) - 90° - 90° - \tan^{-1} 0\cdot5$$
$$= 180(2m+1) - 180 - 26\cdot6°$$
$$= -26\cdot6° \ (323\cdot4°).$$

For the lower pole the angle of departure is $+26\cdot6°$. To complete the roots-loci it is necessary to work away from the precisely known region in small search areas, the angle and/or the magnitude conditions being satisfied at each point by trial and error location. With the knowledge so far we can draw the roots-loci to a well-defined pattern however. The initial sketch and this may be compared. Note that full accurate plotting requires extensive computation.

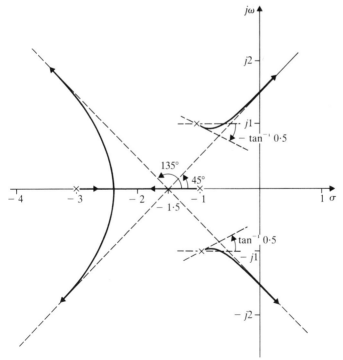

Fig. 4.33 Roots loci plot (compare Fig. 4.32)

The accurate plotting may require and benefit from the use of a computer to evaluate sufficient points on the loci. The use of 'pocket calculators' enables rapid solution of trial and error calculations for the angle and magnitude conditions, perhaps in preference to such aspects as the Louth-Hurwitz method for the imaginary axis crossing points. However, the above rules lead to a logical progression to the shape of the roots loci plot and are recommended in gaining familiarity with the root locus method.

4.4 Summary

In earlier chapters we have considered the dynamics of general systems without placing much emphasis on their structure. Now, single loop feedback control has been introduced and characteristics of feedback systems in terms of their time (transient) response have been considered. Early control theory has been based on the single input–single output concept in which the effect of one input is referred to a particular single output. Linearity and the algebra of Laplace transforms has been retained and the response of simple, first and second order, systems with feedback may then easily be established. For more complex higher order systems the response from a finite input may more readily become unbounded, i.e. the system becomes unstable. It is then necessary to restrict the range of certain parameters of the system to ensure that instability is not reached. In the 'time domain' the conditions for stability/instability may be studied by the Routh–Hurwitz and roots loci methods as well as by full solution of the dynamic equations. The underlying consideration is the presence of poles of the closed loop transfer function in the right (positive) half of the complex plane.

Chapter 5
Single input–single output feedback control: frequency response

5.1 Introduction

This chapter is a continuation of the material of Chapter 4 in that it extends the discussion of the single loop control system. The techniques presented here are based on the use of the frequency response of the general single input–single output system as first presented in Chapter 3. A major consideration will remain that of stability but a more quantitative interpretation will be sought, as distinct from the absolute stability or instability which has dominated so far. The methods of frequency response representation are the polar (Nyquist) plot and the log magnitude and phase plots as functions of the exciting frequency (Bode). To consider the use of these plots in the areas of feedback, stability and dynamic system improvement, it is necessary to extend our understanding beyond that of the 'representation' of a cycling forced system.

5.2 The Nyquist criterion

The use of the polar plot to represent frequency response was introduced in section 3.2. In this section we shall be particularly concerned with the polar plots for feedback systems and their use in the study of system stability. The polar or Nyquist plot indicates not only absolute stability but it enables the degree of stability (relative stability) and steady state errors to be assessed. It also indicates by its general shape how system performance can be improved. In Chapter 3 a function $G(j\omega)$ was plotted directly in the complex plane. To relate the polar plot more directly to the basis of system dynamics and to the poles and zeros of the various transfer functions, we shall look at the source of the plot more closely, in particular how its shape and position are related to the closed loop poles.

For the standard closed loop with forward transfer function $G(s)$

and feedback transfer function $H(s)$ the closed loop transfer function is

$$\frac{C(s)}{R(s)} = \frac{G(s)}{1 + G(s)H(s)}.$$

The denominator, $F(s) = 1 + G(s)H(s)$, may be expressed in terms of its own poles and zeros and if $F(s)$ is equated to zero then we have the characteristic equation

$$F(s) = 1 + G(s)H(s)$$
$$= \frac{(s - z_1')(s - z_2') \ldots (s - z_v')}{s^n(s - p_1')(s - p_2') \ldots (s - p_u')}$$
$$= 0. \tag{5.1}$$

s is the complex variable $\sigma + j\omega$ having a real part σ and imaginary part $j\omega$. The zeros, z_i', of $F(s)$ are the roots of the characteristic equation, equation (5.1). These z_i' are also of course the poles of the closed loop transfer function $C(s)/R(s)$ so that for a stable closed loop system none of these must have a positive real part. The poles of $F(s)$, p_i' and those at the origin from s^n, are also the poles of the open loop transfer function $G(s)H(s)$. If any of these have a positive real part, i.e. lie in the right half of the complex s-plane then the open loop system is unstable but the closed loop system is still stable if the zeros of $F(s)$ are all in the left half of the plane. This is the important feature of the feedback system and the Nyquist Stability Criterion is based on determining the position of the zeros of $F(s)$ (i.e. the closed loop poles) from a consideration of the open loop poles and zeros, i.e. of $G(s)H(s)$. These relationships are summarized in Table 5.1.

Table 5.1   Relationship between system zeros and poles

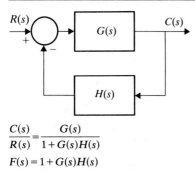

$$\frac{C(s)}{R(s)} = \frac{G(s)}{1 + G(s)H(s)}$$

$$F(s) = 1 + G(s)H(s)$$

(i)   The zeros of $F(s)$ are the roots of the characteristic equation $F(s) = 0$

(ii)   The zeros of $F(s)$ are the poles of the closed loop transfer function $\dfrac{C(s)}{R(s)}$

(iii)   The poles of $F(s)$ are the poles of the open loop transfer function $G(s)H(s)$

Nyquist's criterion may be developed by a rigorous analytical analysis (Nyquist 1932) but it may also be shown by a qualitative argument based on the ideas of mapping from one plane to another.

Mapping and transformation

A major concept which is required is that of the 'enclosure' of a region in a plane. A region in a plane, say the complex *s*-plane, is said to be enclosed only if it is on the right side of a closed path when the path is traced in a prescribed direction. (An alternative definition of enclosure, i.e. taking the left-hand side, is just as acceptable but that given here is the most widely used.) Thus in Fig. 5.1 although both points *A* and *B* are encircled by a closed path only point *A* is said to be enclosed by the path around it. The region enclosed by each part is on the side of the path which is shaded. Thus point *A* is enclosed both by the closed path (*a*) and by the closed path (*b*). *B* is enclosed by neither. The taking of points, curves or regions in one plane, say the *s*-plane, and the plotting of the corresponding points of a function, say *F*(*s*), in another plane is called mapping. For every value or point in the *s*-plane there will be one value or point in the *F*(*s*) plane. Thus the mapping of a point will give a point, and the mapping of curves and regions will give curves and regions in the plane correspondingly. For an analytic function *F*(*s*), i.e. one for which the derivative exists, the mapping is conformal. This is a mapping which preserves both the size and the sense of angles. Orientation may, however, change and some excellent illustrative examples appear in the literature (see Ogata (1970) and Fig. 5.2).

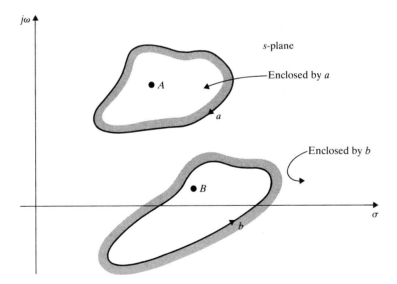

Fig. 5.1 Enclosed regions

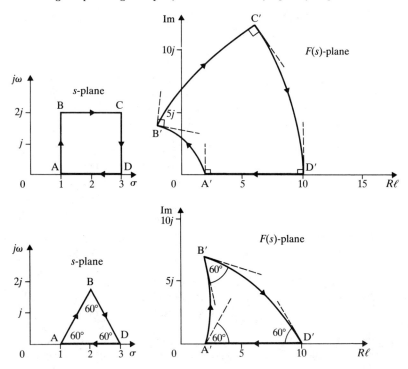

Fig. 5.2 Examples of conformal mapping from the s-plane to the $F(s)$ plane for $F(s) = s^2 + 1$

Mapping from the s-plane to the $F(s)$ plane is a one to one process but the reverse is not generally true, i.e. if $F(s)$ is multivalued in s so that $F(s) = 0$ has a number of roots in s. For example if $F(s)$ is $(s^2 + 1)$ then for $s = 1$, $F(s)$ is equal to 2. But for $F(s) = 2$ we have $s^2 = 1$ and s is equal to plus one or minus one.

Let us examine the simple examples of mapping from the s-plane to the $F(s)$ plane when the function $F(s)$ has solely a simple zero or a simple pole. For a simple zero

$$F(s) = s - z_1' \quad \text{with} \quad z_1' \text{ real.}$$

This mapping is shown in Fig. 5.3.

The function $F(s) = s - z_1'$ will map any point, say s_1, in the s-plane into a point, F_1 in the $F(s)$ plane, where $F_1 = s_1 - z_1'$. If we consider the particular point $s = z_1'$, i.e. the zero of $F(s)$, then

$$F(z_1') = s - z_1'$$
$$= z_1' - z_1'$$
$$= 0$$

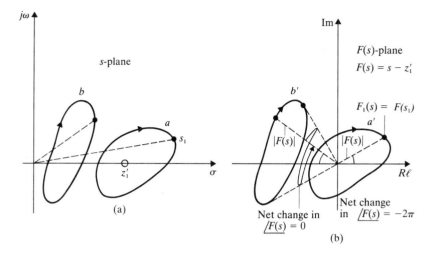

Fig. 5.3 Mapping from the s-plane (a) to the $F(s)$ plane (b) for $F(s) = s - z_1'$

i.e. the zero z_1' maps into the origin of $F(s)$. Any closed contour, (a), which encircles the s-plane zero (z_1') must encircle the $F(s)$ plane origin when mapped into the $F(s)$ plane and the direction of rotation is retained. Any closed contour which does not encircle the zero in the s-plane will also not encircle the origin in the $F(s)$ plane. The net change in the phase (angle) of $F(s)$ as s takes on all the values along the contour (a) as it traverses it in the direction shown is 2π radians, the modulus, $|F(s)|$, sweeping through a full 2π clockwise. As (b) is traversed it can be seen that the net change in angle in $F(s)$ is zero as $F(s)$ sweeps along the mapping (b').

If instead of a zero we consider $F(s)$ to have a simple pole, e.g. $F(s) = (s - p_1')^{-1}$, then the argument is similar. Any value, e.g. s_1, will map into the $F(s)$ plane at $F_1(s) = (s_1 - p_1')^{-1}$. The pole, i.e. $s = p_1'$ does not itself map into the $F(s)$ origin but a closed contour about the pole will map into a closed contour about the origin but with the direction reversed. For example, if $p_1' = 2 + j0$, the point $s = 1 + j1$ maps into the $F(s)$ plane at the point $F(s) = (1 + j - 2)^{-1} = \frac{1}{2}(-1 - j)$. Thus any point P in the s-plane which is enclosed by a closed path as shown in Fig. 5.4 will map into the $F(s)$ plane as shown, Q. The point Q in the $F(s)$ plane is 'enclosed' according to the enclosure definition above but it is not encircled.

Let us now start from a fuller, more general open loop transfer function, say

$$G(s)H(s) = \frac{K(s - z_1)}{s(s - p_1)(s - p_2)}.$$

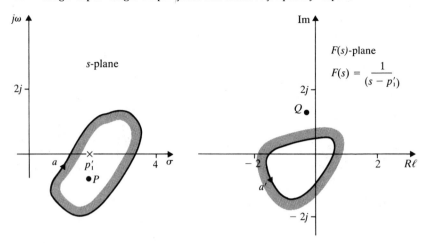

Fig. 5.4   Mapping of a contour from the s-plane to the $F(s)$ plane, $F(s) = (s - p_1')^{-1}$

The form of the closed-loop denominator will then be

$$F(s) = 1 + G(s)H(s)$$
$$= \frac{s(s - p_1)(s - p_2) + K(s - z_1)}{s(s - p_1)(s - p_2)}$$
$$= \frac{(s - z_1')(s - z_2')(s - z_3')}{s(s - p_1)(s - p_2)}$$

where the z_i' are the roots of the characteristic equation, the closed loop poles. These z_i' will be functions of the gain K, as we saw also in the study of the roots loci. Thus for a specific K, the roots z_1', z_2', z_3' will have specific values, e.g. Fig. 5.5.

We now turn our attention to the zeros and poles of $F(s)$ and consider contours drawn in turn around the zero z_2', the pole p_1 and the pair z_1', and p_2 (Fig. 5.6). The factors of $F(s)$ may be expressed in terms of their modulus and argument so that for some $s = s_1$

$$(s_1 - z_1') = |s_1 - z_1'| \,\underline{/s_1 - z_1'}$$
$$= A_1 e^{j\theta_1}$$

where A and θ are defined by this relationship and Fig. 5.6.

To see that this is so in more detail remember that both s_1 and z_1' may be complex numbers, each having a real and imaginary part. If, (Fig. 5.7),

$$s_1 = \sigma_s + j\omega_s$$

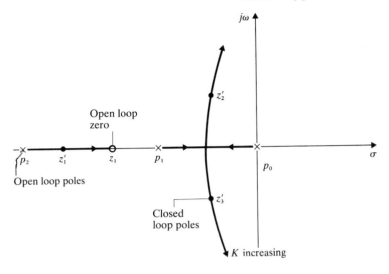

Fig. 5.5 Possible closed loop poles z'_1, z'_2, z'_3 for a given gain K and their loci as K changes

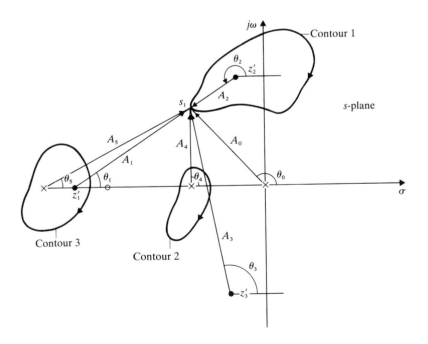

Fig. 5.6 Computation of $F(s)$

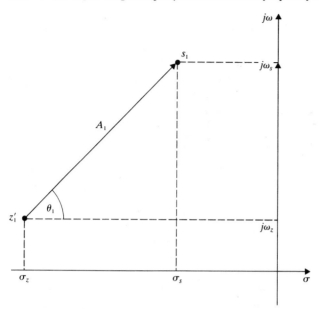

Fig. 5.7   Relationship between points in the complex plane

and
$$z_1' = \sigma_z + j\omega_z$$
then
$$s_1 - z_1' = (\sigma_s - \sigma_z) + j(\omega_s - \omega_z)$$

It is seen from Fig. 5.7 that
$$\omega_s - \omega_z = A_1 \sin \theta_1$$
$$\sigma_s - \sigma_z = A_1 \cos \theta_1$$
so that
$$s_1 - z_1' = A_1 (\cos \theta_1 + j \sin \theta_1).$$

But from de Moivres theorem
$$e^{j\theta} = \cos \theta + j \sin \theta$$
so that
$$(s_1 - z_1') = A_1 e^{j\theta_1}$$

Then, using this notation for each factor of $F(s)$,
$$F(s) = \frac{A_1 e^{j\theta_1} . A_2 e^{j\theta_2} . A_3 e^{j\theta_3}}{A_0 e^{j\theta_0} . A_4 e^{j\theta_4} . A_5 e^{j\theta_5}}$$
$$= \left| \frac{A_1 A_2 A_3}{A_0 A_4 A_5} \right| (\underline{/\theta_1} + \underline{/\theta_2} + \underline{/\theta_3} - \underline{/\theta_0} - \underline{/\theta_4} - \underline{/\theta_5}).$$

As the point s_1 is taken around contour 1 in the complex plane in the direction shown this new expression for $F(s)$ can be used to produce the conformal mapping of the contour into the $F(s)$ plane. For the contour 1 as drawn we may tabulate the net angle contribution to $F(s)$ as the full traverse around contour 1 is completed, bearing in mind the above consideration for simple pole and zero plottings. These angle contributions are given in Table 5.2.

Table 5.2   Angle contributions to $F(s)$ for a full traverse of contour 1

Pole or zero	Angle contribution (positive angles anticlockwise)
z_1'	0
z_2'	-2π
z_3'	0
p_1	0
p_2	0
p_{origin}	0
Total	-2π

Thus the mapping of contour 1 into the $F(s)$ plane gives a contour which encircles the origin as shown (Fig. 5.8), the total angle swept through about the origin by $|F(s)|$ being -2π. Considering contours 2 and 3 similarly we see that the total angle contributions to $F(s)$ are 2π and zero respectively (Tables 5.3 and 5.4, and Fig. 5.8).

With the knowledge gained from these observations we may consider the Nyquist stability criterion itself.

The Nyquist path and stability criterion

From the above section the following general principles result. For a single-valued function $F(s)$, (i) for each point in a specified region in the s-plane there is a corresponding point in the $F(s)$ plane and (ii) the mapping results in the transformation of a general s-plane locus to a

Table 5.3   Angle contributions to $F(s)$ from contour 2

Pole or zero	Angle contribution
z_1'	0
z_2'	0
z_3'	0
p_1	2π
p_2	0
p_{origin}	0
Total	2π

Table 5.4 Angle contributions to $F(s)$ from contour 3

Pole or zero	Angle contribution
z_1'	-2π
z_2'	0
z_3'	0
p_1	0
p_2	2π
p_{origin}	0
Total	0

locus in the $F(s)$ plane. We deduce as well that (iii) for a closed path in the s-plane the corresponding mapped path in the $F(s)$ plane will be closed and will encircle the origin as many times as the difference between the number of the zeros and the number of the poles of $F(s)$ that are encircled by the s-plane path. Thus if

> Z = number of zeros of $F(s)$ encircled by the s-plane path in the s-plane
>
> P = number of poles of $F(s)$ encircled by the s-plane path in the s-plane (i.e. open loop poles).
>
> N = number of encirclements of the origin made by the $F(s)$ path in the $F(s)$-plane

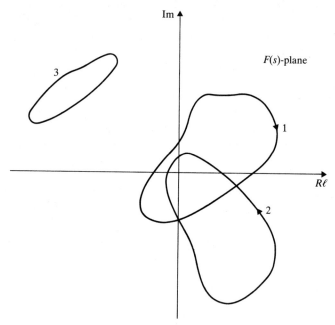

Fig. 5.8 Encirclements of the origin of $F(s)$ plane

then

$$N = Z - P. \tag{5.2}$$

If N is positive the $F(s)$ path encircles the origin N times and the $F(s)$-plane path and the s-plane path are in the same direction. If N is negative the direction of encirclement in the $F(s)$ plane is opposite to that in the s-plane. The significance is that if the s-plane closed path is suitably defined this principle can be used for the determination of stability, i.e. to find if any roots of the characteristic equation $F(s) = 0$ (poles of the closed loop transfer function) lie in the right half of the s-plane. This we do by use of the path defined in Fig. 5.9, the Nyquist path. Alternative conventions exist, the path being drawn either clockwise or anti-clockwise. (The alternatives do not usually create confusion provided a consistent view is taken in any problems.) Consistent with our chosen definition of enclosure the path is shown clockwise and encloses the entire right half of the s-plane but is deviated around any poles of $F(s)$ (open loop poles) which fall on the imaginary axis. The mapping of this path into the $F(s)$ plane is the Nyquist plot of the system.

Now that the special Nyquist path is specified the stability of the closed loop system is determined by first plotting the path of $F(s)$ as s takes on values along the Nyquist path, and then observing the nature

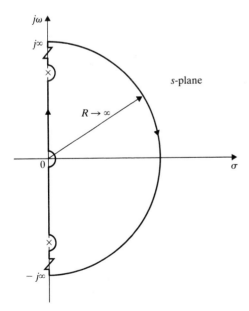

Fig. 5.9 The Nyquist path

of the $F(s)$ path with respect to the origin in the $F(s)$-plane. We shall see that because of the symmetry of the s-plane and because the whole of the right half of the s-plane is enclosed by the locus, we need only consider during use that initial part of the path from the origin region $s = 0 + j0$ to $s = 0 + j\infty$. Because $F(j\omega)$ is so strongly related to the open loop frequency transfer function $G(j\omega)H(j\omega)$, this first part of the Nyquist plot is equivalent to the open loop polar plot already considered in Chapter 3. The 'frequency' ω appears again as a parameter along the curve. However, consider initially the full region in the s-plane as shown and equation (5.2).

(i) $P = 0$. If the poles of $F(s)$ which are the open loop poles, are all located in the left half of the complex s-plane or on the imaginary axis then the Nyquist path does not enclose them so for this contour P is zero. This means that the open loop system is stable. If also the Nyquist plot of $F(s)$ does not encircle the origin of the $F(s)$ plane then N is zero. In this case, given P is zero with N also zero equation (5.2) gives

$$Z = N + P$$
$$= 0.$$

But for the closed loop system to be stable we know already that there must be no closed loop pole, i.e. no zero of $F(s)$, in the right half plane. Thus for stability we require that

$$Z = 0$$

and if P is also zero then from equation (5.2) the Nyquist plot of $F(s)$ must not enclose the origin in the $F(s)$ plane. That is N must also be zero. If Z is not equal to zero, i.e. there are closed loop poles in the right half plane, then this instability will be shown by the $F(s)$ locus enclosing its origin, the number of times it encloses the origin being equal to Z.

(ii) $P \neq 0$. If P is not zero, i.e. there are open loop poles (poles of $F(s)$) in the right half plane, then the open loop response is unstable. Since we must ensure for a closed loop system to be stable that $Z(F(s)$ zeros, closed loop poles in the right half plane) must be zero then for a stable system

$$Z = 0$$

and

$$N = -P.$$

That is, to show stability the Nyquist plot of $F(s)$ must now encircle the origin P times but in the reverse direction. Thus although the origin is encircled it is still not enclosed for a stable system. For the majority of cases, when $G(s)$ and $H(s)$ are stable functions, P is zero and hence for stability N must also be zero.

Although the criterion is established on the function $F(s)$ and the origin in the $F(s)$ plane, this function is always derived from the open loop function $G(s)H(s)$ and the $G(s)H(s)$ plane is normally used. If $H(s)$ is unity then we simply use the $G(s)$ plane. This relationship between $F(s)$ and $G(s)H(s)$ is depicted in Fig. 5.10 where the transformation or mapping in this case is equivalent solely to a sideways shift of the plot or to a shift of the imaginary axis, the origin of the $F(s)$ plane corresponding to the point $(-1, j0)$ in the $G(s)H(s)$ plane, i.e.

$$G(s)H(s) = F(s) - 1.$$

The Nyquist stability criterion can be restated in terms of the plot in the $G(s)H(s)$ plane.

For a stable closed loop system the Nyquist plot of $G(s)H(s)$ should encircle the $(-1, j0)$ point as many times as there are poles of $G(s)H(s)$ in the right half of the s-plane. The encirclements, with the chosen convention, will be in a counterclockwise direction. Since in general the open loop transfer function $G(s)H(s)$ may itself be stable, i.e. P is zero, the Nyquist plot of $G(s)H(s)$ for a stable system will not then enclose the new critical point $(-1, j0)$ at all.

In Chapter 3 the general plotting of polar plots was covered. This procedure is followed for the Nyquist path but of the full Nyquist path shown in Fig. 5.9 it is necessary only to consider the range of values of s from 0 to $0 + j\infty$. Since our physical systems give open loop functions $G(s)H(s)$ with the denominator of higher order than the numerator, as s increases $G(s)H(s)$ maps into its origin and the small semicircle around the s-plane origin maps into a finite value or tends to infinity in the $G(s)H(s)$ plane. In the later case the mapping sweeps a semicircle of infinite radius about the origin. By symmetry the Nyquist plot for values of s from zero to $0 - j\infty$ yields a symmetrical mirror image about the real axis to that obtained for values of s from zero to $0 + j\infty$. The

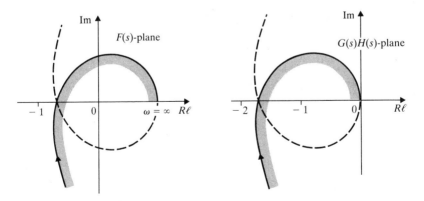

Fig. 5.10 Equivalent plots in the $F(s)$ and $G(s)H(s)$-planes

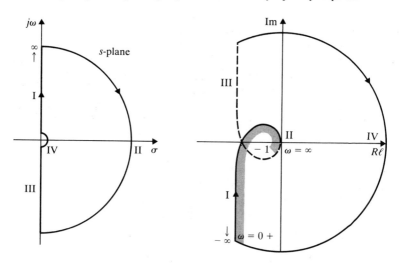

Fig. 5.11 Equivalent regions of the s-plane and $G(s)H(s)$-plane for a possible third order system, i.e. $G(s)H(s)$ third order in s

Nyquist plot is thus reduced normally to using the range s equals zero to s equals $j\infty$ only, and as such it becomes synonymous with the frequency response polar plot for the open loop system. Using the definitions of this chapter, the region to the right of the $G(s)H(s)$ plot as ω is increased from zero to infinity is said to be enclosed by the plot (Fig. 5.11). As drawn, the figure shows that section I of the plot passes above the critical point, i.e. the critical point is to the right and hence enclosed by this section of the plot. This alone is sufficient to indicate that the closed loop system is unstable. The full plot encloses the critical point twice and hence, in the absence of open loop zeros in the right half of the s-plane, we know that there must be two zeros of $F(s)$, poles of the closed loop transfer function, in the right half of the s-plane. In further diagrams only section I will generally be shown.

Effects of additional open loop poles and zeros

In Chapter 4 the effect of additional open loop poles and zeros on the root locus and stability were discussed. The corresponding changes for the Nyquist plot are shown here.

(a) *The addition of poles.* The addition of open loop poles increases the possibility of instability. Figure 5.12 shows that closed loop instability is not possible (with non-positive open loop zeros and negative feedback) with first and second order open loop functions $G(s)H(s)$. At infinite frequency in all cases the modulus of $G(s)H(s)$ tends to zero but the addition of a finite pole, with a negative real part results in the $G(s)H(s)$ plot going around a further $\pi/2$ radians in the

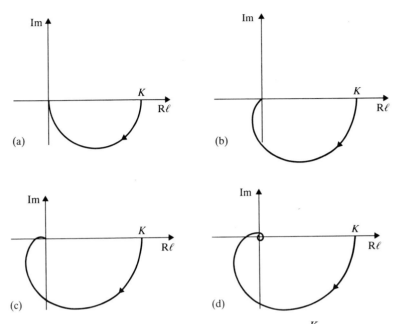

Fig. 5.12 General form of Nyquist plots for $G(s)H(s) = (a)\dfrac{K}{1+T_1s}$,

(b) $\dfrac{K}{(1+T_1s)(1+T_2s)}$, (c) $\dfrac{K}{(1+T_1s)(1+T_2s)(1+T_3s)}$, (d) $\dfrac{K}{(1+T_1s)\ldots(1+T_ns)}$

clockwise direction. The approach to the origin is asymptotic to either the real or imaginary axis depending on system order.

The notation of Fig. 5.12 is used in order that the steady state gain ($\omega = 0$) is constant at K. The $T_1,\ldots,T_n$ are the corresponding time constants the poles being located in the s-plane at $-1/T_1$, $-1/T_2$ etc. In the case of (c) and (d) the point at which the plot cuts the real negative axis determines the stability or instability of the closed loop system.

If the open loop poles added are poles at the origin, i.e. factors of s^{-1} in $G(s)H(s)$, then both the starting ($\omega = 0$) and finishing ($\omega \to \infty$) parts of the plot are rotated by $\pi/2$ radians for each pole added. The plot itself only occurs in the same number of quadrants as it did before the poles at the origin were added, i.e. one quadrant if first order, two if second order and so on. The shape is modified however (Fig. 5.13). Note that the total order of $G(s)H(s)$ still determines the 'finishing' quadrant of the plot.

(b) *The addition of zeros.* The addition of open loop zeros can have a significant effect on stability but the effect is dependent on the relative magnitudes of the system parameters. However a system which is unstable may be stabilized by adding zeros to the open loop transfer function, e.g. Fig. 5.14.

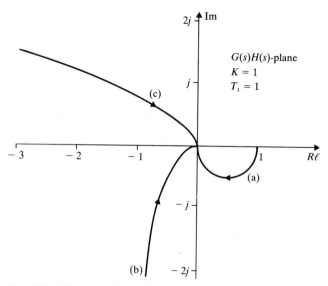

Fig. 5.13 Effect of poles at the origin,

$$G(s)H(s) = \text{(a)}\ \frac{K}{1+T_1s}, \quad \text{(b)}\ \frac{K}{s(1+T_1s)}, \quad \text{(c)}\ \frac{K}{s^2(1+T_1s)}$$

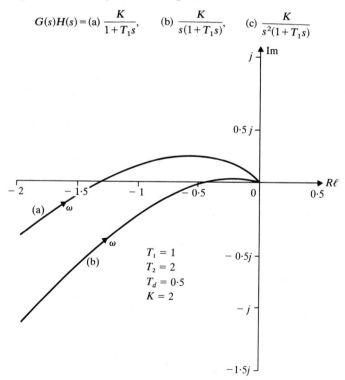

Fig. 5.14 The effect of adding a zero (e.g. derivative action) to the open loop.

$$G(s)H(s) = \text{(a)}\ \frac{K}{s(1+T_1s)(1+T_2s)}, \quad \text{(b)}\ \frac{K(1+T_ds)}{s(1+T_1s)(1+T_2s)}$$

The use of the Nyquist plot for showing the relative stability of systems will be taken up after we have reconsidered the Bode plot. The simultaneous addition of both poles and zeros enables one to rotate and change the shape of the Nyquist plot considerably. That is the high and low frequency characteristics of the system may be changed to a large extent independently and this is also discussed further later.

The inverse Nyquist diagram

The Nyquist criterion may be applied with some modification to the inverse polar plot. The inverse plot is that of the function $[G(s)H(s)]^{-1}$. As poles predominate in number in the open loop functions over zeros the combination of transfer functions and their subsequent polar plot construction may be quite laborious (Fig. 3.29). Using the inverse plot may be more rapid in such cases although in general its use is not so widespread as is that of the direct polar plot. Dividing $F(s)$ by $G(s)H(s)$ we see that a similar function $F'(s)$ is given

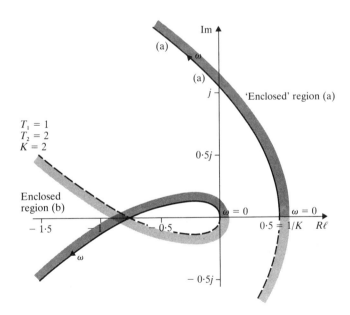

Fig. 5.15 Inverse Nyquist plots,

$$[G(s)H(s)]^{-1} = \text{(a)} \; \frac{(1+T_1 s)(1+T_2 s)}{K}, \qquad \text{(b)} \; \frac{s(1+T_1 s)(1+T_2 s)}{K}$$

by

$$F'(s) = \frac{1 + G(s)H(s)}{G(s)H(s)}$$

$$= \frac{1}{G(s)H(s)} + 1.$$

The zeros of $F'(s)$ are the same as those of $F(s)$ and hence are the poles of the closed loop transfer function. The poles of $F'(s)$ are not equal to the poles of $F(s)$ but are equal to the poles of $[G(s)H(s)]^{-1}$, the zeros of $G(s)H(s)$. Reasoning along the lines for the Nyquist criterion and the normal polar plot leads to the conclusion that for a stable system the net number of rotations of $F'(s)$ about the origin must be clockwise (on the chosen convention) and equal to the number of poles P' of $[G(s)H(s)]^{-1}$ (zeros of $G(s)H(s)$) that lie in the right half of the s-plane. As discussed above P' will normally be zero for physical systems so that then for stability the net rotations about the origin in the $F'(s)$ plane must be zero. Similarly to before we plot in the $[G(s)H(s)]^{-1}$ plane rather than in the $F'(s)$ plane so that the critical point is once again at $(-1, j0)$. Figure 5.14 shows examples of stable and unstable closed loop systems using the inverse plots of the open loop functions.

5.3 Stability and the Bode plots

The Bode plot was introduced as a general graphical method for representing the frequency response of a system in Chapter 3. The frequency response of a complex system is easily built up from the simple first and second order factors of the more complex function. In addition to just showing the dynamic gain of the system and the phase lag introduced the Bode plot, like the other methods, illustrates the conditions for instability to occur in closed loop systems.

In the Nyquist analysis it was seen that instability occurred in the closed loop system if the polar plot of the open loop frequency response went beyond the critical point $(-1, j0)$. That is, if the amplitude of $G(s)H(s)$ exceeds unity when the phase angle is $-180°$. This result was reached purely by consideration of the response of a system in terms of the poles and zeros in its analytical expression, and the need for the poles of the closed loop transfer function to have negative real parts for stability to be maintained. Thus the form of the closed loop response, i.e. whether stable and unstable, when the system is excited in some (non-oscillatory) manner, e.g. pulse, step, random input, has been related to the frequency response of the open loop system. The critical point of the Nyquist plot $(-1, j0)$, is equivalent on the Bode plot to the two conditions, log magnitude ($\mathcal{L}m$) of

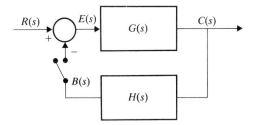

Fig. 5.16 Open loop system for stability discussion

$G(s)H(s)$ is 20 log 1, i.e. zero, and the phase angle is $-180°$. Consider now a further qualitative argument to illustrate the theory.

In Fig. 5.16 the loop is 'opened' in the feedback as shown. The reference input, e.g. controller set point, is varied sinusoidally and the frequency adjusted until the feedback value B is exactly $180°$ out of phase, lagging, the input. If now the break in the feedback path is reconnected a further $180°$ phase lag is introduced by the negative sign at the summing junction so that the feedback signal is in phase (equivalent to $360°$ out of phase for a steady sinusoidal input) with the reference input. The reference input may now be stopped and all excitation of the system can come from its own feedback. If the gain of the combined forward and feedback paths, i.e. of $G(s)H(s)$, is less than unity at this frequency then the oscillations in the system will die away to zero. If this gain is unity the oscillations within the system and

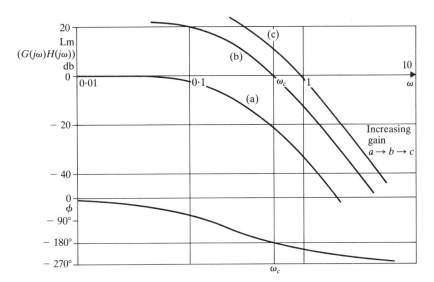

Fig. 5.17 Stability and the Bode diagram, (a) stable, (b) critically stable and (c) unstable closed loop systems showing the effect of increased open loop gain

hence at the output will continue at a steady amplitude. If the gain exceeds unity then the amplitude will continue to increase without limit, the amplitude increasing in magnitude each time that the signal passes around the loop. This means that the closed loop system will be unstable if the open loop frequency response has a gain greater than unity at the frequency giving an open loop phase lag of 180°. As expected this qualitative argument agrees with the earlier mathematical reasoning.

This basic limiting condition for stability is shown in the Bode plot of Fig. 5.17, together with the frequency response for a stable system (*a*) and an unstable system (*c*).

5.4 Relative stability

So far consideration has been concentrated on the absolute stability of a closed loop system. However, we may use the representation of system dynamics to tell us how stable a system is, how much scope exists for the alteration of system parameters, and how stability may be improved. In the discussion of the Nyquist plot some indication of these uses has already been given but to establish a numerical measure of stability the gain margin and phase margin are introduced. These measures of the degree of stability of a system may be illustrated on both the Nyquist and Bode plots and also on the lesser used magnitude-phase plot.

The gain margin

The gain margin and phase margin only really exist where there is a possibility of both stable and unstable behaviour resulting from a plant or system as parameters, e.g. controller gain, are varied. Thus we are concerned mainly with those systems whose frequency responses show that at some frequency a phase lag of 180° is produced by the open loop. The Nyquist plot (a polar plot of the open loop function) will have a form near the origin which can be illustrated by Fig. 5.18. If the variable gain factor of the open loop, K, is increased the Nyquist plot may move from a position indicating stability, through the critical point $(-1, j0)$ to a position indicating instability. The frequency on any plot at which the phase lag becomes 180° is the phase crossover frequency, ω_c. The gain margin is defined as the reciprocal of the magnitude $|G(j\omega_c)H(j\omega_c)|$ at the phase crossover frequency. Expressed in terms of decibels this is

$$\text{Gain margin } (G.M.), \text{ db} = 20 \log \frac{1}{|G(s)H(s)|_{s=\omega_c}}$$

$$= -20 \log |G(s)H(s)|_{s=\omega_c}$$

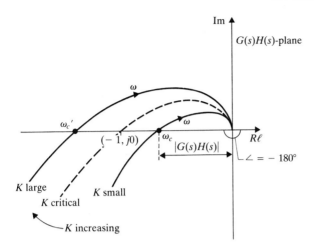

Fig. 5.18 Basic plot for consideration of relative stability

Thus if the absolute value of gain margin is greater than unity then its expression in decibels is positive, and if less than unity then it is negative when expressed in db. Thus a positive gain margin in db means that the system is stable and a negative one that the system is unstable. The gain margin (db) is thus the amount by which the gain can be allowed to increase before the system becomes unstable (Fig. 5.21). For stable first and second order systems the gain margin is obviously infinite since this negative section of the real axis is never cut by the open loop plot. As the plot passes through $(-1, j0)$ the gain margin is unity, zero db.

The numerical value of the gain margin alone may be insufficient to indicate the relative stabilities of two systems. Figure 5.19 indicates two systems with the same gain margin but the stability of *B* may more readily be upset than that of *A* by small changes in system parameters.

The phase margin

The gain cross over frequency is the frequency at which $|G(s)H(s)|$, the magnitude of the open loop transfer function, is unity. The phase margin is the amount by which the open loop phase lag may increase at the gain cross over frequency, before instability is reached. Thus in terms of the Nyquist plot it may be defined as the angle through which the Nyquist locus must be rotated so that the unity magnitude point passes through the critical point $(-1, j0)$ (Fig. 5.20). The phase margin is 180° plus the open loop phase angle ϕ i.e.

Phase margin, $\gamma = 180° + \phi$

$$= 180° - \text{open loop phase lag.}$$

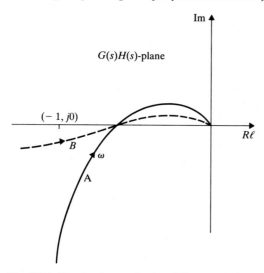

Fig. 5.19 Equal gain margins but different 'relative stabilities'

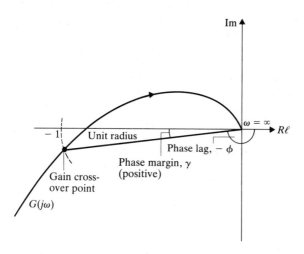

Fig. 5.20 The phase margin from the Nyquist plot

(Remember that phase lag is the negative of the phase angle which is taken as positive for a phase advance.) Thus the phase margin must be positive for stability.

The phase and gain margins may be shown on the Bode and magnitude-phase diagrams with equal ease. The equivalent representations are given in Fig. 5.21 for the general stable and unstable cases. Note that a positive gain margin in db (i.e. *G.M.*) is always associated with a possible phase margin and vice versa. Also, all that has been said

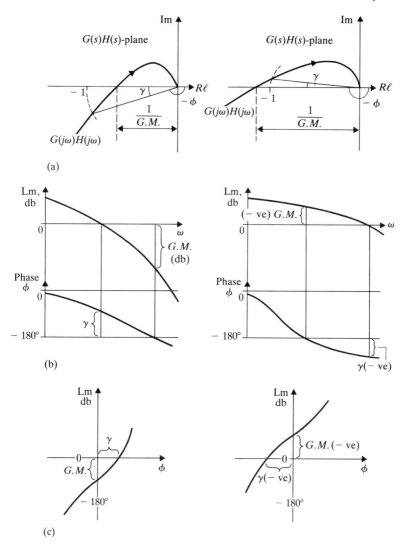

Fig. 5.21 Equivalent representations of gain and phase margins, (a) polar (Nyquist), (b) Bode, and (c) magnitude phase diagrams

in this section is true for a minimum phase system only, that is one having no open loop poles or zeros in the right half s-plane.

For a conditionally stable system, stability may be ensured either by increasing or decreasing the open loop gain. The gain margin then has little meaning (Fig. 5.22). If the critical point is on the section A then we have stability, if on B we have instability. By having certain requirements specified for the phase and gain margins we can tolerate

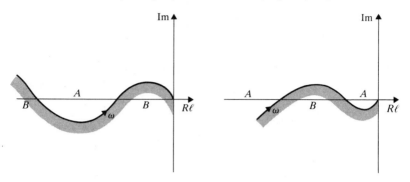

Fig. 5.22 Conditionally stable systems, Nyquist plots

some variation in the system parameters or component behaviour without a severe risk of instability occurring. The actual recommendations depend on particular plant circumstances, see Chapter 6, but a phase margin between 30 and 60° and a gain margin of more than 6 db will usually be sufficient (e.g. Oldenburger 1956; Gould 1969). Procedures to obtain these margins, e.g. by altering the gain K or by adding additional (compensation) control elements, will be outlined below.

Example The open loop transfer function of a third order system is

$$G(s) = \frac{K}{s^3 + 6s^2 + 11s + 6}.$$

Determine the gain and phase margins using the Nyquist and Bode plots for the closed loop when K is unity. Hence determine the limiting value of K to maintain stability. We can check that $K = 1$ will give us a stable system. The characteristic equation is $s^3 + 6s^2 + 11s + 7 = 0$ and using Routh–Hurwitz we could see if this is stable and we could also use this method as in examples in Chapter 4 to find the limiting gain. The open loop function factorizes as

$$G(s)H(s) = \frac{1}{(s+3)(s+2)(s+1)}$$

and it is easiest to use either the Bode plot as a graphical method or the inverse Nyquist. Because of the wider use of the Bode plot we shall use this method. The Bode plot is constructed using the rules of Chapter 3. Rearrange $G(s)H(s)$ as

$$G(s)H(s) = \frac{0 \cdot 167}{(1 + 0 \cdot 33s)(1 + 0 \cdot 5s)(1 + s)}$$

to give

$$\mathscr{L}m[G(j\omega)H(j\omega)] = 20\log_{10}\frac{1}{\sqrt{(1+0.33^2\omega^2)}} + 20\log_{10}\frac{1}{\sqrt{(1+0.5^2\omega^2)}}$$

$$+ 20\log_{10}\frac{1}{\sqrt{(1+\omega^2)}} + 20\log_{10}0.167$$

$$= 20\log_{10}0.167 - 20\log_{10}\sqrt{(1+0.33^2\omega^2)}$$

$$- 20\log_{10}\sqrt{(1+0.5^2\omega^2)} - 20\log_{10}\sqrt{(1+\omega^2)}.$$

The phase angle is

$$\underline{/G(j\omega)H(j\omega)} = -\tan^{-1}0.33\omega - \tan^{-1}0.5\omega - \tan^{-1}\omega.$$

The Bode plot is constructed initially in terms of its asymptotes. The corner frequencies are at $\omega = 3, 2, 1$ respectively. From the Bode plot the gain margin and phase margin are 35.5 db and, because the gain never exceeds unity, no finite value respectively. These are shown on the Nyquist plot (Fig. 5.24); the gain margin is 59.5 in absolute terms. Thus the gain K can be increased to 59.5 before instability is reached.

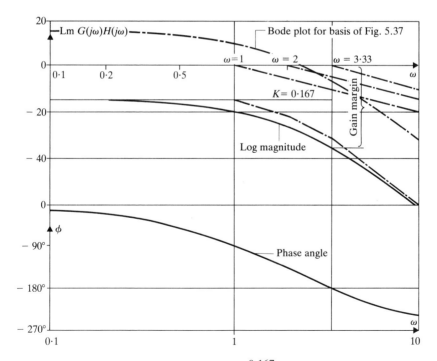

Fig. 5.23 Bode plot for $G(j\omega)H(j\omega) = \dfrac{0.167}{(1+0.33s)(1+0.5s)(1+s)}$

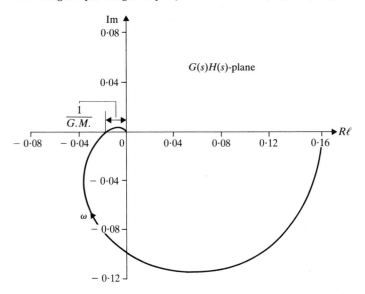

Fig. 5.24 Nyquist plot equivalent of Fig. 5.23

Relative stability and conformal mapping

If an experimental frequency response is obtainable then further information can be obtained by conformal mapping. The relationship between the position of the Nyquist plot and the position of the closed loop poles in the complex plane is further illustrated by this mapping from the s-plane to the $G(s)H(s)$ plane. It is possible to then estimate

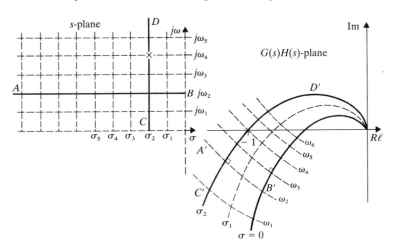

Fig. 5.25 Conformal mapping of the s-plane grid into the $G(s)H(s)$ plane

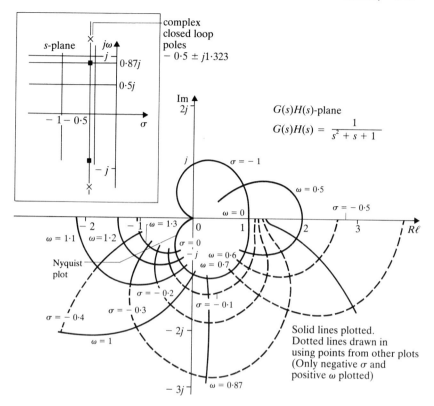

Fig. 5.26 (a) Mapping of second order system function (closed loop poles at $-0.5 \pm j1.323$)

the values of those closed loop poles closest to the imaginary axis, i.e. the dominant poles. (An alternative mapping is shown in West (1962).) The grid of the s-plane maps into the $G(s)H(s)$ plane according to the general pattern of Fig. 5.25. The actual position of the lines mapped into the $G(s)H(s)$ plane depending of course on the open loop function. Now the poles of the closed loop function, i.e. the roots of $1 + G(s)H(s) = 0$ must map into the point $G(s)H(s) = -1$, to satisfy their defining equations, that is, into the critical point $(-1, j0)$. The mapping that we get for two distinct cases for systems with two closed loop poles is shown in Fig. 5.26(a) and (b). It can be seen that the closer the closed loop poles are to the imaginary axis the closer the Nyquist plot (i.e. the mapping for $s = j\omega$) runs to the critical point.

Example The experimental determination of the open loop frequency response of a linear dynamic system is given in Fig. 5.27. Use this result to estimate the dominant closed loop poles, i.e. those poles with real parts closest to the imaginary axis and which control the

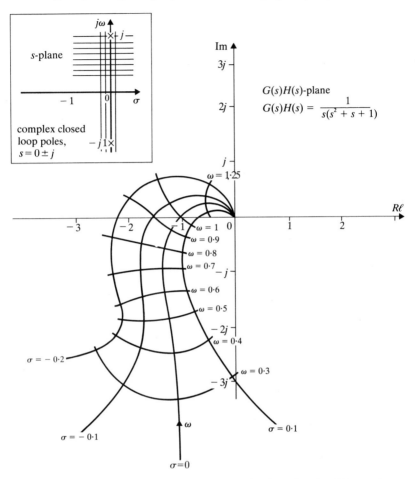

Fig. 5.26 (b) Mapping of third order system function (complex closed loop poles at $s = 0 \pm j$)

longer time behaviour of the system. What is the damping factor associated with these values and (as the system is underdamped) how rapidly do oscillations in the closed loop system output decay if it is disturbed momentarily from a steady condition? Using the principle of conformal mapping we may build up a grid on this frequency response corresponding to the grid of the $(\sigma, j\omega)$ complex plane.

We have seen that the constant real (σ) and imaginary $(j\omega)$ lines from the s-plane map into a grid in the open loop transfer function $G(s)$ plane. If the $(-1, j0)$ point in the $G(s)$ plane is found at the intersection of a constant σ and a constant ω curve mapped into that plane then these σ and ω values are those of the closed loop poles $\sigma \pm j\omega$. Since we move in our grid structure from $\sigma = 0$ to increasing negative values of σ these poles will also be the dominant poles.

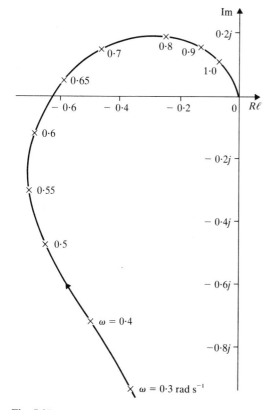

Fig. 5.27 Open loop experimental frequency response

If four points forming a square in the *s*-plane are mapped into the *G(s)* plane then this gives a 'curvilinear square', in which the sums of the lengths of the two pairs of opposite sides are equal. By taking as our *s*-plane square small increments in σ and ω we may map this grid into the *G(s)* plane by a series of curvilinear squares (Fig. 5.28). From the construction it is seen that the point $(-0\cdot07, j0\cdot59)$ maps (approximately) into the $(-1, j0)$ point. Thus the dominant closed loop poles are

$$s = -0\cdot07 \pm j0\cdot59$$

and the quadratic factor in the closed loop denominator is $(s+0\cdot07 - j0\cdot59) \cdot (s+0\cdot07+j0\cdot59)$. Factors of this nature give time response terms of the form

$$c = k e^{-0\cdot07t} \sin (0\cdot59t + \beta)$$

The time between oscillatory peaks is given by

$$2\pi = 0\cdot59t$$

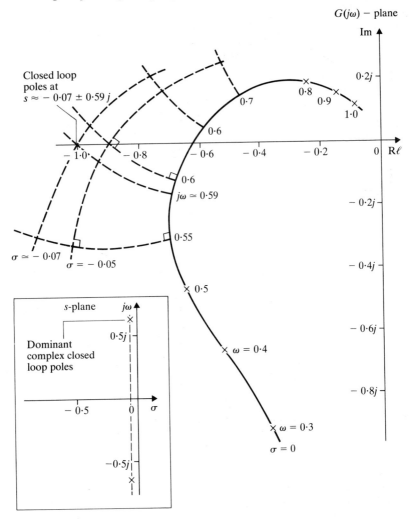

Fig. 5.28 Mapping of s-plane grid on to the given frequency response (mapping drawn 'by eye')

or

$$t = 11 \text{ (sec)}.$$

The magnitude ratio between successive peaks is

$$\frac{c_n}{c_{n+1}} = \frac{e^{-0.07t}}{e^{(-0.07t-0.07\times11)}}$$

$$= e^{11\times0.07}$$

$$= 2.2.$$

The quadratic factor expands to $(s+0.07)^2+0.59^2$ or $s^2+0.14s+0.35$ to give a damping factor ζ of

$$\zeta = \frac{0.14}{2\sqrt{0.35}} = 0.12.$$

Throughout the sections of this chapter so far the general open loop expression $G(s)H(s)$ has been used. Some simplification in notation could have been given by using the unity feedback loop with $H(s) = 1$ and working in the $G(s)$ plane since it is possible to reduce any feedback system to a transfer function in series with a unity feedback system (Fig. 3.8). The frequency response of the resulting system can be easily established using the Bode plot. In the following section we shall start by considering a unity feedback system.

5.5 Closed loop frequency response

In the earlier sections and Chapter 4 we have considered closed loop stability but where frequency response has been used to determine this stability it has been the frequency response of the open loop system which has been used. Let us now consider the frequency response of the closed loop systems and the relationship of this to the chosen stability criteria.

For the unity feedback system the open loop transfer function is simply $G(s)$, the closed loop transfer function is

$$\frac{C(s)}{R(s)} = \frac{G(s)}{1+G(s)}$$

and the error-to-input transfer function is

$$\frac{E(s)}{R(s)} = \frac{1}{1+G(s)}.$$

In the frequency domain we obtain the response by substituting $j\omega$ for s and the closed loop frequency response is given by

$$\frac{C(j\omega)}{R(j\omega)} = \frac{G(j\omega)}{1+G(j\omega)}$$

$$= \frac{G(j\omega)}{F(j\omega)}$$

$$= \frac{|G(j\omega)|\ e^{j\phi}}{|F(j\omega)|\ e^{j\lambda}}$$

where the phase angles ϕ and λ are also functions of ω. Rearranging leads to

$$\frac{C(j\omega)}{R(j\omega)} = \frac{|G(j\omega)|}{|F(j\omega)|} e^{j(\phi-\lambda)}$$

$$= \frac{|G(j\omega)|}{|F(j\omega)|} e^{j\alpha} \tag{5.3}$$

and similarly

$$\frac{E(j\omega)}{R(j\omega)} = \frac{1}{|F(j\omega)|} e^{-j\lambda}. \tag{5.4}$$

Figure 5.29 illustrates graphically the relationship

$$F(j\omega) = 1 + G(j\omega)$$

where $G(j\omega)$ may be taken directly from the Nyquist plot. That is, knowing only the Nyquist plot, or open loop frequency response, the amplitude $|F(j\omega)|$ and phase angle α may be determined. Using equation (5.3) the closed loop frequency response may then be determined also. Similarly measurement or calculation of λ enables us to establish the error-input relationship, equation (5.4). The magnitude of the closed loop response is found from $|G(j\omega)|/|F(j\omega)|$ and is referred to as *M*. The angle α is the phase angle between input and output for the closed loop system. The magnitude of the error $|E(j\omega)|/|R(j\omega)|$ is found also and we see that the further the distance from the point $(-1, j0)$ to a point on the $G(j\omega)$ locus for a given frequency the smaller is the error amplitude.

In the following let us denote the closed loop frequency response,

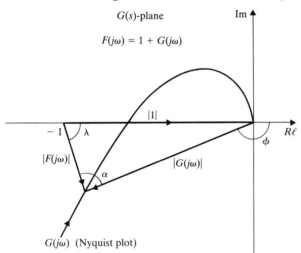

Fig. 5.29 Closed loop frequency response and the Nyquist plot (ϕ, λ and α are shown here as lags)

equation (5.3), by the condensed form,

$$\frac{C(j\omega)}{R(j\omega)} = Me^{j\alpha} \tag{5.5}$$

In Chapter 3 it was seen that resonance peaks occur in second and higher order systems. The use of contours of constant M and constant α drawn in the complex plane yields both a rapid means of determining these resonance peaks and frequencies from the polar plot of the open loop transfer function and also a method of obtaining the closed loop transfer function more rapidly than from its direct polar plot. The M and α contour method is developed for a unity feedback system and some modification is required for non-unity feedback systems.

Constant M loci (circles)

The complex function $G(j\omega)$ has real and imaginary parts so that we may write

$$G(j\omega) = x + jy$$

and the closed loop amplitude is

$$M = \frac{|G(j\omega)|}{|1 + G(j\omega)|}$$

$$= \frac{|x + jy|}{|1 + x + jy|} = \sqrt{\left\{ \frac{x^2 + y^2}{(1+x)^2 + y^2} \right\}}.$$

Squaring and rearranging enables us to write

$$\left(x - \frac{M^2}{1 - M^2} \right)^2 + y^2 = \left(\frac{M}{1 - M^2} \right)^2. \tag{5.6}$$

For a constant value of M, equation (5.6) is the equation of a circle of radius $M/(1 - M^2)$ with its centre at $x = M^2/(1 - M^2)$, $y = 0$. Equation (5.6) is the equation of the constant M loci (or M circles) in the complex plane (Fig. 5.30).

The pattern of the loci is established by the following observations:
(a) As $M \to 0$, the centre of the M circles tends to $x = 0$, $y = j0$ and the radius to zero, i.e. the circle tends to the origin
(b) For $M < 1$, all centres lie to the right of the imaginary axis
(c) For $M = 1$, the radius extends to infinity and the locus becomes the straight line $x = -\frac{1}{2}$. This is the condition $|C(j\omega)| = |R(j\omega)|$
(d) For $M > 1$, all centres lie to the left of $x = -1$
(e) As $M \to \infty$, the radius tends to zero and the centres tend to $x = -1$, $y = j0$
(f) All centres are on the real axis.
If the open loop $G(j\omega)$ plot is plotted on the same axes then the point where this plot intersects a constant M loci gives the value of the

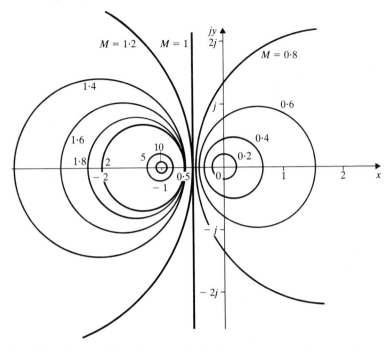

Fig. 5.30 Constant M circles in the (x, jy) plane

closed loop magnification M at the fequency on the $G(j\omega)$ plots at that point (Fig. 5.31(a)). That is, from the constant M loci which are completely independent of the system equation type or form, and the open loop plot we can determine the closed loop gain directly at discrete frequencies. The gain–frequency relationship can then be plotted (Fig. 5.31(b)). If an M circle forms a tangent to the $G(j\omega)$ plot this circle corresponds to the highest value of gain of the closed loop system, i.e. the resonant peak, M_r. The resonant frequency ω_r is read from the $G(j\omega)$ locus at that point. For the general case, as the open loop gain K is increased the $G(j\omega)$ locus moves nearer to the critical point $(-1, j0)$ and the resonant frequency ω_r approaches the phase crossover frequency, ω_c. As K is increased so that the $G(j\omega)$ locus passes through the critical point peak magnification M_r becomes infinite and ω_r equals ω_c. The system is on the limit of stability. Further increase in gain leads to instability.

Constant α loci (N circles)

If $G(j\omega) = x + jy$ the closed loop phase angle α is given by

$$\alpha = \underline{/\dfrac{x + jy}{1 + x + jy}}$$
$$= \underline{/x + jy} - \underline{/1 + x + jy}.$$

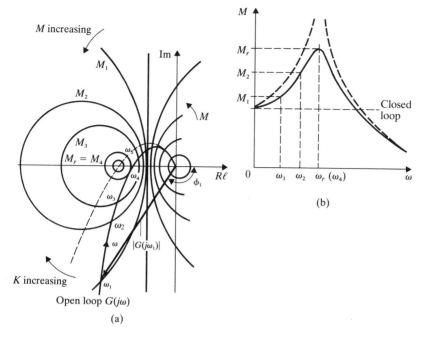

Fig. 5.31 Determination of the closed loop frequency response (b) from the constant M loci and open loop frequency response (a)

Hence

$$\alpha = \tan^{-1}\frac{y}{x} - \tan^{-1}\frac{y}{1+x}$$

$$= \tan^{-1}\frac{y}{x^2 + x + y^2}.$$

If we let $\tan\alpha$ be N and keep α, and hence N, constant, this relationship may be written

$$x^2 + x + y^2 - \frac{y}{N} = 0$$

Addition of $\frac{1}{4} + 1/(2N)^2$ to both sides yields

$$\left(x + \frac{1}{2}\right)^2 + \left(y - \frac{1}{2N}\right)^2 = \frac{1}{4} + \frac{1}{4N^2}$$

$$= \frac{N^2 + 1}{4N^2}. \tag{5.7}$$

For a constant value N this equation represents a circle of radius $\sqrt{\{(N^2+1)/4N^2\}}$ and centre $x = -\frac{1}{2}$, $y = 1/2N$. The family of circles given by equation (5.7) is illustrated in Fig. 5.32, where the parameter on the circles is α. Note that all circles lie with their centres on the line $x = -\frac{1}{2}$ and both positive and negative values of α (and N) are used. Intersection of the $G(j\omega)$ plot with the constant N circles gives the closed loop phase angle α at the frequency of the $G(j\omega)$ plot at the point of intersection. Thus Fig. 5.33 may be drawn in which the phase angle versus frequency relationship for the closed loop is deduced. The combination of Figs. 5.31 and 5.33 gives the Bode plot for the closed loop frequency response.

The use of the N circles is not quite so straightforward since the circles are multivalued in α as

$N = \tan \alpha$

$\quad = \tan (\alpha \pm 180°)$ etc.

The correct value of α is selected by starting at the low frequencies at which we know the phase angle tends to zero and then moving steadily towards the higher frequencies, maintaining a smooth continuous phase curve.

In the case of non-unity feedback $H(s)$ is not unity, then the closed

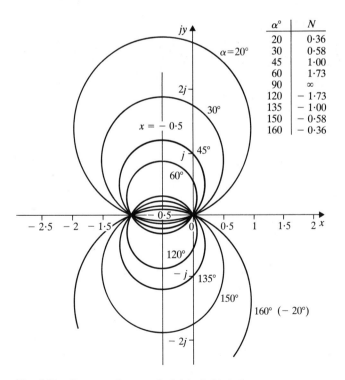

$\alpha°$	N
20	0·36
30	0·58
45	1·00
60	1·73
90	∞
120	$-1·73$
135	$-1·00$
150	$-0·58$
160	$-0·36$

Fig. 5.32 Constant phase angle (α) loci, N circles

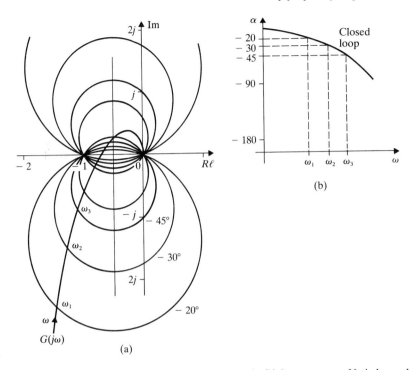

Fig. 5.33 Determination of closed loop phase angle (b) from constant N circles and open loop frequency response (a)

loop transfer function is rearranged,

$$\frac{C(s)}{R(s)} = \frac{G(s)}{1 + G(s)H(s)}$$

$$= \frac{1}{H(s)} \cdot \frac{G(s)H(s)}{1 + G(s)H(s)}$$

$$= \frac{1}{H(s)} \cdot \frac{G_1(s)}{1 + G_1(s)}.$$

Thus the M and N circles may be drawn in conjunction with $G_1(j\omega)$ where $G_1(j\omega)$ is $G(j\omega)H(j\omega)$. The magnitude and phase angles determined in this way must be adjusted by the factor $H(j\omega)$, i.e. $|H(j\omega)|\underline{/H(j\omega)}$. This is most easily done using a Bode plot.

The Nichols chart

To construct the Nichols chart the M and N circles drawn separately in the x, jy plane are redrawn and superimposed on the log-magnitude, phase angle plane. To obtain a point in the gain phase plane a vector is drawn from the origin of the complex plane to a particular point on a

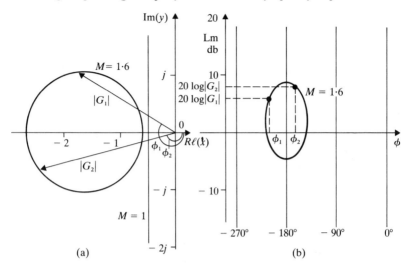

Fig. 5.34 Construction of the Nichols chart: (i) use of constant M circles. (The $\mathscr{L}$m vs ϕ curve shape depends on the axis scaling)

constant M circle. The vector length, converted to db, and the phase angle determine the corresponding point in the gain phase plane (Fig. 5.34).

The method is repeated using the constant N loci (Fig. 5.35), and the two families of curves are superimposed to form the Nichols chart (Fig. 5.36).

Since ϕ is multivalued the charts repeat themselves every 360° of phase angle but the most commonly required section is shown in Fig. 5.36. It will be observed that dominant features are

(a) There is symmetry about $\phi = -180°$

(b) The $M = 1$, (zero db), curve is asymptotic to $\phi = -90°$ and $\phi = -270°$,

(c) The curves for $M < \frac{1}{2}(-6\,\mathrm{db})$ are always negative in the db scale,

(d) $M = \infty$ is the point at 0 db, $\phi = -180°$, and

(e) The curves for $M > 1$ are closed curves inside the limits $\phi = -90°$ and $\phi = -270°$.

The Nichols charts may be used as the individual charts were by superimposing onto them the gain–phase relationship of $G(j\omega)$. This may most readily be obtained from the Bode plot. The intersections give the gain and phase angles at the corresponding frequency on the $G(j\omega)$ plot and, as before, the resonance peak M_r and frequency ω_r are determined by a tangent point. The bandwidth of the closed loop system is given by the frequency of the intersection of the $M = 0.707$

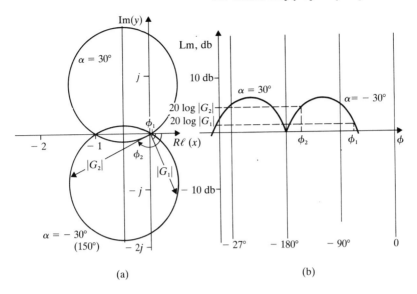

Fig. 5.35 Construction of the Nichols chart: (ii) use of constant N circles

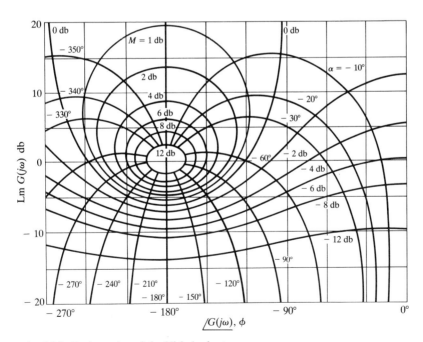

Fig. 5.36 Basic section of the Nichols chart

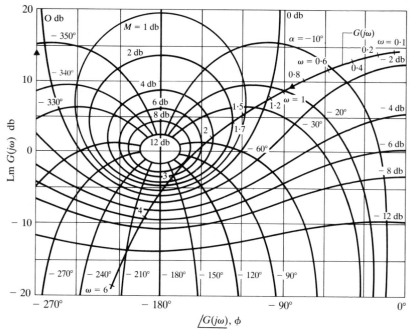

Fig. 5.37 $G(j\omega)$, log magnitude vs phase angle, superimposed on Nichols chart.

$$G(j\omega) = \frac{5}{(1+0\cdot33s)(1+0\cdot5s)(1+s)} \quad \text{(Fig. 5.23)}$$

(-3 db) locus and the $G(j\omega)$ plot. The gain and phase margins may be read directly from the Nichols chart.

Example Using the Bode plot of the above example construct the magnitude phase on the Nichols chart and hence plot the frequency response of the closed loop system. Use $K = 30$.

Values are taken from the open loop Bode plot and transferred to the Nichols chart to give Fig. 5.36.

From Fig. 5.37 the closed loop frequency response is shown in Fig. 5.38.

5.6 Compensation and the frequency response

So far attention has been focussed on the dynamic response of systems with the overriding consideration being that of stability. Changes in controllers and in the system which affect stability also affect the other

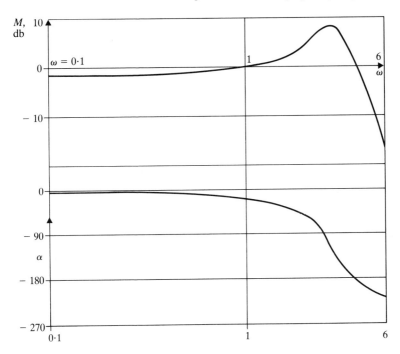

Fig. 5.38 Frequency response, M and α, of the closed loop system as taken from the Nichols chart (Fig. 5.37)

dynamic characteristics of system behaviour, e.g. bandwidth and transient response times. Changes or 'compensation' of the system are required which while satisfying some stated requirements for a system do not distract from, and maybe improve, other general system dynamic and steady state behaviour. For example, although increasing the system gain reduces steady state error, increased oscillations and a greater tendency to instability occur and all of these must be considered together. Thus additional compensation, or modifications are sought to satisfy both the steady state and dynamic requirements. The effects of such compensation may be represented both on the roots-loci diagram or the Nyquist and Bode plots or their derivatives.

Compensation of a control system is dependent on the specifications already given and in this sense little new information is required. These specifications are phase margin, gain margin, resonance peak and bandwidth in the frequency domain and damping ratio, rise time, peak overshoot etc. in the time domain. With single input–single output systems systemmatic trial and error adjustments of the controller interspaced with performance assessment may enable the required dynamics to be obtained. Alternatively it may be possible to change in a similar way some plant parameters but, except at the design stage, this may be unrealistic and far from convenient, time consuming, or impossible. The trial and error method is obviously improved if the changes can be brought about in a model and simulation rather than on the plant itself. Then our graphical methods may be used to the full before any physical implementation is made. However, the limiting qualities of physical components must be given adequate weight.

For single input–single output systems compensation will be principally of three types, series, parallel or feedback and load compensation. These are illustrated in Fig. 5.39, (a), (b) and (c) respectively.

Examples of series and feedback compensation have already been mentioned. Load compensation can be looked on as a new system with tight loop control about the first part, the output of this part then entering a further compensation system. It has some features of a feedforward system (Chapter 6).

Where specifications are given in the time domain the effects of system compensation are shown well by the root locus plots. If direct specification is made in the frequency domain then the polar plot and

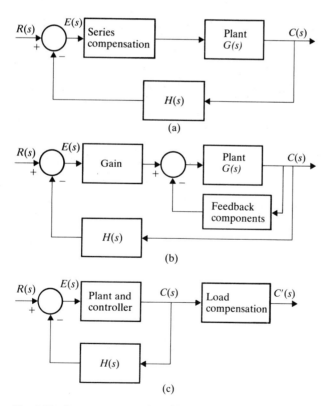

Fig. 5.39 System compensation, (a) series, (b) feedback, (c) load

Bode plot are more suitable vehicles for the investigation. The transient response may then be checked after the closed loop frequency response methods. The performance of control elements themselves may be expressed conveniently by experimentally determined frequency response data. Within the frequency domain the Bode diagrams may be found more suitable as the straight line asymptotes may be used in conjunction with the phase and gain margins, in arriving at the approximate values of compensation required. If a resonance peak value is specified then the polar plot or log magnitude-phase plot may be more useful. The shape of the open loop frequency response indicates more than just plain stability. The low frequency region (say for less than the gain crossover frequency) indicates the steady state behaviour of the closed loop, the region near the critical point frequency indicates relative stability and the high frequency region indicates the complexity and order of the system response. Satisfaction of the steady state, stability and rapid response specifications thus can be seen in terms of the frequency response loci.

Concerning ourselves principally with series compensation the basic additional control modes which are introduced are integral, derivative and integral plus derivative. These are known commonly also, as lag, lead and lag-lead respectively. In addition to these single or additive control modes, series compensation may take a form so its transfer function is of the general nature

$$C_p(s) = \frac{1 + Ts}{1 + \alpha Ts} \cdot K_p$$

or equivalent forms. There is a zero at $-1/T$ and a pole at $-1/\alpha T$. Such transfer functions may be representative of electrical and mechanical controllers, the key features depending on the relative magnitude of the parameters α. In the following discussion let the compensation gain K_p be unity. If α is small (0·05 to 0·1) the resulting compensation is that of a lead network. The function has a positive phase angle, i.e. phase advance is introduced into the system and the phase margin is consequently improved. If α is large (say 10 to 20) then the resulting compensation is that of a lag network. Now T is chosen to be much greater (say a factor of ten) than the largest time constant of the system. The attenuation at high frequency is increased but the gain margin is not greatly affected. The equivalent process control (P.I.D.) modes are the derivative and integral action respectively.

A lag-lead combination effectively produces the lag and lead effects in series, at low frequencies it acts as a lag network and at the higher frequencies as a lead network. This may be seen by looking first at the Bode plots for the separate lag and lead terms and then at the plots for the product of such terms.

The frequency transfer function is

$$C_p(j\omega) = \frac{1 + j\omega T}{1 + j\alpha\omega T}$$

and

$$|C_p(j\omega)| = \sqrt{(1+\omega^2 T^2)} \cdot \frac{1}{\sqrt{(1+\alpha^2 \omega^2 T^2)}}$$

or

$$\mathscr{L}m\,|C_p(j\omega)| = 10 \log (1+\omega^2 T^2) - 10 \log (1+\alpha^2\omega^2 T^2).$$

The phase angle is

$$\underline{/C_p(j\omega)} = \tan^{-1} \omega T - \tan^{-1} \alpha\omega T.$$

Immediately by inspection it can be seen that if α is greater than unity a phase lag results and for α less than unity there is a phase advance. Similarly with α greater than unity the log magnitude will be negative (db) and for α less than unity it will be positive.

The corner frequencies (Chapter 3) are at $\omega = 1/T$ for the zero and at $\omega = 1/\alpha T$ for the pole. The relative positions of these 'corners' will again depend upon whether α is greater or less than unity, Fig. 5.40 and 5.41.

For a lag-lead compensator we may write the form

$$C_p(s) = \left(\frac{1+sT_1}{1+\alpha sT_1}\right)\left(\frac{1+T_2 s}{1+sT_2/\alpha}\right).$$

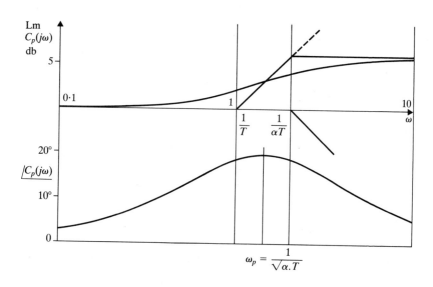

Fig. 5.40 Bode diagram for lead network $(\alpha < 1)$,

$$C_p(j\omega) = \frac{1+Tj\omega}{1+\alpha Tj\omega} \quad (T=1,\ \alpha=0.5)$$

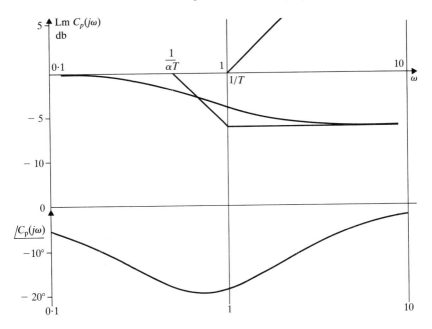

Fig. 5.41 Bode diagram for lag network $(\alpha > 1)$,

$$C_p(j\omega) = \frac{1 + Tj\omega}{1 + \alpha Tj\omega} \quad (T = 1, \alpha = 2)$$

If α is small, say 0.1, then the first term gives us the lead term and the second gives the lag term $(\alpha < 1)$. As arrived at by electrical networks it is usually cheaper to produce a combined lag-lead network than to use lag and lead separate networks in series. An additional gain constant appears also in C_p but this does not affect the arguments here but shifts the gain plot vertically. The time constant T_2 is chosen large compared with T_1, e.g. $T_2 = 10T_1$. The plot for such a combined compensator is given in Fig. 5.42.

The lead portion of C_p adds phase angle and increases the phase margin at the gain cross over frequency. The phase lag portion adds increased attenuation above this frequency. It is possible to improve the gain, giving low frequency and steady state improvements while the high frequency attenuation prevents instability.

To bring the 'P.I.D.' terms into comparison with these 'compensation networks' we may summarize the major features of integral and derivative action and their association with lag and lead modes.

Integral (lag) compensation

This additional mode improves the steady state response by giving effectively a large (infinite) increase in steady state gain. This is associated with some decrease in the natural frequency and bandwidth and an increase in the settling time. However if large noise signals are present integral action is preferable to derivative action.

Derivative (lead) compensation

The essential feature of derivative action is the establishment of additional phase lead, as distinct from the destabilizing lag introduced with integral action. This is accompanied by a moderate increase in gain. The band width is increased and there is a substantial reduction in settling time although, as mentioned above, the performance is poor in the presence of high frequency noise. Because of the phase advance characteristic there is an increase in phase margin, giving additional relative stability.

Integral plus derivative compensation

By using integral and derivative action together, or lag-lead compensation, with suitable weighting on each mode some of the advantages of each basic mode may be included to give both a significant increase in gain and steady state performance (the low frequency end of frequency response) and increased natural frequency, bandwidth and faster transient response and settling times.

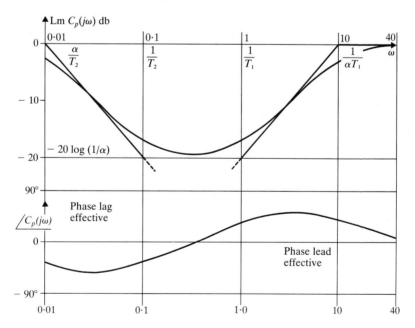

Fig. 5.42 Bode plots for lag-lead network ($\alpha = 0.1$, $T_1 = 1$, $T_2 = 10T_1$)

The application of these compensation techniques may be described in terms of reshaping the polar or logarithmic plot. If a system is stable and has suitable dynamic characteristics but a large steady state error then the low frequency end of the plot must be changed. The required increase in gain can be effected safely by integral action for first order systems but higher order systems suffer a reduction in stability. If a system is stable with a suitable resonance peak but slow response, i.e. a low resonant frequency, the high frequency portion of the polar plot is altered. The addition of derivative action will increase the resonant frequency without changing the resonant peak value. If both dynamic and steady state responses are unsuitable we must alter the high frequency end of the plot to improve the transients and also the low frequency end to improve the steady state error. Stability must be maintained so the curve in the region of the critical point must not be too severely affected. In this case integral plus derivative action will improve the behaviour in both dynamic and steady state criteria. Finally if the system is unstable the first attempt must be at stabilization and so derivative action may first be added and additional considerations given after stabilization.

The frequency response modifications by additional modes are of course also shown on the Bode plot and with its additive qualities coupled with the clear expression of movement of phase and gain margins this is often the easiest way of actually assessing the effect of compensation values.

Example A servomechanism with unit feedback comprises a proportional controller of gain 10 and a plant with transfer function $G(s) = 10/s(1+0\cdot1s)$. Determine the phase margin of this system and a suitable phase compensator to increase the phase margin to about $50°$. Figure 5.43 shows the system.

Fig. 5.43 Servomechanism with series compensation block diagram

The bode plot for the uncompensated system is shown in Fig. 5.44. The phase margin is low although we naturally have stability for this system. The $1/s$ factor in the open loop transfer function ensures zero steady state error, i.e. low frequency behaviour might be expected to be satisfactory, so we look at a phase lead compensator to improve the phase margin. The phase lead compensator is

$$C_p(s) = K_p \cdot \frac{1+sT}{1+\alpha sT}, \quad \alpha < 1$$

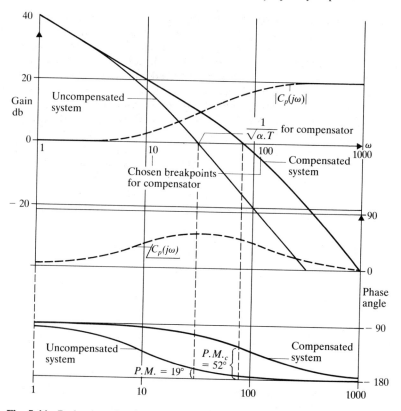

Fig. 5.44 Bode plots showing progression from initial to compensated system

Using a value of $\alpha = 0.1$, gives a maximum phase advance of about 50° (Fig. 5.40). The break points for the compensator are chosen to come either side of the gain cross-over frequency (Fig. 5.44), so that the peak phase advance of the compensator also falls in this region of ω to give a large phase advance and improves the phase margin as desired. With $\alpha = 0.1$, and a phase cross over frequency of about 30 rad/ sec a suitable value of $1/T$ (from Fig. 5.40) would be approximately $30 \times \sqrt{0.1}$, say 10 rad/sec. This value of T is not too critical but some adjustment may be necessary and α may also be adjusted. The break points for the compensator are then $1/T$ and $1/\alpha T$, i.e. at 10 and 100. The gain K_p could be varied but since we already have a large gain and zero steady state error put K_p as unity. The new phase margin of the compensated system becomes 52°.

When feedback compensation is used a compensation system is put into the feedback path. The actuating signal to the plant is now produced by comparison of the input signal with effectively a modified form of the output. As shown in Fig. 5.39(b) the direct major loop of the feedback is retained. This is essential if the correct steady state is ever to be achieved since the compensation addition is normally a rate function, e.g. tachometer system. The design of a feedback compensator is less direct than for a series compensator and may be laborious. In trying to select a series or feedback compensator one needs to consider the signal types to the compensators which are available; the weight of the hardware required, the environment, accessibility and noise associated with possible measurements, the response time, possibility of 'tight loop' control around part of the system and availability of suitable components must also be considered. Feedback compensation may be studied by forming the overall system equation and characteristic equation, and the roots-loci plotted using the gain of this system. Alternatively the roots of the inner loop characteristic equation may first be adjusted and then these roots, the poles of the forward transfer function, be used to draw the roots-loci of the overall system. This effectively treats the transfer function of the plant and the feedback compensation as a joint forward function (Fig. 5.39(b)).

In the frequency domain the inverse polar plot has some advantages over the polar plot and Bode plot when there is feedback compensation. Consider Fig. 5.45. The forward transfer function is

$$G(s) = \frac{G_p(s)}{1 + G_p(s)H(s)}$$
$$= \frac{C(s)}{E(s)}$$

i.e.

$$\frac{1}{G(s)} = \frac{1}{G_p(s)} + H(s)$$
$$= \frac{E(s)}{C(s)}.$$

This addition of phasors is shown in Fig. 5.46(a). Figure 5.46(b) shows

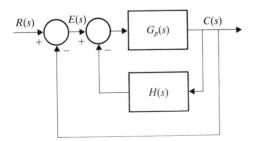

Fig. 5.45 Simple feedback compensation system

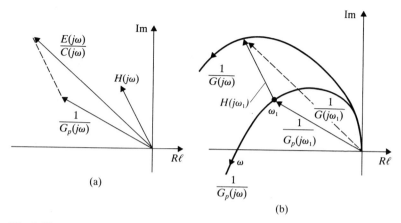

Fig. 5.46 Feedback compensation on the inverse polar plot

the inverse polar plot for both the uncompensated system and the compensated. The compensator $H(s)$ is the difference between the two plots at any given frequency ω.

5.7 Pure delays

Actual plant, especially those involving hydraulic or pneumatic flow in the plant or control system are liable to pure delays. These pure delays or distant velocity lags, mean that the output will not begin to respond to an input until after a definite time interval. They may be either in the feedback loop or in the forward path, and are also called dead time and transportation lags. We have seen in the Laplace transforms that the transfer function of a system containing a pure delay is no longer a rational function of polynomials but the delay introduces the exponential term $e^{-\tau s}$, where τ is the actual time delay. Two simple examples are shown in Fig. 5.47. In each case the time taken for the process material to pass from the point of interest and controller action to the measurement point is d/v,
i.e.

$$\tau = \frac{d}{v}.$$

In the examples shown the delay occurs between completion of the process and its measurement and consequent feedback in a control system so that, if $c(t)$ is the controlled variable and $l(t)$ is its value at the measurement point,

$$l(t) = c(t - \tau)$$

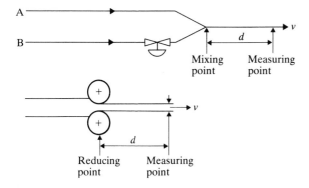

Fig. 5.47 Examples of distance/velocity lags

and

$$L(s) = C(s)e^{-s\tau}$$

The block diagram is thus as Fig. 5.48.

Some of the difficulties presented by the pure delay can be overcome by representing its transcendental transfer function by a rational algebraic function although the latter will be approximate to various degrees. The Routh–Hurwitz criterion, root-locus etc. may then be applied, bearing in mind that the new function is only an approximation. The following are some of the more widely used approximations.

Multiple pole. The exponential function can be expressed by the limit

$$e^{-\tau s} = \lim_{n \to \infty} \left(\frac{1}{1 + \tau s/n} \right)^n.$$

By using a finite number n, $e^{-\tau s}$ is approximated to a multiple pole located at $-n/\tau$ on the negative real axis of the s-plane. At times it may be sufficient to use a low value of n, even $n = 1$, but the approximation even with $n = 3$ is not particularly good (Fig. 5.49).

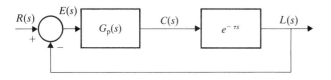

Fig. 5.48 Block diagram of system with pure delay

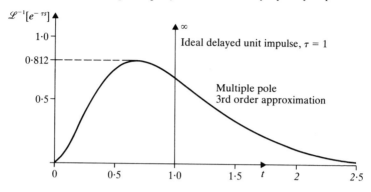

Fig. 5.49 Approximate impulse response of pure delay by multiple pole transform, $n = 3$

Maclauren series. The first few terms of the series expansions for $e^{-\tau s}$ may be used, i.e.

$$e^{-\tau s} = 1 - \tau s + \frac{\tau^2 s^2}{2!} - \ldots$$

$$= (1 + \tau s + \tau^2 s^2/2! + \ldots)^{-1}$$

If τs is small then this may be adequate but for large time lags or consideration far from the origin of the s-plane (short times) this is not suitable. Although this approximation may be useful in establishing a transfer function, investigation of stability subsequently by the Routh–Hurwitz criterion must be counterchecked. Its validity will be determined by the values of s at which the zeros of the characteristic polynomial pass into the right-half plane. Thus the number of terms to be used in the truncated expression cannot be determined until it has been used to determine the region of the s-plane of interest. For a system close to instability many terms may be required.

Padé-approximation. This approximation is more involved in that both numerator and denominator of the transfer function are algebraic functions of s. There are many Padé approximations to various order but the basic criterion is that the maximum number of terms in the Maclauren expansion of the approximation agree with the equivalent order terms in the expansion of the true exponential function. For example if the approximation

$$e^{-\tau s} \simeq \frac{1 + a_1 s + a_2 s^2 + a_3 s^3}{b_0 + b_1 s + b_2 s^2}$$

is used then the coefficients are chosen so that at least the first six terms are equal in the Maclauren expansion of this expansion and of $e^{-\tau s}$. Although widely used in analogue simulation it can be expensive in its use of components and difficulties may occur with non-positive values in its response (Fig. 5.50). The common disadvantages of the approximations is that the required accuracy can only be determined by trial and error, i.e. after first using the approximation, and the higher order but more accurate approximations greatly complicate existing transfer functions.

The major problem caused by pure delays is that of instability. Although the gain of the function $e^{-\tau s}$ is unity the delay causes a phase lag $\tau\omega$ and since the delay is a constant, the phase lag increases without limit as the frequency increases. Even first and second order systems when associated with a pure delay may thus become easily unstable in the closed loop form. Similarly increasing the time delay at a given frequency will cause instability eventually. In the polar plot the term $e^{-j\omega\tau}$ just rotates each point on the $G(j\omega)$ locus by the angle $\omega\tau$ in the clockwise direction. On the Bode plot the gain curve is unchanged and the phase lag increases linearly with frequency. These are both shown in Fig. 5.51 for a simple second order system. As the frequency ω increases and $G(j\omega)$ approaches the origin of the polar plot the plot encircles the origin an infinite number of times. The value of τ which makes the closed loop system go unstable can be found by trial and error using either the polar or Bode plots. From the Bode plot this is relatively easy as we only locate the point A (Fig. 5.51) and find the value of τ which forces the phase curve through this. From the polar plot it is a little more involved. For the chosen system the characteristic equation is

$$1 + \frac{e^{-\tau s}}{s(s+1)} = 0.$$

This is the condition when the path of $G(j\omega)$ passes through the critical point $(-1, j0)$.

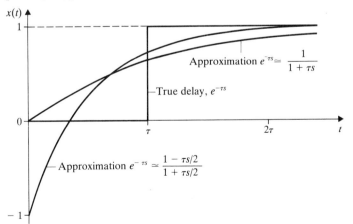

Fig. 5.50 Use of simple Padé approximations to evaluate step response of pure delay τ

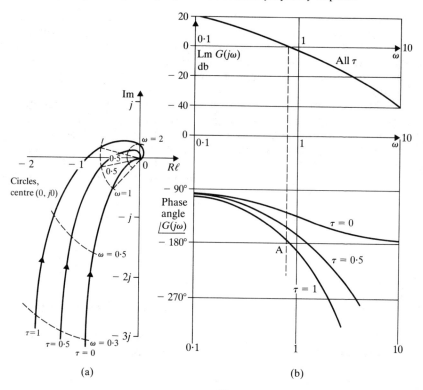

Fig. 5.51 Open loop response for $G(s) = \dfrac{e^{-\tau s}}{s(s+1)}$, (a) Nyquist, (b) Bode plots

Thus

$$\frac{e^{-\tau s}}{s(1+s)} = -1$$

$$= G(s)$$

or

$$\frac{1}{s(1+s)} = -e^{\tau s}$$

$$= G_1(s)$$

so that the poles of the characteristic equation are given by the solution of last relationship also. Hence, corresponding to the critical $(-1, j0)$ point in the $G(s)H(s)$ plane we have the $e^{\tau s}$ locus in the $G_1(s)$ plane for s equal to $j\omega$. As before the complete open loop transfer function is factored so that

$$G(s) = G_1(s)e^{-\tau s}$$

and we work in the $G_1(s)$ plane as distinct from the $G(s)$ plane. (When $\tau = 0$ these obviously are the same functions.) The $e^{j\omega\tau}$ locus is a circle of unit radius as for all values of ω its modulus is unity but the phase angle continues to increase with ω (Fig. 5.52). This circular locus is a locus of critical points each one corresponding to a given value of $\omega\tau$. As the frequency ω increases from zero the critical points trace a set of overlapping circles in the $G_1(s)$ plane (Fig. 5.52). The intersection of the $e^{j\omega\tau}$ locus and the $G_1(s)$ locus corresponds to the intersection in the $G(s)$ plane of $G_1(s)e^{-\tau s}$ with the $(-1, j0)$ point, provided the frequencies on the two loci are the same at intersection. In this example the circular locus has a value $\omega\tau = 0{\cdot}907$ at intersection and the frequency on the $G_1(j\omega)$ plot is $0{\cdot}79$. Thus for this point to represent the stability limit $\tau = 0{\cdot}907/0{\cdot}79$, i.e. $1{\cdot}15$ sec. If τ is greater than this marginal value, the critical point on the unit circle is enclosed by the $G_1(j\omega)$ locus and the system is unstable.

To use the pure delay with the root locus brings much difficulty and suffice it at this time to say that its effect will be shown by the bending of loci towards the right half plane and the subsequent cutting of the imaginary axis.

As well as being a characteristic of many plants the concept of the pure delay is useful in the generation of a non-periodic forcing function, e.g. Fig. 5.53. The function of Fig. 5.53(a) is the sum of the

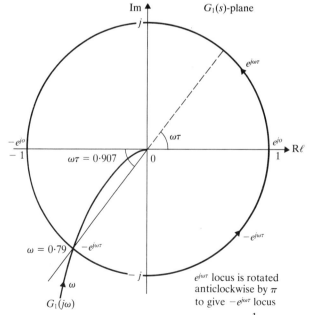

Fig. 5.52 Critical locus in the G_1 plane, $G_1(s) = \dfrac{1}{s(s+1)}$

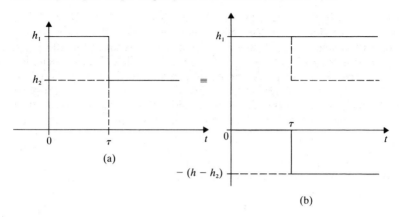

Fig. 5.53 Use of the pure delay to generate a forcing function

inputs $h_1(t)$ and $-(h_1-h_2)(t-\tau)$. Thus the transform form of the input, say $R(s)$ is

$$R(s)=\frac{h_1}{s}-\frac{(h_1-h_2)}{s}e^{-\tau s}$$

and if

$$C(s)=G(s)R(s)$$

then

$$C(s)=\frac{h_1}{s}G(s)-\frac{(h_1-h_2)}{s}e^{-\tau s}G(s).$$

The output $C(s)$ is obtained by inversion; inverting the two halfs of the expression separately to give the response to the input. As the input contains a delay term the output also will contain a delay term and if $G(s)$ is, say, the simple lag function $1/(s+2)$ then

$$C(s)=\frac{h_1}{s(s+2)}-\frac{(h_1-h_2)}{s(s+2)}e^{-\tau s}$$

and

$$c(t)=\frac{h_1}{2}(1-e^{-2t})-\frac{(h_1-h_2)}{2}\{U(t-\tau)-e^{-2(t-\tau)}\}.$$

$U(t-\tau)$ is the delayed step function and $e^{-2(t-\tau)}$ is a decaying exponential. Both by definition are zero for time less than τ. Thus the response looks like Fig. 5.54.

Example   A simple plant, represented by two first order lags, is fitted with a proportional controller of gain k (Fig. 5.55). Because of

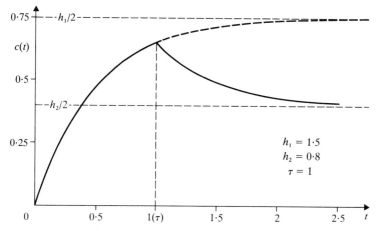

Fig. 5.54 Response from input with delayed component

the offset produced by the limited gain of the controller it is to be replaced by a proportional plus integral action controller. At the same time an alternative measurement element is to be tested but this introduces a pure delay of 2 minutes in the feedback loop (Fig. 5.56). For the new controller and alternative feedback schemes determine the critical value of the gain k when T_i is set to 3 minutes.

Taking the new controller but existing, unity, feedback,

$$G(s)H(s) = \frac{k(1+3s)}{3s(2+s)(1+s)}$$

$$= \frac{k(1+3s)}{6s(1+0\cdot5s)(1+s)}.$$

To evaluate points for the Nyquist plot it is most convenient to evaluate the gain and phase angle from the elements of the overall function so that

$$G(j\omega)H(j\omega) = \frac{k}{6} \cdot \frac{(1+3j\omega)}{j\omega(1+0\cdot5j\omega)(1+j\omega)}$$

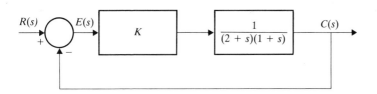

Fig. 5.55 Initial control system

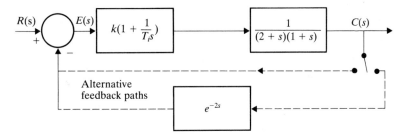

Fig. 5.56 Alternative control schemes

and

$$|G(j\omega)H(j\omega)| = \left|\frac{k}{6}\right| \cdot \frac{|1+3j\omega|}{|j\omega|\,|1+0{\cdot}5j\omega|\,|1+j\omega|}$$

$$= \frac{k}{6} \cdot \frac{\sqrt{(1+9\omega^2)}}{\omega\sqrt{(1+0{\cdot}25\omega^2)}\cdot\sqrt{(1+\omega^2)}}$$

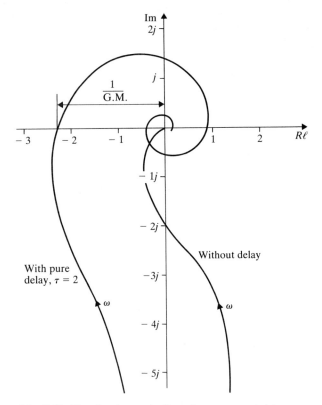

Fig. 5.57 Nyquist plots and effect of measurement delay

with

$$\underline{/G(j\omega)H(j\omega)} = \underline{/k/6} + \underline{/1+3j\omega} - \underline{/j\omega}$$
$$- \underline{/1+0{\cdot}5j\omega} - \underline{/1+j\omega}$$
$$= 0 + \tan^{-1}3\omega - \pi/2 - \tan^{-1}0{\cdot}5\omega - \tan^{-1}\omega.$$

If $G(j\omega)H(j\omega)$ is plotted for $k/6$ equal to unity, the intersection of the Nyquist plot with the real axis will give the gain margin when $k/6$ is unity, i.e. when k is six. Hence the critical gain will be

$k_{crit} = 6 \times$ gain margin from Fig. 5.57

When the pure delay is added the gain $|G(j\omega)H(j\omega)|$ is unchanged as $|e^{-2j\omega}|$ is unity, but there is an additional phase angle, a lag, given by

$$\underline{/e^{-2j\omega}} = -2\omega.$$

The Nyquist plot at any point evaluated at a frequency ω is thus rotated by 2ω radians in the clockwise (lag) direction. Both plots are shown in Fig. 5.57. Without the measurement delay the plot fails to intersect the real axis away from the origin with this integral action time, (i.e. the gain margin is infinite, and the system is stable for all positive k.

With the additional measurement delay stability is reduced, the gain margin from the plot is now $1/2{\cdot}3$ and the critical gain is

$k_{crit}(with\ delay) = 6 \times 1/2{\cdot}3 = 2{\cdot}6.$

It will be seen in Chapter 6 that full solution of the dynamics of feedback systems with delay is more difficult.

5.8 Summary

In addition to the use of the Routh–Hurwitz and roots-loci method of Chapter 4 the frequency response methods using the Nyquist and Bode plots may be used for closed loop stability studies.

Knowledge of absolute stability is insufficient alone, and a measure of the relative stability may be found in the form of phase and gain margins of the system. Stability of the closed loop system may be investigated using plots of the open loop frequency response.

The control modes additional to proportional action, i.e. derivative and integral in particular affect both the dynamic and steady state behaviour of the system. These additional modes may conveniently be determined and discussed in terms of their effects on the Nyquist and Bode plots. A further addition to a system in the form of a pure delay is a common occurrence in process plant, frequently appearing as a direct distance velocity lag. This is difficult to handle in system simulation and may be approximated to by a number of series expansions and ratios. The approximate form is then included, with limitations, in the simulation of open or closed loop systems, including stability studies.

Chapter 6
Further aspects of simple controllers

6.1 Introduction

In Chapters 4 and 5 attention has been focussed on control systems with a simple feedback loop. The conditions for stability and acceptable performance have been established from an analytical basis. In this chapter the establishment of controller settings for the series compensation on the basis of experimental testing will be discussed and then feedforward control considered as an alternative to feedback control. This leads on to the combination of controllers in cascade control and in parallel. This raises in turn problems of interaction between multiple inputs and outputs and we shall use this as a preview of the more general multivariable control problems of Chapter 7.

6.2 Experimental determination of controller settings

For an existing plant for which a transfer function is unknown it is necessary to adjust controller settings on the knowledge gained by observation of the plant. This is particularly so on a comparatively 'simple' plant where the input–output relationship of a controller and plant can be observed. It may be possible to carry out full testing to establish the frequency response curves of the open loop system and use these as a basis for controller-setting determination, or a few test settings may prove good enough. However, much initial selection of controller action (without full frequency response testing) and especially in process systems, has been based largely on the work of Ziegler and Nichols where we are dealing with simple control loops. It will be seen that these empirically determined rules can be explained on the analysis already covered.

The methods of Ziegler and Nichols (1942) and of Coon (1956a, b) are not 'optimal' as that word is now interpreted but the function of the methods is to establish good controller settings on existing plant or at the design stage in simple loops. From these initial settings subsequent plant performance may lead to closer tuning of the controller to improve the performance. The methods are appropriate for one, two, or three term controllers having proportional, derivative, and integral action.

Continuous cycling method

Values for the controller settings are established in this case by testing the closed loop system using proportional action only. For an installed three term controller the integral action time T_i is set to its highest value (infinity) and the derivative action time T_d set to its lowest value (zero) so that only the proportional mode is left effective. Starting with low values of the proportional gain, k_p, this gain is increased until continuous cycling at a constant amplitude occurs in the plant output. The plant plus controller is now at the limit of stability. The period of oscillation is the ultimate period P_u and the value of the gain is the ultimate gain k_u. By observing the output at each value of k_p the ultimate value is reached in a few tests (Fig. 6.1(a)).

Based on these results Ziegler and Nichols suggested the following settings for the various controller modes:
Proportional

$$k_p = 0.5 \, k_u$$

Proportional plus integral

$$k_p = 0.45 \, k_u$$

$$T_i = \frac{P_u}{1.2}$$

Proportional plus integral plus derivative

$$k_p = 0.6 \, k_u$$

$$T_i = P_u/2$$

$$T_d = P_u/8.$$

These settings were suggested as giving a decay ratio of about $\frac{1}{4}$ (a setting time of between three and four oscillations), a period of oscillation close to the ultimate and hence fairly fast, and a reasonable overshoot (Fig. 6.1(b)). Note that introduction of integral action is accompanied by a lowering of the gain k_p. This is in keeping with the extra gain and phase lag introduced by the integral action and its effect on stability. For proportional action alone a gain margin of two is proposed and this is approximately maintained for the new integral plus proportional action settings. The phase advance introduced with the derivative action has a stabilizing effect and the gain k_p can be increased because of this. A reduction in integral action time is recommended at the same time.

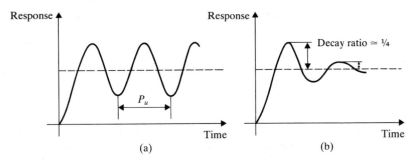

Fig. 6.1   (a) Continuous cycling and (b) desired response following a change in set point

These same values could be used during design and simulation of a system as they depend only on the value of the gain producing proportional control instability and on the frequency of cycling at this point. These are determined readily from the Bode plot, given the plant transfer function is known, during a design or simulation study.

Damped oscillation method

In place of increasing the proportional gain until the sustained oscillations are reached, which may be hazardous in some plant and extremely unpopular in others, the gain is increased until the closed loop decay ratio is $\frac{1}{4}$, say k'_p. The corresponding period P, greater than P_u is used as a basis for T_i and T_d so that for the three term controller

$$k_p = k'_p$$
$$T_i = P/1 \cdot 5$$
$$T_d = P/6.$$

With these conservative settings testing is repeated and k_p subsequently reset.

Reaction curve methods

The open loop response is now the basis of testing, a step change is made to the process with the feedback loop opened, say between the controller and the control variable, e.g. a change in valve-stem position might be made. The controller settings are then based on the maximum slope N of the response curve and the effective delay, L (Fig. 6.2).

For an input Δ (say p.s.i.), Ziegler and Nichols proposed the controller settings,
Proportional

$$k_p = \Delta/NL \quad \text{(e.g. psi/°C)}$$

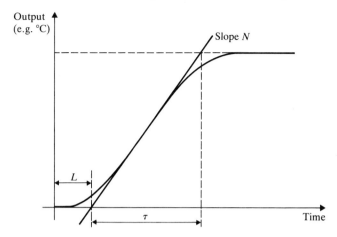

Fig. 6.2 Open loop process reaction curve

Proportional plus integral

$$k_p = 0.9\Delta/NL$$
$$T_i = L/0.3$$

Proportional plus integral plus derivative

$$k_p = 1.2\,\Delta/NL$$
$$T_i = L/0.5$$
$$T_d = 0.5\,L$$

If the change, Δ, is expressed as a fraction of its full scale, and the output plot as a fraction of its full deviation, the gain k_p becomes a dimensionless quantity. The effective lag is greater than any pure lag in the system, and the gain is generally conservative. With one dominant time constant in the system the behaviour tends to first order and lag L tends to zero so that the k_p value increases, in agreement with what is expected as the restriction on k_p is removed generally for first order systems.

More detailed expressions are given by Coon who generalized many results and expressed them in terms of the reaction curve also. However, the reaction curve was expressed approximately by the transfer function of a pure delay and first order lag, $G(s) = Ke^{-Ls}(1+\tau s)^{-1}$. Using minimum offset, minimum integral error and a decay ratio of about $\frac{1}{4}$, the recommended controller settings are:
Proportional

$$k_p = \frac{\tau}{KL}\left(1 + \frac{L}{3\tau}\right)$$

Proportional plus integral

$$k_p = \frac{\tau}{KL}\left(0.9 + \frac{L}{12\tau}\right)$$

$$T_i = L\left(\frac{30 + 3L/\tau}{9 + 20L/\tau}\right)$$

Proportional plus derivative

$$k_p = \frac{\tau}{KL}\left(1.25 + \frac{L}{6\tau}\right)$$

$$T_d = L\left(\frac{6 - 2L/\tau}{22 + 3L/\tau}\right)$$

Proportional plus integral plus derivative

$$k_p = \frac{\tau}{KL}\left(1.33 + \frac{L}{4\tau}\right)$$

$$T_i = L\left(\frac{32 + 6L/\tau}{13 + 8L/\tau}\right)$$

$$T_d = L\left(\frac{4}{11 + 2L/\tau}\right).$$

The effect of the additional modes is not so immediately obvious in these more involved expressions but once again these settings may be considered as starting points as they are based on the approximation to the real reaction curve and limited, though extensive, experiment.

It is apparent that the different recommendations cannot all give precisely the same answer but, where the minimum or maximum performance criterion, e.g. for process system responses, is generally flat with regard to the control variables, this is not critical and detailed optimization on the controller settings is not warranted. In general the settings are aimed at giving a gain margin of about two and a phase margin around 30° for basically stable process systems rising to three and 45° respectively for fast servo systems.

Minimum error integrals

As well as the variation in selecting the controller settings there are alternatives in measuring the effectiveness of the chosen control for process and other plant. The optimum settings may be regarded as those which give a minimum value to some measure of the error following a change in set point. Thus although integral action eliminates final error it has a time-integral error value, but if offset is not eliminated then the integral error is infinite. However, by working to within a limited region of the required output we may still use integral error functions. Process control quality may be measured by, among others,

(i) integral of the square of the error $\int e^2 dt$, which is mathematically suitable and weights large errors;
(ii) integral of time and absolute error product, $\int t|e| dt$, which is suitable for weighting heavily errors of longer duration;
(iii) the integral of the error magnitude $\int |e| dt$ which is suitable if the 'loss' due to any error is approximately proportional to its magnitude and to its duration.

The actual criterion used may have little effect on the selection of the controller, in many cases there being less than 10 per cent difference between the controller settings as determined by the three methods (Wills 1962, Harriot 1964). In addition the type of disturbance expected to enter a plant has some effect on the best controller settings to use. Although this is apparently in conflict with the idea of stability and performance depending on the plant and not on the input, it must be remembered that actual plant are nonlinear and we must guard against uncertainty of parameters. Thus controllers selected on the response to a step input may need further tuning in normal operation, and if cyclic disturbances occur then a phase angle criterion, say a 30° phase margin, is more suitable than an error integral criterion based on a step input. Also if there is considerable noise present in the input then a low gain should be initially used in general and derivative action may prove unsuitable. Studies using the above integral criteria (Jackson 1958) show that the resulting controllers are not too dissimilar from the Ziegler and Nichols settings but that there is dependence on the relative size of the time constants of the system. The best settings for step response and frequency response conditions are significantly different when the range of time constants is large. The number of equal or near equal time constants also affects the amount of integral action required. If there are many time constants of approximately equal value then the lag introduced by integral action is small in comparison with that of the system and hence a lower integral action time, more integral action, may be used.

The location of the major plant disturbances will affect controller settings and pure time delays also introduce problems, it being difficult to get good control with two or three term controllers on account of the large lags introduced by the delay. Controllers containing time delays and predictor elements have been suggested but complexity of the controller is obviously increased (Lupto and Oglesby 1961).

Example Compare the settings obtained for a plant which gives the reaction curve shown, using Ziegler–Nichols and Coon recommended values, for P, P + I, P.I.D. controllers. All settings correspond closely between the two methods and show the movement of gain and integral action which we might expect from the general theory. We see

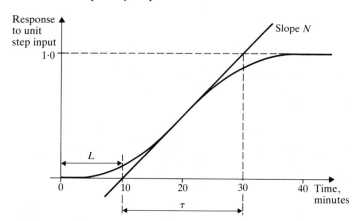

Fig. 6.3 Reaction curve method

Table 6.1 Controller settings

Controller modes	Ziegler–Nichols values		Coon values	
Proportional	$k_p = 2 \cdot 0$		$k_p = 2 \cdot 33$	
P + I	$k_p = 1 \cdot 8$,	$T_i = 33$ min	$k_p = 1 \cdot 88$,	$T_i = 33$ min
P + I + D	$k_p = 2 \cdot 4$,	$T_i = 20$ min	$k_p = 2 \cdot 9$,	$T_i = 21$ min
		$T_d = 5$ min		$T_d = 3$ min

that, for a typical order of reaction curve times, the agreement is such that either criterion could be used for determining initial settings for a controller.

6.3 Cascade control

In Chapter 3 the reduction of complex block diagrams was discussed yet, in the main, simple feedback systems have been considered which have required little or no reduction. A further stage of complexity (slight) is reached when considering cascade control. By measuring process variables as close as we can to where external disturbances enter the plant we stand a better chance of improving the control of the plant. However, to maintain overall control and allow for change in the setpoint of the desired output we need overall feedback. Thus we are concerned with the effect of two (or more) inputs to the plant. In such a case two controls may be combined in a cascade control system which utilizes two controllers, the setpoint of one being set by the output of the other. The object is to improve control by reducing the

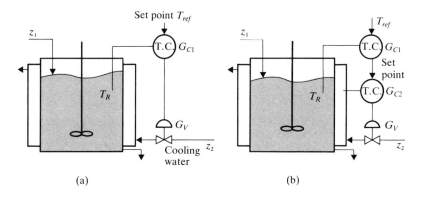

Fig. 6.4 Single loop (a) and cascade control (b)

effect of a disturbance by measurement near to its source. This disturbance may be in a utility or power source to the plant, such as cooling water, fuel flow, steam, or electric power. For example, Fig. 6.4 shows a jacket-cooled, stirred reactor with two possible control configurations. In (a) control is effected solely by the feedback loop measuring the reactor temperature and controlling the cooling water. Any variation in cooling water temperature or supply pressure will affect the jacket temperature also but this has to pass 'through' the process vessel itself before being corrected for via the reactor temperature. By controlling the jacket temperature directly and setting the setpoint of this controller with the output of the controller which is 'measuring' the reactor temperature a speed-up in correcting for supply disturbances should result. (In some chemical processes the additional dynamics of the extra measurement may be important but for simplicity here they are considered small.) By drawing the block diagrams and considering the overall transfer functions for the linearized equations for the two cases we may make a direct comparison (Figs. 6.5, 6.6). The block diagrams shown are already simplified to some extent, since, for example, the form of the transfer function relating the change in reactor input z_1 to reactor temperature T_R will

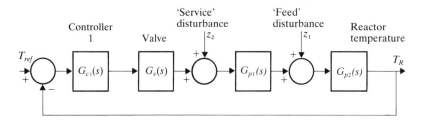

Fig. 6.5 Block diagram for single loop control

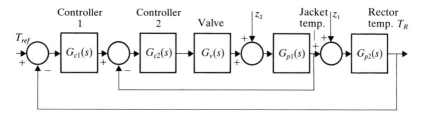

Fig. 6.6   Cascade control block diagram

differ from that relating jacket temperature changes and reactor temperature. However, insertion of the additional functions adds little to
clarify the situation and makes the diagrams more complex without
changing their essential nature. To realize the advantages of cascade
control the inner loop should cope with the major service disturbances
z_2 and act much faster than the outer loop. The closed loop transfer
functions relating the reactor temperature T_R to a change in cooling
water conditions z_2 are for the two cases respectively:

Single loop

$$\frac{T_R(s)}{z_2(s)} = \frac{G_{p1}(s)G_{p2}(s)}{1 + G_{c1}(s)G_v(s)G_{p1}(s)G_{p2}(s)}$$

and for the cascade system

$$\frac{T_R(s)}{z_2(s)} = \frac{G_{p1}(s)G_{p2}(s)}{1 + G_{c2}(s)G_v(s)G_{p1}(s) + G_{c2}(s)G_v(s)G_{c1}(s)G_{p2}(s)G_{p1}(s)}.$$

(These may be readily established by redrawing the block diagrams to
show z_2 in the conventional input position with T_R as output.) That the
second controller $C2$ has a significant effect on the dynamics is seen by
its double appearance in the denominator of the closed loop transfer
function. Investigation (Gould 1969) of the characteristic equation
shows that the crossover frequency is greatly increased for the overall
system so that the stability is increased as well as the dynamics being
improved.

The secondary controller $C2$ will normally be only a proportional or
proportional plus integral controller. The inner loop controller settings
are first established and the principal controller then tuned for the
overall system using, for example, the earlier methods of this chapter.

Associated in some respects with the cascade controllers is a ratio
controller. In Fig. 6.7 a flow controller controls the flow in one stream
B based on measurements in both streams. One of these measurements will determine the setpoint of the controller so that a constant
ratio between the two is maintained. As flow in A increases the
setpoint for B is increased so that flow in B increases also to keep the
ratio between the flows unchanged.

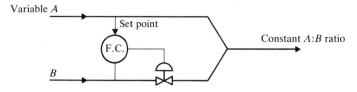

Fig. 6.7 Ratio controller

6.4 Feedforward control

Where the disturbances to the plant may be measured, and this is by no means always the case, and the transfer function between the plant output and the disturbance is known then, in theory, we can adopt a feedforward strategy as opposed to that of feedback. If, for example, the feed into the reactor of Fig. 6.4 is subject to uncontrollable fluctuations then it may be possible to measure these fluctuations and control the temperature in the jacket so as to counteract their effect, and in this way keep the reactor temperature itself constant. By counteracting such effects before they materialize perfect control of the reactor temperature would be possible. Pure feedforward control of the example reactor would have the block diagram of Fig. 6.8 (omitting z_2 for the moment). To entirely counteract the effect of z_1 the transfer function of the additional controller (and of its measurement lags) $G_f(s)$ must be given by the equation

$$G_f(s)G_v(s)G_{p1}(s) = 1$$

so that the output from the summer is zero, i.e.

$$G_f(s) = \frac{1}{G_v(s)G_{p1}(s)}.$$

However, as well as requiring exact measurement and knowledge of the transfer functions $G_v(s)$ and $G_{p1}(s)$ we must be able to implement $G_f(s)$ exactly. This $G_f(s)$ will contain high orders of s, i.e. high order derivative action which cannot be achieved and which in the presence of noise is undesirable anyway and would lead to quite impossible practical control action. Pure feedforward is also ruled out if there is a

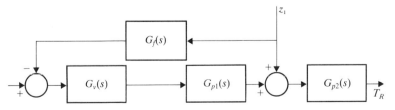

Fig. 6.8 Pure feedforward control for the disturbance z_1

pure delay in $G_v(s)G_{p1}(s)$ as $G_f(s)$ then contains the factor $e^{\tau s}$, which indicates a prediction or very early measurement before the disturbance z_1 reaches the plant. Thus, although some advantage may be gained by a simpler feedforward controller it is not a fully satisfactory solution. When coupled with feedback control it can, however, substantially reduce the effect of the disturbance, the normal feedback controller being left to complete the task (Fig. 6.9). Note that if z_1 is measured right at entry of the plant then $G_f(s)G_v(s)G_{p1}(s)$ must represent a very rapid response if the effect of measuring z_1 is going to be of any use at all. If commercial controllers are to be used then only an approximate $G_f(s)$ will be possible anyway. Obviously the cascade control of Fig. 6.6 and the feedforward control of Fig. 6.9 can be combined.

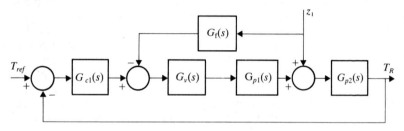

Fig. 6.9 Feed forward with feedback control

Example Draw the block diagram for the control system and plant shown in Fig. 6.10(a) and establish the overall closed loop transfer function relating a change in solvent flow to effluent composition. Has the controller $C1$ improved the control of the system? Compare Fig. 6.10(a) with the pure cascade control (Fig. 6.10(b)). Consider the valves to be fast acting.

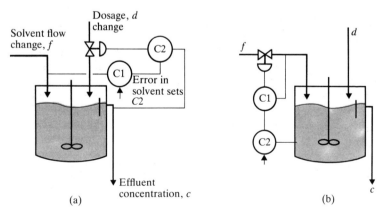

Fig. 6.10 Alternative cascaded controllers

The full block diagram (Fig. 6.11) is obtained from the system equations. These equations will supply the individual transfer functions, i.e.

$$G_{p1}(s) = \frac{C(s)}{D(s)}$$

$$G_{p2}(s) = \frac{C(s)}{F(s)}$$

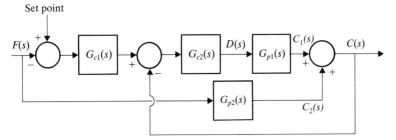

Fig. 6.11 Block diagram for Fig. 6.10(a)

and the controller transfer functions will depend on the chosen controller modes and parameters. Figure 6.11 may be redrawn as Fig. 6.12. Note that a change in solvent flow requires a change in dosage flow of the same sign if the effluent concentration is to be 'unchanged'. In its new form the block diagram more easily yields the overall transfer function which is sought, and shows the control system to contain some feedforward characteristics. One more block diagram reduction is useful (Fig. 6.13): so that

$$\frac{C(s)}{F(s)} = \frac{G_{c2}(s)G_{p1}(s)}{1 + G_{c2}(s)G_{p1}(s)} \cdot \frac{G_{c1}(s)G_{c2}(s)G_{p1}(s) + G_{p2}(s)}{G_{c2}(s)G_{p1}(s)}$$

$$= \frac{G_{c1}(s)G_{c2}(s)G_{p1}(s) + G_{p2}(s)}{1 + G_{c2}(s)G_{p1}(s)}.$$

The controller $C1$ may improve the performance of the system in a feedforward roll. In fact treating $C1$ as a feedforward controller full dynamic compensation is possible if

$$G_{c1}(s) = \frac{-G_{p2}(s)}{G_{c2}(s)G_{p1}(s)}$$

so that the numerator of the closed loop transfer function is zero. For the cascade system of Fig. 6.10(b) with fixed dosage d, but with the solvent flow subject to controlled changes and uncontrolled (additive) disturbances, e.g. due to pressure fluctuation, we obtain Fig. 6.14. This is a flow controller with the setpoint coming from the composition controller. Redrawing this we get Fig. 6.15, and this reduces to Fig. 6.16.

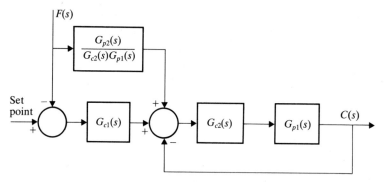

Fig. 6.12   More useful configuration of block diagram Fig. 6.11

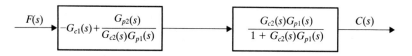

Fig. 6.13   Further block diagram reduction

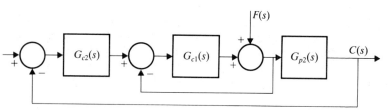

Fig. 6.14   Block diagram for Fig. 6.10(b)

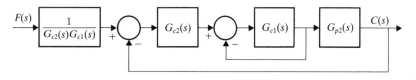

Fig. 6.15   First stage in diagram reduction (Fig. 6.14)

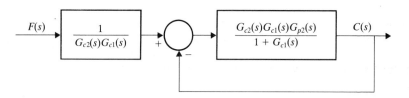

Fig. 6.16   Second stage in diagram reduction

It is now easily seen that

$$\frac{C(s)}{F(s)} = \frac{1}{G_{c2}(s)G_{c1}(s)} \cdot \frac{G_{c2}(s)G_{c1}(s)G_{p2}(s)}{1 + G_{c1}(s) + G_{c2}(s)G_{c1}(s)G_{p2}(s)}$$

$$= \frac{G_{p2}(s)}{1 + G_{c1}(s)(1 + G_{c2}(s)G_{p2}(s))}.$$

It can be seen also that the two controllers in this arrangement both influence the speed of response of the system. Controller $C1$ (in the inner loop) in particular dominates the response but full dynamic cancellation is not possible.

6.5 Interaction in control systems

Process interaction

In cascade and feedforward control the success of the systems depends on the combination of one controller with another to give a required output of one variable. Also these remain systems with one desired reference input although we also consider additional external disturbances. Because of our linearization and superposition the external disturbances give little more conceptual difficulty than does the loop without these disturbances.

The next logical step might be to question what happens when we use two controllers on one plant but with the purpose of controlling two distinct outputs by manipulation of two distinct inputs. For example, it might be required to control the temperature and concentration of a product by acting upon two inputs, say a steam rate and reactant flow. Can we be sure that increasing steam rate to raise the product temperature might not also affect the concentration because of changing the temperature-dependent reaction rate? Similarly, increasing the reactant to adjust composition of the product may, by heat of reaction, affect the product temperature. This sort of result is known as interaction and is always a possibility with those multivariable systems having multiple inputs and outputs. The general state space method of dealing with multivariable systems will be examined in Chapter 7. However, for a low order system we can examine the problem of interaction, and the design of non-interacting controllers, on the basis of single loop theory and the transfer function matrix of section 3.4. The two input–two output case can in theory be extended to higher orders but the number of controllers required to remove the interaction goes up as the square of the system order.

If the two controlled outputs are c_1 and c_2, the reference values r_1 and r_2 and the plant inputs of the control variables are m_1 and m_2 then interaction is shown by Fig. 6.17. Each control variable m_1 and m_2 affects both outputs c_1, c_2 through a plant transfer function. The effects

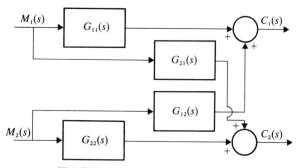

Fig. 6.17   Interaction within the plant

on the outputs are additive (due to linearization). Note the order of subscripts on the $G_{ji}(s)$. As a basis for argument consider that the control variable m_1 is used principally to control c_1 and m_2 principally to control c_2. The full block diagram becomes that of Fig. 6.18, in which $G_m(s)$ is a feedback (measurement) transfer function, $G_v(s)$ an actuator (e.g. valve) transfer function and $G_c(s)$ a controller transfer function.

Because of the interaction within the system the tuning of one of the feedback loops influences the performance of the other and to obtain reasonable control on both loops it may be necessary to use the individual controllers at other than their best settings, e.g. at a lower gain. If one of the loops is much more rapid than the other then individual tuning is possible, otherwise iterative tuning on both loops, using simulation, may be a lengthy alternative. The addition of two

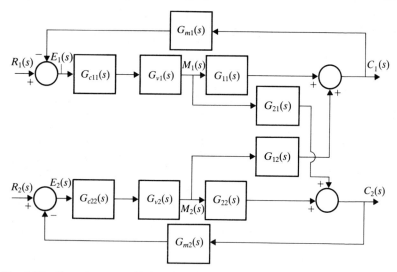

Fig. 6.18   Control system with in-plant interaction

further controllers can remove the interaction effect but it will be seen that this may suffer from further undesirable effects.

The analysis of the system in Fig. 6.18 provides a useful introduction to the matrix algebra of feedback systems as well as to the interaction problem. It must be remembered, however, that in this chapter our matrices are comprised of inputs and outputs as transformed variables or are transfer functions. In the state space methods of Chapter 7, we have state variables and out primary equations are principally in the time domain.

From Fig. 6.18

$$C_1(s) = G_{11}(s)M_1(s) + G_{12}(s)M_2(s)$$
$$C_2(s) = G_{21}(s)M_1(s) + G_{22}(s)M_2(s)$$

or in vector-matrix notation

$$\mathbf{C}(s) = \mathbf{G}_p(s)\mathbf{M}(s) \tag{6.1}$$

where $\mathbf{G}_p(s)$ is the plant matrix $\begin{bmatrix} G_{11}(s) & G_{12}(s) \\ G_{21}(s) & G_{22}(s) \end{bmatrix}$, $\mathbf{C}(s) = \begin{bmatrix} C_1(s) \\ C_2(s) \end{bmatrix}$, $\mathbf{M}(s) = \begin{bmatrix} M_1(s) \\ M_2(s) \end{bmatrix}$. The control system introduces the additional equations

$$M_1(s) = G_{v1}(s)G_{c11}(s)E_1(s)$$
$$M_2(s) = G_{v2}(s)G_{c22}(s)E_2(s)$$

and

$$E_1(s) = R_1(s) - G_{m1}(s)C_1(s)$$
$$E_2(s) = R_2(s) - G_{m2}(s)C_2(s)$$

or in matrix form

$$\mathbf{M}(s) = \mathbf{G}_v(s)\mathbf{G}_c\mathbf{E}(s)$$
$$\mathbf{E}(s) = \mathbf{R}(s) - \mathbf{G}_m(s)\mathbf{C}(s) \tag{6.2}$$

where

$$\mathbf{G}_v(s) = \begin{bmatrix} G_{v1}(s) & 0 \\ 0 & G_{v2}(s) \end{bmatrix}, \quad \mathbf{G}_c(s) = \begin{bmatrix} G_{c11}(s) & 0 \\ 0 & G_{c22}(s) \end{bmatrix}$$
$$\mathbf{G}_m = \begin{bmatrix} G_{m1}(s) & 0 \\ 0 & G_{m2}(s) \end{bmatrix}.$$

From equations (6.1) and (6.2) the forward transfer matrix is given by

$$\mathbf{C}(s) = \mathbf{G}_p(s)\mathbf{G}_v(s)\mathbf{G}_c(s)\mathbf{E}(s)$$
$$= \mathbf{G}(s)\mathbf{E}(s) \tag{6.3}$$

and the overall transfer matrix is given by

$$[\mathbf{I} + \mathbf{G}_p(s)\mathbf{G}_v(s)\mathbf{G}_c(s)\mathbf{G}_m(s)]\mathbf{C}(s) = \mathbf{G}_p(s)\mathbf{G}_v(s)\mathbf{G}_c(s)\mathbf{R}(s)$$

or

$$\mathbf{C}(s) = [\mathbf{I} + \mathbf{G}_p(s)\mathbf{G}_v(s)\mathbf{G}_c(s)\mathbf{G}_m(s)]^{-1}\mathbf{G}_p(s)\mathbf{G}_v(s)\mathbf{G}_c(s)\mathbf{R}(s). \qquad (6.4)$$

Note the similarity of this in form to the single input–single output case already considered and which of course may be obtained as a special case of equation (6.4), viz. $C(s) = [1 + G(s)H(s)]^{-1}G(s)R(s)$. Writing equation (6.4) again using $\mathbf{G}(s)$ for $\mathbf{G}_p(s)\mathbf{G}_v(s)\mathbf{G}_c(s)$

$$\mathbf{C}(s) = [\mathbf{I} + \mathbf{G}(s)\mathbf{G}_m(s)]^{-1}\mathbf{G}(s)\mathbf{R}(s). \qquad (6.4\text{a})$$

The degree of interaction in the overall system equation (6.4(a)) is assessed by the off-diagonal elements of the matrix $[\mathbf{I} + \mathbf{G}(s)\mathbf{G}_m(s)]^{-1}\mathbf{G}(s)$ in particular. Interaction causes a first order effect in these terms but also a second order effect in the diagonal elements. Thus, the effect of interaction, $G_{21}(s)$, of m_1 on c_2 is first order but it also has a lesser effect on the reaction of c_1 itself to m_1 (Gould 1969).

Non-interacting controllers

To remove all interaction effects for changes in set points r_1, r_2 additional cross-controllers are introduced, (Fig. 6.19). The number of measurements remains the same but two controllers use each error signal, and although the number of actuators remains the same their action is a function of the sum of two controller outputs. The purpose of the extra controllers is to reduce the off-diagonal terms of the overall transfer matrix to zero so that for changes in setpoints r_1 and r_2

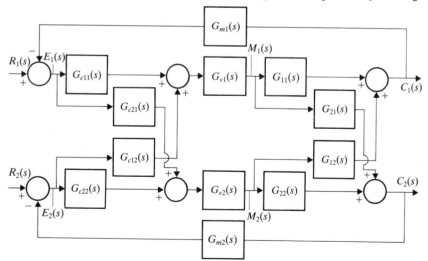

Fig. 6.19 Addition of cross controllers

respectively only c_1 or c_2, will change. The condition that $[\mathbf{I} + \mathbf{G}(s)\mathbf{G}_m(s)]^{-1}\mathbf{G}(s)$ be diagonal is not so imposing as it may at first seem since $\mathbf{I}$ is diagonal, $\mathbf{G}_m(s)$ is diagonal and if $\mathbf{G}(s)$ is made diagonal then this whole matrix is also diagonal. With the extra controllers,

$$\mathbf{G}_c(s) = \begin{bmatrix} G_{c11}(s) & G_{c12}(s) \\ G_{c21}(s) & G_{c22}(s) \end{bmatrix}$$

and

$$\mathbf{G}(s) = \begin{bmatrix} G_{11}(s)G_{v1}(s)G_{c11}(s)\left\{1 + \dfrac{G_{12}(s)G_{v2}(s)G_{c21}(s)}{G_{11}(s)G_{v1}(s)G_{c11}(s)}\right\} & G_{11}(s)G_{v1}(s)G_{c12}(s) + G_{12}(s)G_{v2}(s)G_{c22}(s) \\ \hline G_{21}(s)G_{v1}(s)G_{c11}(s) + G_{22}(s)G_{v2}(s)G_{c21}(s) & G_{22}(s)G_{v2}(s)G_{c22}(s)\left\{1 + \dfrac{G_{21}(s)G_{v1}(s)G_{c12}(s)}{G_{22}(s)G_{v2}(s)G_{c22}(s)}\right\} \end{bmatrix}$$

Writing $\mathbf{G}(s)$ in this way emphasizes the contribution of the plant interaction $G_{12}(s)$ and $G_{21}(s)$. Equating the off-diagonals to zero gives the required transfer function for the additional controllers,

$$\left.\begin{aligned} G_{c12}(s) &= \frac{-G_{12}(s)G_{v2}(s)G_{c22}(s)}{G_{11}(s)G_{v1}(s)} \\ G_{c21}(s) &= \frac{-G_{21}(s)G_{v1}(s)G_{c11}(s)}{G_{22}(s)G_{v2}(s)}. \end{aligned}\right\} \tag{6.5}$$

Substitution of the $G_{c12}(s)$, $G_{c21}(s)$ expressions into the diagonal elements leaves terms which are dependent solely on the plant parameters and the primary controllers.

In establishing the cross-controllers cancellations of factors in the individual transfer functions may occur. This means that poles present in the initial system will not be present in the expression for the noninteracting system and external disturbances may not be controllable. Thus, the use of this method is in establishing a one to one input–output relationship for r_1, c_1 and r_2, c_2 but the lack of controllability of external disturbances may rule out the method. Also, the number of controllers has doubled from two to four and the general rule is that noninteracting controllers increase the number of controllers as the square of the order of the system, e.g. nine controllers are needed for a three input–three output system if interaction is present from all inputs to all outputs. The form of the controller established by equation (6.5) may not be realizable, e.g. higher derivatives may be required, a factor common with feedforward control which relies on similar analysis. Although the full cancellation of dynamic interaction may thus not be possible, good steady state results may be achieved. This enables changes in the steady desired output to be effected by simple manned supervision (in the two output case) by the sole adjusting of just two set points.

Example   Evaluate the effect of the interaction term on each of the outputs for the control system in Fig. 6.20 and determine the additional controller(s) required to eliminate the interaction effect.

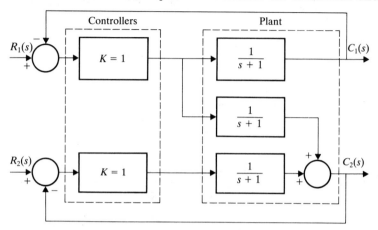

Fig. 6.20   Simple system with interaction

The plant matrix $\mathbf{G}_p(s)$ is

$$\begin{bmatrix} \dfrac{1}{s+1} & 0 \\ \dfrac{1}{s+1} & \dfrac{1}{s+1} \end{bmatrix},$$

the controller matrix $\mathbf{G}_c(s)$ is

$$\begin{bmatrix} 1 & 0 \\ 0 & 1 \end{bmatrix},$$

i.e. $\mathbf{I}$, and the actuator and measurement matrices $\mathbf{G}_v(s)$ and $\mathbf{G}_m(s)$ are the identity matrix also. Thus, from equation (6.4)

$$\begin{bmatrix} C_1(s) \\ C_2(s) \end{bmatrix} =$$

$$\begin{bmatrix} \mathbf{I} + \begin{bmatrix} \dfrac{1}{s+1} & 0 \\ \dfrac{1}{s+1} & \dfrac{1}{s+1} \end{bmatrix} \begin{bmatrix} 1 & 0 \\ 0 & 1 \end{bmatrix} \end{bmatrix}^{-1} \begin{bmatrix} \dfrac{1}{s+1} & 0 \\ \dfrac{1}{s+1} & \dfrac{1}{s+1} \end{bmatrix} \begin{bmatrix} 1 & 0 \\ 0 & 1 \end{bmatrix} \begin{bmatrix} R_1(s) \\ R_2(s) \end{bmatrix}.$$

To evaluate the interaction it is necessary to compute the matrix, giving,

$$\begin{bmatrix} C_1(s) \\ C_2(s) \end{bmatrix} = \begin{bmatrix} \dfrac{1}{s+2} & 0 \\ \dfrac{s+1}{(s+2)^2} & \dfrac{1}{s+2} \end{bmatrix} \begin{bmatrix} R_1(s) \\ R_2(s) \end{bmatrix}.$$

The individual transfer functions are thus

(i)

$$C_1(s) = \frac{1}{s+2} R_1(s)$$

$$= \frac{1/(s+1)}{1 + 1/(s+1)} R_1(s),$$

the normal closed loop which we expect as the lower half of the system can have no effect on the upper part, and

(ii)

$$C_2(s) = \frac{s+1}{(s+2)^2} R_1(s) + \frac{1}{s+2} R_2(s).$$

As there is no interaction from the lower half to the upper half there is no feedback to the lower via the upper part of any signal originating in the lower part. Thus the transfer function relating $C_2(s)$ and $R_2(s)$ is unchanged also from its single loop response. However, there is an interaction term from the upper reference input $R_1(s)$. For a unit step input in both r_1 and r_2

$$C_2(s) = \frac{s+1}{s(s+2)^2} + \frac{1}{s(s+2)}$$

$$= \frac{1}{4s} - \frac{1}{4(s+2)} + \frac{1}{2(s+2)^2} + \frac{1}{s(s+2)}$$

therefore

$$c_2(t) = \tfrac{1}{4}(1 - e^{-2t} + 2te^{-2t}) + \tfrac{1}{2}(1 - e^{-2t}).$$

Note that although the interaction term has the same time constant as the main process term, and hence we can expect considerable influence from it, its steady state effect is only one half of that of the direct effect from r_2.

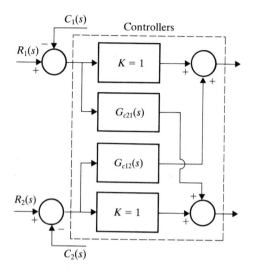

Fig. 6.21 The addition of corss controllers to remove interaction

To remove the interaction we add cross-controllers (Fig. 6.21). We need to have $\mathbf{G}_p(s)\mathbf{G}_c(s)$ diagonal. Thus

$$
\begin{bmatrix} \dfrac{1}{s+1} & 0 \\ \dfrac{1}{s+1} & \dfrac{1}{s+1} \end{bmatrix}
\begin{bmatrix} 1 & G_{c12}(s) \\ G_{c21}(s) & 1 \end{bmatrix}
$$

is diagonal, i.e.

$$
\begin{bmatrix} \dfrac{1}{s+1} & \dfrac{G_{c12}(s)}{s+1} \\ \dfrac{1}{s+1}+\dfrac{G_{c21}(s)}{s+1} & \dfrac{G_{c12}(s)}{s+1}+\dfrac{1}{s+1} \end{bmatrix}
$$

is diagonal, so that $G_{c12}(s)=0$ and $G_{c21}=-1$. We therefore need only one additional controller when only one interaction path is present and this acts in the 'same direction' as the interaction (in the nature of a feedforward controller).

6.6 The general transfer function matrix

The comparatively detailed block diagram of Figs. 6.18 and 6.19 may be reduced considerably and this is indicated by the equation (6.4(a)). To establish more similarity with the single input–single output case

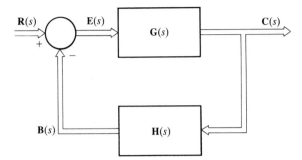

Fig. 6.22 General multi-input multi-output system

replace $\mathbf{G}_m(s)$ by $\mathbf{H}(s)$. The multi-input multi-output system is then represented by Fig. 6.22. Then

$$\mathbf{B}(s) = \mathbf{H}(s)\mathbf{C}(s)$$
$$\mathbf{E}(s) = \mathbf{R}(s) - \mathbf{B}(s)$$

and

$$\mathbf{C}(s) = \mathbf{G}(s)\mathbf{E}(s)$$

giving for the closed loop

$$\mathbf{C}(s) = \mathbf{G}(s)[\mathbf{R}(s) - \mathbf{H}(s)\mathbf{C}(s)]$$

i.e.

$$\mathbf{C}(s) = [\mathbf{I} + \mathbf{G}(s)\mathbf{H}(s)]^{-1}\mathbf{G}(s)\mathbf{R}(s) \qquad (6.6)$$

and this condensed form may be used for establishing conditions of non-interaction (cf. 6.4(a)). External disturbance vectors $\mathbf{Z}(s)$ may be added and the normal block diagram and vector-matrix algebra followed similarly. The individual elements of each matrix are expressions in s of the normal transfer function form.

6.7 Distributed parameter systems

In the discussion of modelling, the level of complexity next to the simplest lumped models was the maximum gradient or distributed parameter model in which one special co-ordinate of the system was dominant. The dynamic model in these cases is comprised of partial differential equations with time and the space co-ordinate as the independent variables. Like the lumped parameter models these equations may be either linear or nonlinear. Although systems which are conveniently modelled in this way are numerous in chemical and physical processes, e.g. heat exchangers, tubular reactors, packed columns, the difficulty in analysis arises because the inversion of the transfer functions requires inversion of irrational, transcendental, functions which can be expressed only by an infinite number of roots.

Digital simulation is more appropriate than analogue simulation but general analytical solutions are usually unavailable. The choice between a lumped and distributed parameter model for a distributed system depends on their relative accuracy and computational difficulty. The accuracy of the distributed model may be better but its actual accurate solution is very expensive, time-consuming or virtually impossible if many components are involved in the model.

The only form of distributed system which has been readily brought into linear control theory is the pure delay. This has been discussed in Chapter 5 and even for this simplest form considerable difficulty may evolve when there is feedback in the system. The detailed discussion of distributed parameter systems becomes extensive and requires the study of individual cases to build up a picture near to completion (Friedly 1972; Douglas 1972; Cohen and Coon 1953). In particular, distributed systems with feedback are of special difficulty and these comprise most of the cases in control situations. However, comparatively little information is available for such systems compared with that for the cases of open loop distributed parameter systems which are fairly extensively listed.

The 'empirical' techniques of Coon and Ziegler and Nichols recognize implicitly the distributed nature of a process by effectively approximating it to a pure delay and a simple lag. The precise nature of the process need not then be known. Even if the first simplifying assumption of a linear model is made, how suitable are the methods of the Nyquist, Bode and root locus plots that are so successful with lumped linear systems? With the simple delay we have seen that the Bode and Nyquist plots may be successfully used with no additional difficulty. With the fuller partial differential equation descriptions the Bode plot will show oscillations which with only small errors can lead to incorrect controller selection based on gain and phase margins. It may be easier to evaluate the stability limits on the Nyquist plot. Despite the full complexity of the distributed parameter solution the system behaviour may be influenced principally by a dominant pair of roots and once these have been identified they may be used in a root locus study. However a more general approach is complex.

In addition to the numerical and analytical problems the choice of measuring element positions in a distributed system may also be involved. This will be governed by system dynamics and stability considerations and repeated simulation will help in this matter.

Example   As an example of a distributed parameter system with a feedback loop consider the simplest situation of Fig. 6.23 in which the plant is a pure unit delay. The signal which will form the output flows through the plant of effective cross section A from $z = 0$ to $z = 1$ with unit velocity. The model equation for the process is then the simple partial differential equation obtained by a mass balance on the element

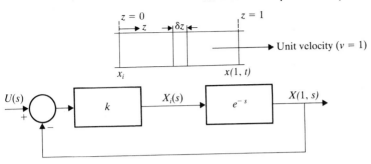

Fig. 6.23 Unity feedback and simple distributed system (pure delay)

of length δz:

$$Av\left\{\left(x+\frac{\partial x}{\partial z}\cdot\delta z\right)-x\right\}\delta t-A\delta z\frac{\partial x}{\partial t}\delta t=0$$

i.e. with $v=1$

$$\frac{\partial x}{\partial t}+\frac{\partial x}{\partial z}=0 \tag{6.7}$$

with $x=x_i$ at $z=0$. The outlet signal $x(1,t)$ is sent back to the plant via the controller and the input to the plant x_i is thus given by

$$x_i(t)=k(u(t)-x(1,t)) \tag{6.8}$$

ignoring any delays in the feedback path.

The transfer function relating the outlet $X(1,s)$ to the input $U(s)$ is obtained by taking Laplace transforms of the two equations (6.7) and 6.8):

$$sX(s)+\frac{dX(s)}{dz}=0$$

$$X_i(s)=k(U(s)-X(1,s)).$$

Note that transforming the partial differential equation (6.7) leaves us with an ordinary differential equation and this is so for all partial differential equations with two independent variables. This differential equation is linear and has separable variables to give

$$-\frac{dX(s)}{sX(s)}=dz$$

or

$$\frac{-1}{s}\cdot\ln X(s)=z+constant$$

where s is a constant during the integration. From the boundary condition $z = 0$, $x = x_i$ the constant of integration is $-\dfrac{1}{s} \ln X_i(s)$ which by our second relationship, the transform of equation (6.8), becomes $-\dfrac{1}{s} \ln \{k(U(s) - X(1, s))\}$ so that the full integration is

$$z = -\frac{1}{s} \ln \left\{ \frac{X(s)}{k(U(s) - X(1, s))} \right\}.$$

At $z = 1$, $X(s) = X(1, s)$ and on rearranging to remove the logarithmic term,

$$e^{-s} = \frac{X(1, s)}{kU(s) - kX(1, s)}$$

so that

$$\frac{X(1, s)}{U(s)} = \frac{ke^{-s}}{1 + ke^{-s}}, \quad \text{i.e.} \quad \frac{G(s)}{1 + G(s)} \quad \text{as previously.}$$

This is a most unpromising looking transfer function yet the basic 'plant' is only a simple delay. For a unit step input $U(s) = 1/s$ and the solution is the slowly converging series (Friedly 1972)

$$x(1, t) = kU(t - 1)\left\{ \frac{1}{1 + k} \right.$$

$$+ \frac{2}{k^{t-1}} \sum_{j=1}^{\infty} \left. \frac{\ln k \cos \{(2j-1)\pi(t-1)\} + (2j-1)\pi \sin \{(2j-1)\pi(t-1)\}}{\ln^2 k + (2j-1)^2 \pi^2} \right\}$$

where $U(t) = 0$, $t < 0$; 1, $t \geq 0$, the Heaviside unit step function. Only if $k < 1$ does the solution converge to a stable solution so that for $k > 1$, the system is unstable. In comparison, if the distributed system had been approximated to a simple lumped representation,

$$\tau \frac{dx}{dt} + x = x_i,$$

the closed loop transfer function would have been stable for all k, being

$$\frac{X(1, s)}{U(s)} = \frac{k}{\tau s + 1 + k}$$

with the explicit non-series solution for a step input, $U(s) = 1/s$

$$x(1, t) = \frac{k}{1 + k} (1 - e^{-(1+k)t/\tau})$$

The distributed system output may be generated by considering waves

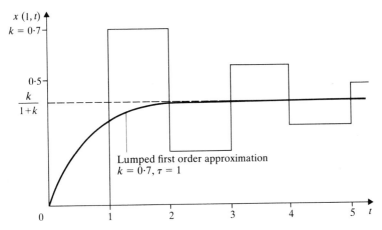

Fig. 6.24 Response of simple distributed system (Fig. 6.23) with feedback to a unit step input

passing around the loop. The response is shown in Fig. 6.24 for a unit step input. Both representations give the same steady state solution. For further discussion of distributed systems see also Chapter 2.

Example The difficulty of obtaining an analytical expression for a distributed system, partial differential equation, representation is illustrated further by taking up again the heat transfer system of Chapter 2. The plant comprises a single pass heat exchanger with a uniform jacket temperature T_j. The temperature of the fluid leaving the tube side of the exchanger is T_L. The change in fluid exit temperature (i.e. change in T_L) is θ_L resulting from a change T'_j in the jacket (shell side) temperature. The open loop transfer function relating θ_L to T'_j has been shown to be

$$\frac{\theta_L(s)}{T'_j(s)} = \frac{1}{1 + \tau s}(1 - e^{-(1+\tau s)L/\alpha})$$

where τ, L, α are parameters of the system and a time solution has been obtained. As earlier in this chapter the transfer function is distinguished by the presence of a term e^{-ks}. If θ_L is controlled, by using feedback of this deviation to change T'_j, then the full feedback closed loop expression will be difficult to invert to obtain a time solution. With simple proportional control the closed loop system is as Fig. 6.25, where other dynamics, e.g. in measuring elements, have been assumed rapid compared with those of the jacket to tube-content heat transfer. The closed loop transfer function is

$$\frac{\theta_L(s)}{\theta_{L\,ref}(\theta)} = \frac{K(1 - e^{-(1+\tau s)L/\alpha})}{1 + \tau s + K(1 - e^{-(1+\tau s)L/\alpha})}.$$

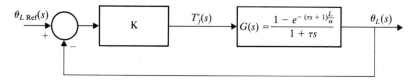

Fig. 6.25   Distributed heat exchange system with feedback and proportional control

In Chapter 1 the initial value theroem,

$$\lim_{t \to 0} f(t) = \lim_{s \to \infty} sF(s)$$

and the final value theorem

$$\lim_{t \to \infty} f(t) = \lim_{s \to 0} sF(s)$$

were given and we may use these to determine the initial and final, steady state, response to a step input in $\theta_{L\,ref}$. Taking a unit step input $\theta_L(s)$ becomes

$$\theta_L(s) = \frac{1}{s} \cdot \frac{K(1 - e^{-(1+\tau s)L/\alpha})}{1 + \tau s + K(1 - e^{-(1+\tau s)L/\alpha})}$$

and

$$\theta_L(t)_{t \to 0} = 0$$

with

$$\theta_L(t)_{t \to \infty} = \frac{K(1 - e^{-L/\alpha})}{1 + K(1 - e^{-L/\alpha})}.$$
$$< 1$$

Thus, as with 'lumped' systems, offset is observed with proportional control only, decreasing as K increases.

The system dynamics are better observed via the open loop, using Nyquist or Bode plots, than they are by solution of the closed loop equations. The open loop frequency response becomes

$$G(j\omega) = \frac{\theta_L(j\omega)}{\theta_{L\,ref}(j\omega)}$$

$$= \frac{K(1 - j\omega\tau)}{(1 + \tau^2\omega^2)}(1 - e^{-L/\alpha} \cdot e^{-L\tau/\alpha \cdot j\omega})$$

$$= \frac{K(1 - j\omega\tau)}{1 + \tau^2\omega^2}\left\{1 - e^{-L/\alpha}\left(\cos\frac{L\tau\omega}{\alpha} - j\sin\frac{L\tau\omega}{\alpha}\right)\right\}.$$

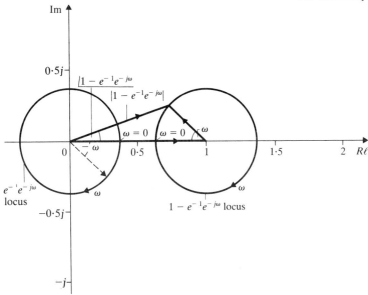

Fig. 6.26 Nyquist plot for $1 - e^{-1}e^{-j\omega}$

Let τ, L, α all be unity for simplicity so that

$$G(j\omega) = \frac{K(1-j\omega)}{1+\omega^2}\{1 - e^{-1}(\cos\omega - j\sin\omega)\}.$$

The amplitude term is, following the normal frequency response rules,

$$|G(j\omega)| = \frac{K}{\sqrt{(1+\omega^2)}} \cdot |1 - e^{-1}(\cos\omega - j\sin\omega)|.$$

As ω increases the term $1 - e^{-1}(\cos\omega - j\sin\omega)$, i.e. $1 - e^{-1}e^{-j\omega}$, follows the values in Fig. 6.26. Thus the amplitude and phase angle of this term both oscillate between finite limits. When combined with the terms $K(1+j\omega)^{-1}$ the Bode plots have the form of Fig. 6.27. It is seen, as expected from the final value theorem result, that this system is stable for all gains despite the delays in the system.

6.8 Summary

Although controller settings can be determined by the Bode plot etc. this requires that either the plant transfer function is known or else that a full frequency response test has been possible. On existing plants, or for simulation, plant controller settings may be made according to the empirically established methods of Ziegler and Nichols or

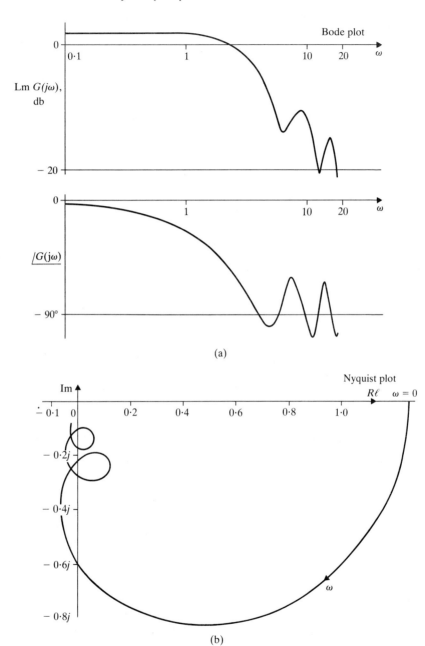

Fig. 6.27   Frequency response of distributed system ($K = 2$): (a) Bode plot (b) Nyquist plot

Coon for simple loops. Advantages may be gained over these simple feedback loops by the use of controllers in series, cascade control, in which the setpoint of one controller is set by the output of another, or by feedforward control. For the latter to be effective the system dynamics must be known and in practice only approximate feedforward control is possible, it being necessary to retain the feedback loop as well. In both these cases two controllers are used to produce a single final output. The simplest multivariable control problem is the two input–two output system with two controllers. This can be treated easily by the transfer function matrix but interaction may arise in the system requiring the addition of further controllers to remove the interaction. The non-interacting system suffers from difficulties similar to those experienced by feedforward control.

Problems arise in treating distributed parameter systems by the lumped parameter methods because of the transcendental functions and the infinite number of roots in the characteristic equation. However, some approximations may be made (and a few exact solutions are obtainable) but most cases have to be considered independently.

Part 3

Control – B – Multivariable, nonlinear and sampled control systems

Chapter 7
Multivariable control systems

7.1 Introduction

Although in Chapter 6 multivariable systems were considered in terms of the transfer function matrix most of the remainder of our work will be in the time domain and will use the state space method. It is therefore convenient to recap briefly on some of the material of Chapter 2 and section 3.4 for the general description of multi-input multi-output systems and of the use of state variables.

As well as the transfer matrix description of the system dynamics the state space model was developed,

$$\dot{\mathbf{x}} = \mathbf{A}\mathbf{x} + \mathbf{B}\mathbf{u}$$

where we consider $\mathbf{A}$ and $\mathbf{B}$ as time invariant, $\mathbf{x}$ the state vector and $\mathbf{u}$ the input vector. $\mathbf{A}$ is the system matrix and $\mathbf{B}$ the distribution matrix. For the general linear case the solution of the state equation is

$$\mathbf{x}(t) = \mathbf{\Phi}(t)\mathbf{x}(0) + \int_0^t \mathbf{\Phi}(t - \tau)\mathbf{B}\mathbf{u}(\tau)\, d\tau,$$

and for the time-invariant case the transition matrix $\mathbf{\Phi}(t)$ is $e^{\mathbf{A}t}$ so that

$$\mathbf{x}(t) = e^{\mathbf{A}t}\mathbf{x}(0) + \int_0^t e^{\mathbf{A}(t-\tau)}\mathbf{B}\mathbf{u}(\tau)\, d\tau.$$

We shall be concerned with this latter case, the transition matrix $e^{\mathbf{A}t}$ being defined as $\mathbf{I} + \mathbf{A}t + \dfrac{\mathbf{A}^2 t^2}{2!} + \ldots$ and being equal also to $\mathcal{L}^{-1}[(s\mathbf{I} - \mathbf{A})^{-1}]$. Then $\int_0^t e^{\mathbf{A}(t-\tau)}\mathbf{B}\mathbf{u}(\tau)\, d\tau$ is equal to $\mathcal{L}^{-1}[(s\mathbf{I} - \mathbf{A})^{-1}\mathbf{B}\mathbf{u}(s)]$. The speed of response of the system is governed by its eigenvalues λ_i, the roots of $|\lambda\mathbf{I} - \mathbf{A}| = 0$, the characteristic equation. Eigenvalues with negative real parts give rise to stable systems but positive real parts lead to instability. The eigenvalues are synonymous with the poles of the transfer function characteristic equation used in Chapter 4.

The solution to the dynamic equation, $x(t)$, may be expressed in terms of its eigenvectors and eigenvalues, the eigenvectors being the solution to the equation $\lambda_i x = Ax$ for each distinct λ_i. The eigenvector e_i requires a boundary condition for its full specification and the solution x may be written

$$x = \sum_{i=1}^{n} e_i e^{\lambda_i t}.$$

By diagonalizing the matrix A by the canonical transformation new state variables are derived so that the dynamics of each new state variable can be expressed solely in terms of that variable and one eigenvalue, plus the effect of the forcing vector u.

With this brief review of the earlier chapters let us now turn to control systems in particular. All of the above summary is applicable to any input–output system, whether or not it is a control system with feedback. In this chapter we shall look at the assessment of controllability, observability and stability of multivariable systems and the use of feedback.

7.2 Controllability and observability

When dealing with single input–single output systems it has not appeared necessary to ask can we control or can we observe the system fully – except in so far as stability is concerned. Having strict conditions for controllability and observability of multivariable systems is more important although the rigid definitions are comparatively recent (e.g. Kalman 1963; Gilbert 1963). All of the following is restricted to linear time-invariant systems. In addition to being able to tell if a system is controllable we need also to be sure that we can observe all our chosen variables which express the system dynamics.

A system is controllable if a control vector $u(t)$ can be found which will enable us to force the system from an arbitrary initial state $x(0)$ to some arbitrary finite state $x(t_f)$ in a finite time t_f. The problem of finding the required control remains but the controllability or otherwise can first be established according to this criterion. If the state vector x and output vector y are the same then state controllability and output controllability are naturally also the same. Usually the output vector is of lower order than the state and output controllability will be easier to satisfy. However, control of the state will be taken to infer control also of the output when they are related by $y = Cx$. Both state and output can be tested separately for controllability.

In section 3.4 the canonical form of the state equation was established. The state equation $x = Ax + Bu$ is expressed in the canonical form by using the transformation $x = Ez$ where E is the matrix whose columns are the eigenvectors of $\mathbf{A}$. Then

$$\dot{\mathbf{z}} = \mathbf{E}^{-1}\mathbf{A}\mathbf{E}\mathbf{z} + \mathbf{E}^{-1}\mathbf{B}\mathbf{u}$$

and using the diagonal matrix $\Lambda(=\mathbf{E}^{-1}\mathbf{A}\mathbf{E})$ whose elements are the eigenvalues λ_i of $\mathbf{A}$

$$\dot{\mathbf{z}} = \Lambda\mathbf{z} + \mathbf{E}^{-1}\mathbf{B}\mathbf{u}$$
$$= \Lambda\mathbf{z} + \mathbf{R}\mathbf{u} \tag{7.1}$$

where

$$\mathbf{R} = \mathbf{E}^{-1}\mathbf{B}.$$

The input or control vector $\mathbf{u}$ is of the same or lower order than $\mathbf{x}$ and $\mathbf{z}$. If controllability can be shown for $\mathbf{z}$ then because of the linear simple transformation between $\mathbf{z}$ and $\mathbf{x}$ this is also true for $\mathbf{x}$ and vice versa. Writing equation (7.1) out fully $(m \leqslant n)$

$$\begin{bmatrix} \dot{z}_1 \\ \dot{z}_2 \\ \cdot \\ \cdot \\ \cdot \\ \dot{z}_n \end{bmatrix} = \begin{bmatrix} \lambda_1 & 0 & \cdots & 0 \\ & \lambda_2 & & \\ & & \cdot & \\ & & & \cdot \\ & & & \lambda_n \end{bmatrix} \begin{bmatrix} z_1 \\ z_2 \\ \cdot \\ \cdot \\ \cdot \\ z_n \end{bmatrix} + \begin{bmatrix} r_{11} & \cdots & r_{1m} \\ r_{21} & \cdots & r_{2m} \\ \cdot & & \\ \cdot & & \\ \cdot & & \\ r_{n1} & \cdots & r_{nm} \end{bmatrix} \begin{bmatrix} u_1 \\ u_2 \\ \cdot \\ \cdot \\ \cdot \\ u_m \end{bmatrix} \tag{7.2}$$

As there is no coupling between the variables z_i, z_j etc., control of any z_i can only be through the control vector $\mathbf{u}$ by way of the matrix $\mathbf{R}$. Thus if any row of $\mathbf{R}$ is all zeros the corresponding z_i cannot be controlled, i.e. if all eigenvalues are distinct then it is necessary that no row of $\mathbf{E}^{-1}\mathbf{B}$ is all zeros. If a row is all zeros then obviously the corresponding variable cannot be controlled, whatever the control vector. Given that this first condition is satisfied, however, this only tells us that we can influence the variable by the control vector. It is still necessary to show that we can select the control to take all the state variables between two arbitrary, different, states. We can argue that the most difficult practical case (Elgerd 1967) would be to use a single control signal u to effect the desired state transition. If it is possible with the single element control then it is also possible with a more versatile multi-signal control vector. The scalar control u also facilitates the analysis. In this case $\mathbf{u} = u$ and $\mathbf{R}$ is the column vector $[r_1 \ r_2 \ldots r_n]^T$ so that for z_i we have

$$\dot{z}_i = \lambda_i z_i + r_i u.$$

Simple integration between the initial and final states yields

$$z_i(t_f) = e^{\lambda_i t_f} z_i(0) + e^{\lambda_i t_f} \int_0^{t_f} e^{-\lambda_i t} r_i u(t) \, dt$$

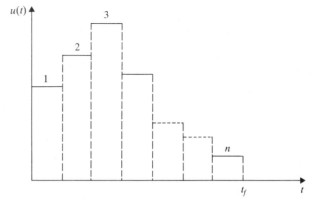

Fig. 7.1   Change in u with time to satisfy controllability requirement

so that for all $z_1 \ldots z_i \ldots z_n$ to be controllable $u(t)$ must satisfy all n equations of this form for arbitrary values of $z_i(0)$, $z_i(t_f)$ and t_f itself. To do this $\int_0^{t_f} e^{-\lambda_i t} u(t)\, dt$ requires to take on n specified arbitrary values, one for each z_i, for only one $u(t)$. Note that $u(t)$ is a function of time and by giving $u(t)$ n values over n discrete time intervals between zero time and the finite time t_f (Fig. 7.1), we can obtain the n values required of the control integral. In fact there exists an infinite number of ways of selecting $u(t)$. In the same way a higher order control vector $\mathbf{u}(t)$ may be selected provided the eigenvalues are distinct and no rows of $\mathbf{E}^{-1}\mathbf{B}$ are all zero. This semi-qualitative analysis may be effected more rigidly leading to a condition expressed in terms of the matrices $\mathbf{A}$ and $\mathbf{B}$. This is that the composite matrix

$$[\mathbf{B} \quad \mathbf{AB} \quad \ldots \quad \mathbf{A}^{n-1}\mathbf{B}]$$

must be of rank n for controllability. If the eigenvalues are not distinct the conditions require fuller detailing but this text will stay with distinct eigenvalues. The possible control has been taken as unbounded.

In the same way that it is necessary to ensure that a system of state variables can be controlled by a lesser number, if needed, of control elements so it is necessary to know if the values of all state variables can be determined by a lesser number of outputs, i.e. by the output vector $\mathbf{y} = \mathbf{Cx}$. That is, can we observe all state variables $\mathbf{x}$ from the measurements $\mathbf{y}$ which are possible. The vector $\mathbf{y}$ of course may be of the same order as $\mathbf{x}$.

Observability may be defined by saying that a system is completely observable if the measurements $\mathbf{y}$, over a finite time, contain the information which completely identifies the state $\mathbf{x}$. Using the canonical transformation $\mathbf{x} = \mathbf{Ez}$ the output vector becomes

$$\mathbf{y} = \mathbf{CEz} = \mathbf{Sz}, \text{ say}$$

and

$$y_1 = s_{11}z_1 + s_{12}z_2 + \ldots s_{1n}z_n$$
$$y_2 = s_{21}z_1 + s_{22}z_2 + \ldots s_{2n}z_n$$

.
.
.

$$y_p = s_{p1}z_1 + s_{p2}z_2 + \ldots s_{pn}z_n.$$

For all z_i to be represented in at least one of the measurements, y_j, it is therefore necessary that none of the columns of **S** is all zeros. If a column is all zeros then that z_i cannot be observed as it does not appear in the $y_j - z_i$ relationships. It does of course continue to be a state variable of the system and may even for example act in an unstable manner. The addition of more measurements may satisfy the observability requirement and in the same way further controls may satisfy the controllability requirement. As with controllability the conditions for full observability may be stated in terms of a matrix rank. In this case the matrix

$$[\mathbf{C}^T \quad \mathbf{A}^T\mathbf{C}^T \quad \ldots \quad (\mathbf{A}^T)^{n-1}\mathbf{C}^T]$$

must be of rank n.

It can be seen that some state variables may be uncontrollable and some may be unobservable. These may or may not be the same so that in general the state variables of a system fall into the four classes of Fig. 7.2.

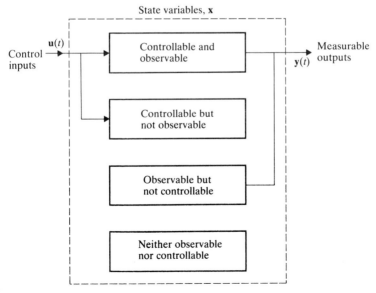

Fig. 7.2 Classification of state variables by controllability and observability criteria

Only one class of variable directly links the inputs, control and the measurable outputs. Thus, as these form the basis of transfer function analysis it can be appreciated that such a function may incompletely show the behaviour of a plant. A transfer function matrix $\mathbf{G}(s)$ determined by experiment, e.g. input–output frequency response, will only enable a model of the process to be deduced if the process is both completely observable and controllable.

Although the property of observability can be established the direct inverse relationship $\mathbf{x} = \mathbf{C}^{-1}\mathbf{y}$ can only be used if $\mathbf{x}$ and $\mathbf{y}$ are of the same order. In general and if possible we measure those state variables in which we are primarily interested but if further information about other state variables is required then it is necessary to construct an 'observer' which enables the reconstruction of the full state vector to be made (Luenberger 1971; Kwakernaak and Sivan 1972). Alternatively, if full simulation is possible the other state variables may be determined and measured in this way.

Example   Using the matrix conditions for observability and state controllability check whether the state is fully controllable and observable for the following systems,

$$\dot{\mathbf{x}} = \mathbf{A}\mathbf{x} + \mathbf{B}u, \qquad \mathbf{y} = \mathbf{C}\mathbf{x},$$

where

(i) $\mathbf{A} = \begin{bmatrix} 2 & 1 \\ -1 & 2 \end{bmatrix} \quad \mathbf{B} = \begin{bmatrix} 1 \\ 1 \end{bmatrix} \quad \mathbf{C} = \begin{bmatrix} 1 & 1 \end{bmatrix}$

(ii) $\mathbf{A} = \begin{bmatrix} 2 & 0 \\ 1 & 1 \end{bmatrix} \quad \mathbf{B} = \begin{bmatrix} 1 \\ 1 \end{bmatrix} \quad \mathbf{C} = \begin{bmatrix} 2 & 1 \end{bmatrix}$

(iii) $\mathbf{A} = \begin{bmatrix} 1 & 4 \\ 1 & 1 \end{bmatrix} \quad \mathbf{B} = \begin{bmatrix} 6 \\ 3 \end{bmatrix} \quad \mathbf{C} = \begin{bmatrix} 1 & 2 \end{bmatrix}.$

Consider each system in turn:

(i) $\begin{bmatrix} \dot{x}_1 \\ \dot{x}_2 \end{bmatrix} = \begin{bmatrix} 2 & 1 \\ -1 & 2 \end{bmatrix} \begin{bmatrix} x_1 \\ x_2 \end{bmatrix} + \begin{bmatrix} 1 \\ 1 \end{bmatrix} u$

$$[\mathbf{B} \quad \mathbf{A}\mathbf{B} \ldots \mathbf{A}^{n-1}\mathbf{B}] = [\mathbf{B} \quad \mathbf{A}\mathbf{B}]$$

$$= \begin{bmatrix} 1 & 3 \\ 1 & 1 \end{bmatrix}.$$

This matrix is non-singular, its determinant is non-zero, and hence its rank is equal to its order $n \ (=2)$. The system therefore is state controllable.

Examine the rank of $[\mathbf{C}^T \quad \mathbf{A}^T\mathbf{C}^T \quad \ldots \quad (\mathbf{A}^T)^{n-1}\mathbf{C}^T], \quad n = 2$

$$[\mathbf{C}^T \quad \mathbf{A}^T\mathbf{C}^T] = \begin{bmatrix} 1 & 1 \\ 1 & 3 \end{bmatrix}.$$

Again this is of rank 2 and the system is observable. In this case then the system is both observable and controllable.

(ii) $[\mathbf{B} \quad \mathbf{AB}] = \begin{bmatrix} 1 & 2 \\ 1 & 2 \end{bmatrix}$.

The determinant of the matrix is zero, and hence the rank is not 2 and the system is not controllable. However,

$[\mathbf{C}^T \quad \mathbf{A}^T\mathbf{C}^T] = \begin{bmatrix} 2 & 5 \\ 1 & 1 \end{bmatrix}$

and the system is observable. Thus the system is not controllable but is observable.

(iii) $[\mathbf{B} \quad \mathbf{AB}] = \begin{bmatrix} 6 & 18 \\ 3 & 9 \end{bmatrix}$.

This is a singular matrix so the system is not controllable.

$[\mathbf{C}^T \quad \mathbf{A}^T\mathbf{C}^T] = \begin{bmatrix} 1 & 3 \\ 2 & 6 \end{bmatrix}$.

This is also singular so the system is neither controllable nor observable.

7.3 Observers

In the use of feedback control so far, and in the following sections, it is assumed that we can measure all the information which we require. For the use of state feedback, section 7.5, a knowledge of the system states is required. In the absence of direct measurement or long calculation we obtain approximations to these missing state variable values by use of an 'observer' or 'state reconstructor'. Only a brief introduction is proposed here as although this method has been of benefit in precise areas like the aerospace industry it has not been so successful in other areas where expenditure and model accuracy is not so high. This is because in order to estimate those state variables not directly observable we rely on a precise model of the system. Lack of precision in the model leads to gross errors in state estimation. Although outlined here in a 'deterministic' way the use of observers is of major importance when extended for use in 'stochastic' systems where there exists uncertainty about true state variable values because of noise (unwanted signals) entering both the plant and the measurements.

The heart of the observer is an accurate model of the system and around this model a feedback loop is added. The aim is to subject this model to the same input as the plant itself and to make the model's output follow accurately the measured output of the plant. The two

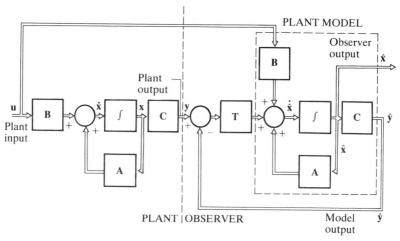

Fig. 7.3 The relationship between the observer and the plant

outputs are brought together by applying a gain matrix (**T**) to the errors between the true (plant) and simulated (model) outputs and this error vector is then fed to the plant model (Fig. 7.3). If (i) the model is accurate, (ii) the inputs **u** are the same and (iii) the outputs are the same then the intermediate state variables must also be the same. Two factors immediately appear as important: firstly the model must be accurate, and secondly since inputs and outputs of the plant may be varying with time the dynamics of the observer must be much faster than those of the plant itself if the model outputs are to follow closely the time output. The ˆ notation is used to denote the estimated values. Note that the output or product of the observer is the estimate $\hat{\mathbf{x}}$ of the state vector **x**. An observer may be used 'on its own' or incorporated in a feedback system (section 7.6). When used in conjunction with a controller the observer and controller operate as an entity.

Consider now how the arrangement of Fig. 7.3 gives the approximation (estimate) of the states. The plant equations are, as usual for the linear case,

$$\dot{\mathbf{x}} = \mathbf{A}\mathbf{x} + \mathbf{B}\mathbf{u}$$
$$\mathbf{y} = \mathbf{C}\mathbf{x}.$$

The observer equations are

$$\dot{\hat{\mathbf{x}}} = \mathbf{A}\hat{\mathbf{x}} + \mathbf{B}\mathbf{u} + \mathbf{T}(\mathbf{y} - \hat{\mathbf{y}})$$
$$\hat{\mathbf{y}} = \mathbf{C}\hat{\mathbf{x}} \tag{7.3}$$

so that elimination of $\hat{\mathbf{y}}$ yields

$$\dot{\hat{\mathbf{x}}} = (\mathbf{A} - \mathbf{T}\mathbf{C})\hat{\mathbf{x}} + \mathbf{B}\mathbf{u} + \mathbf{T}\mathbf{y}. \tag{7.4}$$

The error vector between the estimate of the state and the true state is $\hat{\mathbf{x}} - \mathbf{x}$ and the rate of change of this error is $\dot{\hat{\mathbf{x}}} - \dot{\mathbf{x}}$. Combining the plant state equations and equation (7.4) yields

$$\dot{\hat{\mathbf{x}}} - \dot{\mathbf{x}} = (\mathbf{A} - \mathbf{TC})\hat{\mathbf{x}} + \mathbf{Bu} + \mathbf{TCx} - \mathbf{Ax} - \mathbf{Bu}$$
$$= (\mathbf{A} - \mathbf{TC})(\hat{\mathbf{x}} - \mathbf{x})$$

or

$$\dot{\mathbf{e}} = (\mathbf{A} - \mathbf{TC})\mathbf{e}. \tag{7.5}$$

This equation expresses the error dynamics $\dot{\mathbf{e}}$ of the estimate in terms of the system matrix $\mathbf{A}$, the measurement matrix $\mathbf{C}$ and the observer matrix $\mathbf{T}$ only. The error dynamics are independent' of the state $\mathbf{x}$ and the input $\mathbf{u}$.

Comparing equation (7.5) with the general unforced equation

$$\dot{\mathbf{x}} = \mathbf{Ax}$$

which we have already encountered, we see that if $\mathbf{T}$ is chosen so that $(\mathbf{A} - \mathbf{TC})$ is a stable system matrix then $\mathbf{e}$ tends to zero, i.e. $\hat{\mathbf{x}}$ tends to $\mathbf{x}$ with increasing time, independently of $\mathbf{x}$ and $\mathbf{u}$. The observer design is thus basically one of ensuring that $(\mathbf{A} - \mathbf{TC})$ is stable and that the dynamics are sufficiently fast so that the estimate $\hat{\mathbf{x}}$ attains the value $\mathbf{x}$ (as near as is desired) rapidly. It must be once more emphasized that this is only possible if the $\mathbf{A}$ matrix of the model and the $\mathbf{A}$ matrix of the plant are the same to a high degree of accuracy. Normally we evaluate $\mathbf{T}$ for a chosen system such that the dynamics of the $(\mathbf{A} - \mathbf{TC})$ matrix are an order of magnitude faster than the system matrix $\mathbf{A}$. Then the initial estimate of $\mathbf{x}$, $\hat{\mathbf{x}}(0)$, need not be very good if the observer dynamics are such that $\hat{\mathbf{x}}(t)$ rapidly approaches the true state $\mathbf{x}(t)$.

The determination of a suitable matrix $\mathbf{T}$ is illustrated in the following. Notice again how the fundamental ideas of simple system dynamics, speed of response, eigenvalues etc. are basic to more advanced material as here and in subsequent sections of this chapter.

Example A second order system with state variables x_1 and x_2 is represented by the linear equations

$$\begin{bmatrix} \dot{x}_1 \\ \dot{x}_2 \end{bmatrix} = \begin{bmatrix} -3 & 1 \\ 0 & -1 \end{bmatrix} \begin{bmatrix} x_1 \\ x_2 \end{bmatrix} + \begin{bmatrix} 0 \\ 1 \end{bmatrix} u$$

$$y = \begin{bmatrix} 1 & 0 \end{bmatrix} \begin{bmatrix} x_1 \\ x_2 \end{bmatrix}.$$

What is the observer matrix $\mathbf{T}$ of a suitable observer which enables the full estimate of the state to be obtained?

A suitable $\mathbf{T}$ must result in estimate error dynamics which are much faster than the system dynamics. Therefore first determine the eigenvalues of the system matrix. With the **A** matrix given the eigen-values are $\lambda = -3$, $\lambda = -1$. For the estimate error to decrease rapidly and so allow accurate following of the system state we should choose eigenvalues for the $(\mathbf{A} - \mathbf{TC})$ matrix which are all considerably larger than the dominant system eigenvalues. Since we do not know without full solution the relative effects of each mode of the system let each of the $(\mathbf{A} - \mathbf{TC})$ eigenvalues be set as -15. A standard procedure may now be followed.

Since **C** is a 1×2 row vector **T** must be 2×1 column vector if the product **TC** is to be a 2×2 matrix the same as **A** to enable $\mathbf{A} - \mathbf{TC}$ to be formed. Then

$$\mathbf{A} - \mathbf{TC} = \begin{bmatrix} -3 & 1 \\ 0 & -1 \end{bmatrix} - \begin{bmatrix} t_1 \\ t_2 \end{bmatrix} \begin{bmatrix} 1 & 0 \end{bmatrix}$$

$$= \begin{bmatrix} -3 - t_1 & 1 \\ -t_2 & -1 \end{bmatrix}.$$

The characteristic polynomial of the matrix is $|\lambda \mathbf{I} - (\mathbf{A} - \mathbf{TC})|$ and the roots of the characteristic equation are the chosen eigenvalues.

$$|\lambda \mathbf{I} - (\mathbf{A} - \mathbf{TC})| = \begin{vmatrix} \lambda + 3 + t_1 & -1 \\ t_2 & \lambda + 1 \end{vmatrix}$$

$$= \lambda^2 + \lambda(4 + t_1) + (3 + t_1 + t_2).$$

This must equal $(\lambda + 15)(\lambda + 15)$, i.e.

$$\lambda^2 + 30\lambda + 225 \equiv \lambda^2 + \lambda(4 + t_1) + (3 + t_1 + t_2)$$

so that on equating coefficients

$$4 + t_1 = 30$$
$$3 + t_1 + t_2 = 225$$

and

$$t_1 = 26$$
$$t_2 = 196.$$

Using these values in the **T** matrix will give the desired performance to the observer.

The error dynamics are given by equation (7.5) or on substitution for **A**, **T**, and **C**,

$$\dot{\mathbf{e}} = \begin{bmatrix} \dot{e}_1 \\ \dot{e}_2 \end{bmatrix} = \begin{bmatrix} \dot{\hat{x}}_1 - \dot{x}_1 \\ \dot{\hat{x}}_2 - \dot{x}_2 \end{bmatrix}$$

$$= \begin{bmatrix} -29 & 1 \\ -196 & -1 \end{bmatrix} \begin{bmatrix} e_1 \\ e_2 \end{bmatrix}$$

Remembering that the solution of $\dot{\mathbf{x}} = \mathbf{A}\mathbf{x}$ is

$$\mathcal{L}^{-1}[(s\mathbf{I} - \mathbf{A})^{-1}]\mathbf{x}(0)$$

we obtain

$$\begin{bmatrix} e_1 \\ e_2 \end{bmatrix} = \mathcal{L}^{-1} \left\{ \frac{1}{(s+15)^2} \begin{bmatrix} s+1 & 1 \\ -196 & s+29 \end{bmatrix} \right\} \begin{bmatrix} e_1(0) \\ e_2(0) \end{bmatrix}.$$

Supposing our initial errors are $e_1(0) = 1$, $e_2(0) = 2$ then

$$e_1(t) = \mathcal{L}^{-1} \frac{s+3}{(s+15)^2}$$

$$= (1 - 12t)e^{-15t}$$

$$e_2(t) = \mathcal{L}^{-1} \frac{2s - 138}{(s+15)^2}$$

$$= 2(1 - 84t)e^{-15t}.$$

From these two expressions it is seen that the error, i.e. the amount by which the state estimate differs from the true state, falls rapidly (Fig. 7.4).

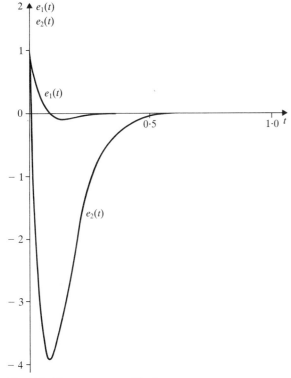

Fig. 7.4 Error response with observer

7.4 Stability of multivariable systems

In the treatment of single input–single output control systems much emphasis was placed on stability and the need for criteria to determine stability. This is a requirement of multivariable systems also and the wide, if laborious in many cases, application of Liapunov's direct method will be introduced. Liapunov's method is applicable, if with some difficulty, to both nonlinear and linear systems. The development will be general but attention will be concentrated on the continuous lumped linear systems which we have concentrated on so far. In this later case stability can also be assessed via the transfer function matrix of the closed loop and this, using the Routh–Hurwitz criterion, will be the easier method. The major benefit from the use of Liapunov's method is for nonlinear systems.

In either case the basis of stability analysis is the determination of the positions of the eigenvalues in the complex plane. As was shown in the general system dynamics of section 3.4 an eigenvalue with a positive real part leads to instability, i.e. an ever-increasing value (possibly oscillatory) for a state variable. In addition to the direct assessment of nonlinear system stability, we may approximate the nonlinear system dynamics by a linear form about some equilibrium point (Chapter 2). The application of linear system stability theory to this approximation then enables us to determine the stability or otherwise of the system when subject only to small perturbations. Stability within a small region of state variable values does not infer that the system is stable under greater excursions from the equilibrium conditions. Treatment in this way is referred to as Liapunov's indirect or first, method.

Stability via the tranfer function matrices

The general transfer function matrix for the closed loop is given by equation (6.6),

$$\mathbf{C}(s) = [\mathbf{I} + \mathbf{G}(s)\mathbf{H}(s)]^{-1}\mathbf{G}(s)\mathbf{R}(s)$$

where $\mathbf{G}(s)$ is the forward transfer function matrix and $\mathbf{H}(s)$ the feed-back path transfer matrix of the lumped, linear, time-invariant system. Write this as

$$\mathbf{C}(s) = \frac{\text{adj}\,[\mathbf{I} + \mathbf{G}(s)\mathbf{H}(s)]}{|\mathbf{I} + \mathbf{G}(s)\mathbf{H}(s)|}\,\mathbf{G}(s)\mathbf{R}(s).$$

Expansion of the expression for each of the n outputs $C_i(s)$ yields n equations, each having the factor $|\mathbf{I} + \mathbf{G}(s)\mathbf{H}(s)|$ in every term. For stability all eigenvalues (poles) of the closed loop must have negative real parts and these eigenvalues are the roots of

$$|\mathbf{I} + \mathbf{G}(s)\mathbf{H}(s)| = 0,$$

the characteristic equation. Thus the roots of this equation must all lie in the left half complex plane and whether this is so or not is most conveniently determined by the use of the Routh–Hurwitz criterion in this equation. This only confirms what might be inferred already in Chapters 3 and 4.

Example Using the general transfer function matrix determine the limiting value of K for stability in the system in Fig. 7.5. The forward transfer matrix $\mathbf{G}(s)$ relates $\mathbf{E}(s)$ and $\mathbf{C}(s)$;

$$\mathbf{C}(s) = \mathbf{G}(s)\mathbf{E}(s)$$

and for the system shown

$$\begin{bmatrix} C_1(s) \\ C_2(s) \end{bmatrix} = \begin{bmatrix} \dfrac{K}{s(s+1)} & \dfrac{1}{s} \\ \dfrac{1}{s+2} & \dfrac{1}{s+1} \end{bmatrix} \begin{bmatrix} E_1(s) \\ E_2(s) \end{bmatrix}.$$

The feedback matrix $\mathbf{H}(s)$ relates the error to the output by

$$\mathbf{E}(s) = \mathbf{R}(s) - \mathbf{H}(s)\mathbf{C}(s)$$

so

$$\begin{bmatrix} E_1(s) \\ E_2(s) \end{bmatrix} = \begin{bmatrix} R_1(s) \\ R_2(s) \end{bmatrix} - \begin{bmatrix} \dfrac{1}{s+2} & 0 \\ 0 & 1 \end{bmatrix} \begin{bmatrix} C_1(s) \\ C_2(s) \end{bmatrix}.$$

Stability is determined by the roots of

$$|\mathbf{I} + \mathbf{G}(s)\mathbf{H}(s)| = 0$$

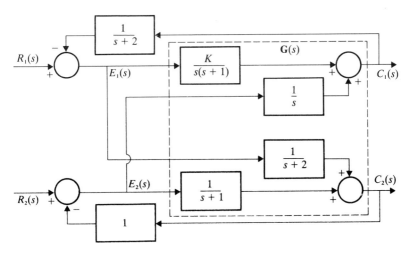

Fig. 7.5 Conditionally stable system

i.e.

$$\left| \begin{bmatrix} 1 & 0 \\ 0 & 1 \end{bmatrix} + \begin{bmatrix} \dfrac{K}{s(s+1)} & \dfrac{1}{s} \\ \dfrac{1}{s+2} & \dfrac{1}{s+1} \end{bmatrix} \begin{bmatrix} \dfrac{1}{s+2} & 0 \\ 0 & 1 \end{bmatrix} \right|$$

$$= \left| \begin{matrix} 1 + \dfrac{K}{s(s+1)(s+2)} & \dfrac{1}{s} \\ \dfrac{1}{(s+2)^2} & 1 + \dfrac{1}{s+1} \end{matrix} \right|$$

$$= 0$$

i.e. multipling out,

$$s^5 + 7s^4 + 18s^3 + (K+19)s^2 + (4K+6)s + (4K-1) = 0.$$

Using the Routh–Hurwitz criterion and array gives four functions in K which must be greater than zero in the first column of the array. For stability we find that $\frac{1}{4} < K < 17$. Even using the array hand computation becomes considerable with the quite simple functions shown and this illustrates the escalation in overall system order brought about by a combination of low order elements.

Stability and the Liapunov direct method

The Liapunov method although of wide applicability will be dealt with here. In contrast to the transfer matrix method it is applied in the time domain in the state space of the system. It is an approach though which does not require an explicit solution to the dynamic equations of the system. An additional reason for considering it with linear systems is that its success in these cases and in systems which are only slightly nonlinear can be assured. Such general assurance is not possible with the more highly nonlinear systems.

The more general method covering, in principle, all systems is called Liapunov's direct or second method, and as mentioned above no effort is made to solve the system equations explicitly. This is as opposed to the indirect method which requires linear approximations of the system equations about the equilibrium point or points and a study by linear theory of the behaviour in the region of the equilibrium points alone, thus indicating the presence or otherwise of stable equilibrium points. (Liapunov 1892; LaSalle and Lefschetz 1961). No attempt will be made to present a rigorous synthesis of the method and only an introduction to the method is intended. We need first to consider more precisely the definitions of stability.

The ideas of stability are made clearer by the two-dimensional diagram of Fig. 7.6. The two-dimensional case is a special case of the general n-dimensional state space whose co-ordinates are the state

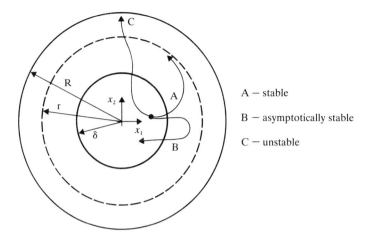

Fig. 7.6 Stability by Liapunov, two-dimensional example

variables $x_1, x_2, \ldots, x_n$. A region may be expressed in terms of the norm (or Euclidean norm) which itself is defined by

$$\|\mathbf{x}\| = \surd(x_1^2 + x_2^2 + \ldots x_n^2)$$

A hyperspherical region (of radius R) is then denoted by the region of space such that

$$\|\mathbf{x}\| \leq R.$$

The consideration of stability may now be considered within the region of radius R. (For linear systems R is not restricted.) Define a further two regions now within R so that $\delta \leq r < R$. Figure 7.6 illustrates this for the two-dimensional case centred on the state space origin. In this case the hypersphere of radius R becomes a circle, and it is seen that

$$R^2 \geq x_1^2 + x_2^2$$

or

$$\|\mathbf{x}\| = \surd(x_1^2 + x_2^2) \leq R.$$

The regions are considered as being about an equilibrium point, 0. By suitable transformation, if necessary, this may be expressed as the origin as shown. Nonlinear systems may have more than one equilibrium point, stable or unstable.

Stability in the sense of Liapunov is now defined by saying that a system is stable if for every radius r there exists a radius $\delta(\leq r)$ so that any trajectory starting within the region of radius δ will always remain with continually increasing time within the radius r. Such a system trajectory is path A.

If in addition to satisfying the Liapunov stability condition the trajectory starting within radius δ converges to the origin with continuing time then the system is asymptotically stable, path *B*. If this holds for all trajectories starting anywhere in the state space then the system is asymptotically stable in the large. For linear systems asymptotic stability and stability in the large are the same. If for some region of radius $r > 0$, the trajectory from some initial state leaves the region of r then the system is unstable, e.g. path *C*.

These definitions of stability do not mean precisely the same as the definition of stability we have used to this point and which corresponds to asymptotic stability only. In addition to these definitions of stability the terms of 'positive definite' etc. must be understood. For a scalar function it is convenient to define these terms here. This scalar may be a function V of the state vector $\mathbf{x}$, i.e. $V(\mathbf{x})$ and its following properties are defined over a specific region for the state space which includes the origin, $\mathbf{x} = \mathbf{0}$.

A scalar function $V(\mathbf{x})$ is positive definite in a region if $V(\mathbf{x}) > 0$ for all non zero states $\mathbf{x}$ in the region and if $V(\mathbf{x}) = 0$ for $\mathbf{x} = \mathbf{0}$.

$V(\mathbf{x})$ is negative definite if $- V(\mathbf{x})$ is positive definite.

$V(\mathbf{x})$ is positive semi-definite if it is positive for all states in the region except at the origin, and at certain other states, at which it is zero.

$V(\mathbf{x})$ is negative semi-definite if $- V(\mathbf{x})$ is positive semidefinite.

The Liapunov stability criterion may be stated in terms of the general autonomous (possibly nonlinear) system defined by the dynamic equation,

$$\dot{\mathbf{x}} = \mathbf{f}(\mathbf{x}, t) \qquad (7.6)$$

with its steady state as the origin,

$$\mathbf{f}(\mathbf{0}, t) = \mathbf{0} \quad \text{for all } t.$$

The state $\mathbf{x}$ is thus the vector of deviation from the steady state. Liapunov proposed a scalar function

$$V = V(\mathbf{x}) \qquad (7.7)$$

having the properties

$$\left.\begin{array}{l} V(\mathbf{x}, t) > 0 \\ V(\mathbf{0}, t) = 0 \end{array}\right\} \text{ for } \|\mathbf{x}\| > 0 \text{ i.e. positive definite,}$$

and with its time derivative having the property

$$\dot{V}(\mathbf{x}, t) < 0 \quad \text{for } \|\mathbf{x}\| > 0, \text{ i.e. negative definite.}$$

If there exists such a function V, a Liapunov function, and it has continuous first partial derivatives, then the equilibrium state at the origin is an asymptotically stable state. If $V(\mathbf{x})$ tends to zero with increasing time even as $\|\mathbf{x}\|$ tends to infinity, then the equilibrium state is asymptotically stable in the large, i.e. in the entire state space.

No guide is given by this definition of $V(\mathbf{x})$ as to how $V(\mathbf{x})$ is formed or even of the form we are looking for. Although the existence of such a function guarantees stability, just because a probable looking function does not satisfy the conditions it does not mean that the system is unstable as there may be other functions which do fall in with the criteria. Thus, although stability is based on the two simple criteria that $V(\mathbf{x}, t)$ must be positive definite and its time derivatives must be negative definite the actual finding of a $V(\mathbf{x}, t)$ may be difficult. In arriving at the time derivative $\dot{V}(\mathbf{x}, t)$ note that

$$\dot{V}(\mathbf{x}, t) = \frac{\partial V}{\partial t} + \frac{\partial V}{\partial x_1} \cdot \dot{x}_1 + \frac{\partial V}{\partial x_2} \cdot \dot{x}_2 + \ldots \frac{\partial V}{\partial x_n} \cdot \dot{x}_n$$

$$= \frac{\partial V}{\partial t} + \dot{\mathbf{x}}^T \begin{bmatrix} \dfrac{\partial V}{\partial x_1} \\ \cdot \\ \cdot \\ \cdot \\ \dfrac{\partial V}{\partial x_n} \end{bmatrix}$$

$$= \frac{\partial V}{\partial t} + \dot{\mathbf{x}}^T \mathbf{grad}\ V. \tag{7.8}$$

If $V(\mathbf{x}, t)$ is not explicit in t then $\dfrac{\partial V}{\partial t} = 0$ and

$$\dot{V}(\mathbf{x}, t) = \dot{\mathbf{x}}^T \mathbf{grad}\ V$$
$$= \mathbf{f}^T(\mathbf{x})\ \mathbf{grad}\ V.$$

Examples of the more direct quadratic forms of V will be given but experience has led to several much more involved expressions (e.g. Friedly 1972).

Linear systems

It is unfortunate in a sense that although a Liapunov function can always be found for a stable linear function the stability or otherwise of such a system can be established in other ways and so the benefit is minimal in the linear case. However, because a stable function can be guaranteed a linear system is the best vehicle on which to illustrate the Liapunov direct method in an introductory manner. In particular consider the autonomous linear time invariant system whose state equation is

$$\dot{\mathbf{x}} = \mathbf{A}\mathbf{x}.$$

The only equilbrium state is at the origin, $\mathbf{x} = \mathbf{0}$. To investigate stability suppose that we try a Liapunov function $V(\mathbf{x})$ where it has the 'quadratic form'

$$V(\mathbf{x}) = \mathbf{x}^T \mathbf{P} \mathbf{x}$$
$$= p_{11}x_1^2 + p_{22}x_2^2 + \ldots p_{nn}x_n^2 + p_{12}x_1x_2 + \ldots \tag{7.9}$$

where $\mathbf{P}$ is a positive definite Hermitian or real symmetrical matrix. The first condition for $V(\mathbf{x})$ is satisfied by this given quadratic form. Note that depending on the p_{ij} not all quadratic forms are positive definite. Forming the time derivative

$$\dot{V}(\mathbf{x}) = \dot{\mathbf{x}}^T \mathbf{P} \mathbf{x} + \mathbf{x}^T \mathbf{P} \dot{\mathbf{x}}$$
$$= (\mathbf{A}\mathbf{x})^T \mathbf{P} \mathbf{x} + \mathbf{x}^T \mathbf{P}(\mathbf{A}\mathbf{x})$$
$$= \mathbf{x}^T \mathbf{A}^T \mathbf{P} \mathbf{x} + \mathbf{x}^T \mathbf{P} \mathbf{A} \mathbf{x}$$
$$= \mathbf{x}^T (\mathbf{A}^T \mathbf{P} + \mathbf{P} \mathbf{A}) \mathbf{x}$$
$$= -\mathbf{x}^T \mathbf{R} \mathbf{x}$$

where

$$\mathbf{R} = -(\mathbf{A}^T \mathbf{P} + \mathbf{P} \mathbf{A}). \tag{7.10}$$

To satisfy the second condition $\mathbf{R}$ must be positive definite, i.e. $-\mathbf{R}$ is negative definite. To check if this is so it is necessary to use Sylvester's criterion (Gantmacher 1960) by which a necessary and sufficient condition for a matrix to be positive definite is that all the determinants of the principle minors are positive, i.e.

$$|r_{11}|, \begin{vmatrix} r_{11} & r_{12} \\ r_{21} & r_{22} \end{vmatrix}, \ldots, \begin{vmatrix} r_{11} & r_{12} & \cdots & r_{1n} \\ r_{21} & r_{22} & \cdots & r_{2n} \\ \cdot & & & \\ \cdot & & & \\ \cdot & & & \\ r_{n1} & & \cdots & r_{nn} \end{vmatrix} > 0.$$

A convenient alternative is to specify first an arbitrary $\mathbf{R}$ matrix which is positive definite and work back through the equation $\mathbf{R} = -(\mathbf{A}^T\mathbf{P} + \mathbf{P}\mathbf{A})$ to determine $\mathbf{P}$ and then to check if $\mathbf{P}$ is positive definite. $\mathbf{R}$ may be selected to be of a simple form as for example $\mathbf{I}$ the identity matrix. Note that any $\mathbf{P}$ and $\mathbf{R}$ matrices where $\mathbf{P}$ is real symmetrical and both $\mathbf{P}$ and $\mathbf{R}$ are positive definite satisfying equation (7.10) in $\mathbf{A}$, give sufficient proof of stability. In addition to stability determination, Liapunov's second method may be used in the transient response analysis of linear and nonlinear systems and in the evaluation of quadratic performances indices, e.g. squared error criteria (Salukvanze 1963). It may be used for discrete time systems and has been the subject of many investigations, especially in its use with nonlinear systems.

Example Two closed loop systems are described by the equations

(i) $\begin{bmatrix} \dot{x}_1 \\ \dot{x}_2 \end{bmatrix} = \begin{bmatrix} -1 & 2 \\ 1 & -1 \end{bmatrix} \begin{bmatrix} x_1 \\ x_2 \end{bmatrix}$.

(ii) $\begin{bmatrix} \dot{x}_1 \\ \dot{x}_2 \end{bmatrix} = \begin{bmatrix} -5 & -4 \\ 2 & 1 \end{bmatrix} \begin{bmatrix} x_1 \\ x_2 \end{bmatrix}$.

Using Liapunov's direct method check their stability and give suitable Liapunov functions. Using the results for linear systems we require a quadratic form Liapunov function $V(\mathbf{x}) = \mathbf{x}^T \mathbf{P} \mathbf{x}$ where $\mathbf{P}$ must be a positive definite symmetric matrix related to another positive definite matrix $\mathbf{R}$ by equation (7.10) i.e.

$$\mathbf{R} = -(\mathbf{A}^T \mathbf{P} + \mathbf{P} \mathbf{A}).$$

By choosing $\mathbf{R}$ to be the identity matrix $\mathbf{I}$, matrix $\mathbf{P}$ may be evaluated via the matrix $\mathbf{A}$. If the system matrix $\mathbf{A}$ contains variable values, such as a gain K, then the matrix $\mathbf{P}$ obtained using the method below would have elements which are functions of K. This K would then be limited by the conditions that the elements of $\mathbf{P}$ are such that $\mathbf{P}$ remains positive definitive.

(i) Returning to this example, select $\mathbf{R} = \mathbf{I}$, then equation (7.10) becomes

$$\begin{bmatrix} -1 & 1 \\ 2 & -1 \end{bmatrix} \begin{bmatrix} p_{11} & p_{12} \\ p_{12} & p_{22} \end{bmatrix} + \begin{bmatrix} p_{11} & p_{12} \\ p_{12} & p_{22} \end{bmatrix} \begin{bmatrix} -1 & 2 \\ 1 & -1 \end{bmatrix} = \begin{bmatrix} -1 & 0 \\ 0 & -1 \end{bmatrix}.$$

Because $\mathbf{P}$ is symmetric it has been possible to write p_{21} as p_{12}. To check on the positive definiteness of $\mathbf{P}$ the elements p_{11}, p_{12}, p_{22} are now determined. Evaluating the matrix product and sum gives

$$\begin{bmatrix} -p_{11} + p_{12} - p_{11} + p_{12} & -p_{12} + p_{22} + 2p_{11} - p_{12} \\ 2p_{11} - p_{12} - p_{12} + p_{22} & 2p_{12} - p_{22} + 2p_{12} - p_{22} \end{bmatrix} = \begin{bmatrix} -1 & 0 \\ 0 & -1 \end{bmatrix}.$$

Equating equivalent elements gives three equations (the two offdiagonal elements are equal)

$$-2p_{11} + 2p_{12} = -1$$
$$2p_{11} - 2p_{12} + p_{22} = 0$$
$$4p_{12} - 2p_{22} = -1$$

so that

$$p_{11} = -\tfrac{1}{4}$$
$$p_{12} = -\tfrac{3}{4}$$
$$p_{22} = -1.$$

Since p_{11} is negative the determinants of all of the principle minors of $\mathbf{P}$ cannot be positive as required since the very first $|p_{11}| = -\tfrac{1}{4}$. Thus the condition for stability is not satisfied and a Liapunov function for this linear system will not exist. This conclusion can be checked by determining the eigenvalues from

$$|\lambda \mathbf{I} - \mathbf{A}| = 0$$

i.e.

$$\begin{vmatrix} \lambda + 1 & -2 \\ -1 & \lambda + 1 \end{vmatrix} = 0$$

or

$$\lambda^2 + 2\lambda - 1 = 0$$

i.e.

$$\lambda_1 = -1 + \sqrt{2}, \qquad \lambda_2 = -1 - \sqrt{2}.$$

The existence of the positive real eigenvalue λ_1, confirms the instability of the system. This procedure and check is only possible of course with linear systems.

(ii) Consider the second system where $\mathbf{A} = \begin{bmatrix} -5 & -4 \\ 2 & 1 \end{bmatrix}$ so that $\mathbf{P}$ is given by

$$\begin{bmatrix} -5 & 2 \\ -4 & 1 \end{bmatrix} \begin{bmatrix} p_{11} & p_{12} \\ p_{12} & p_{22} \end{bmatrix} + \begin{bmatrix} p_{11} & p_{12} \\ p_{12} & p_{22} \end{bmatrix} \begin{bmatrix} -5 & -4 \\ 2 & 1 \end{bmatrix} = \begin{bmatrix} -1 & 0 \\ 0 & -1 \end{bmatrix}$$

i.e.

$$\begin{bmatrix} -5p_{11} + 2p_{12} - 5p_{11} + 2p_{12} & -5p_{12} + 2p_{22} - 4p_{11} + p_{12} \\ -4p_{11} + p_{12} - 5p_{12} + 2p_{22} & -4p_{12} + p_{22} - 4p_{12} + p_{22} \end{bmatrix} = \begin{bmatrix} -1 & 0 \\ 0 & -1 \end{bmatrix}$$

so that

$$-10p_{11} + 4p_{12} = -1$$
$$-4p_{11} - 4p_{12} + 2p_{22} = 0$$
$$-8p_{12} + 2p_{22} = -1$$

and hence

$$p_{11} = \tfrac{1}{3}$$
$$p_{12} = \tfrac{7}{12}$$
$$p_{22} = \tfrac{11}{6}.$$

Checking on the principle minors

$$|p_{11}| = \tfrac{1}{3} > 0$$

$$\begin{vmatrix} p_{11} & p_{12} \\ p_{12} & p_{22} \end{vmatrix} = \begin{vmatrix} \tfrac{1}{3} & \tfrac{7}{12} \\ \tfrac{7}{12} & \tfrac{11}{6} \end{vmatrix}$$

$$= \tfrac{11}{18} - \tfrac{49}{144}$$

$$= \tfrac{39}{144} > 0.$$

Thus **P** is definite positive, the system is stable and we can form a Liapunov function

$$V(\mathbf{x}) = [x_1 \quad x_2] \begin{bmatrix} \frac{1}{3} & \frac{7}{12} \\ \frac{7}{12} & \frac{11}{6} \end{bmatrix} \begin{bmatrix} x_1 \\ x_2 \end{bmatrix}$$

$$= \tfrac{1}{3}x_1^2 + \tfrac{7}{6}x_1 x_2 + \tfrac{11}{6}x_2^2$$

and

$$\dot{V}(\mathbf{x}) = -[x_1 \quad x_2] \begin{bmatrix} 1 & 0 \\ 0 & 1 \end{bmatrix} \begin{bmatrix} x_1 \\ x_2 \end{bmatrix}$$

$$= -(x_1^2 + x_2^2).$$

$V(\mathbf{x})$ is positive definite and $\dot{V}(\mathbf{x})$ is negative definite.

Again a check by eigenvalues gives

$$\begin{vmatrix} \lambda + 5 & 4 \\ -2 & \lambda - 1 \end{vmatrix} = 0$$

i.e.

$$\lambda^2 + 4\lambda + 3 = 0$$

$$\lambda_1 = -3, \qquad \lambda_2 = -1.$$

i.e. we have a stable system.

Nonlinear systems

For the use of Liapunov with linear systems it was seen above that the quadratic form, equation (7.9), was a natural selection, it being possible in the linear case to establish stability or otherwise conclusively. For the nonlinear systems attempts to form a quadratic form of Liapunov function are again the easiest first steps to make, although the failure of our search for such a function does not prove instability. As well as quadratic forms other Liapunov functions are based on integrals of state functions or other special forms but there are no true guidelines for selection in the more difficult cases (e.g. Gurel and Lapidus 1969; Krasovskii 1963).

By way of illustrating the use of a quadratic function to clarify the method, consider the following example.

Example   A third order system is described by the nonlinear state equations

$$\dot{x}_1 = -2x_1 + x_2^2 + 3x_3$$

$$\dot{x}_2 = -x_1 x_2 - 3x_3$$

$$\dot{x}_3 = -x_1 + x_2 - 3x_3^5.$$

Establish by the Liapunov direct method the stability or otherwise of the system.

Inspection shows that there is an equilibrium point, i.e. when $\dot{\mathbf{x}} = \mathbf{0}$, at $x_1 = x_2 = x_3 = 0$. Is this a stable or unstable equilibrium position? In the region of this point we can test for stability by linearizing about the equilibrium position and then checking on the eigenvalues. However, for a full consideration of stability over the whole space of x_1, x_2, x_3 this is not sufficient as the effect of the nonlinear nature of the state equations becomes highly significant for larger values of the x_1 etc. Consider therefore the direct method and take a trial Liapunov function

$$V(x_1, x_2, x_3) = \tfrac{1}{2}(ax_1^2 + bx_2^2 + cx_3^2).$$

That is, we try first a quadratic form of V. Obviously for all $\mathbf{x}$ ($\|\mathbf{x}\| > 0$), V is greater than zero and $V(\mathbf{0}, t)$ is zero so the first requirements of the Liapunov function are satisfied. Now forming the partial derivatives

$$\frac{\partial V}{\partial x_1} = ax_1$$

$$\frac{\partial V}{\partial x_2} = bx_2$$

$$\frac{\partial V}{\partial x_3} = cx_3$$

and

$$\frac{dV}{dt} = \frac{\partial V}{\partial t} + \frac{\partial V}{\partial x_1} \cdot \dot{x}_1 + \frac{\partial V}{\partial x_2} \cdot \dot{x}_2 + \frac{\partial V}{\partial x_3} \cdot \dot{x}_3$$

$$= 0 + ax_1(-2x_1 + x_2^2 + 3x_3) + bx_2(-x_1x_2 - 3x_3)$$
$$+ cx_3(-x_1 + x_2 - 3x_3^5)$$
$$= (-2ax_1^2 - 3cx_3^6) + (a - b)x_1x_2^2 + (3a - c)x_1x_3 - (3b - c)x_2x_3.$$

We can select values of a, b, c in this case so that $\dot{V}$ is negative definite and hence fulfills the Liapunov requirement. Let $a = b = 1$ and $c = 3$ so that

$$\dot{V} = -2x_1^2 - 9x_3^6 < 0, \qquad \mathbf{x} \neq \mathbf{0}$$

Then a suitable (non unique) Liapunov function is

$$V = \tfrac{1}{2}(x_1^2 + x_2^2 + 3x_3^2).$$

The use of the $\tfrac{1}{2}$ term merely simplifies the arithmetic after taking partial differentials.

7.5 State variable feedback in multivariable systems

Having looked at the requirements of controllability, observability and stability of multivariable systems let us return to what has been the basis of the single input–single output control systems – feedback. In the single output case the output is compared with the reference input and the error signal, the difference between these, forms the basis of the actuating signal. If the same system is represented by state space notation feedback operation may again be sought.

The general linear time invariant system equations used so far are

$$\dot{x} = Ax + Bu$$
$$y = Cx$$

where x is the state-vector, u the control vector and y the output vector. These represent the system response to an input u and are open loop in form. To close the loop the control vector u will be a function of the output y (and hence of the state variables x). If a reference input vector is r then the simplest conception of control will be one in which control action is proportioned to the difference between the input r and the output y, i.e. proportional to the error vector $r - y$. This may be shown by the equation,

$$u = K(r - y) \tag{7.11}$$

where in the simplest cases K is a diagonal gain matrix. A block diagram representation is shown in Fig. 7.7. For the regulator problem $r = 0$. However, continuing initially with retaining r and replacing y by Cx,

$$u = K(r - Cx)$$

and

$$\dot{x} = Ax + BK(r - Cx)$$
$$= (A - BKC)x + BKr. \tag{7.12}$$

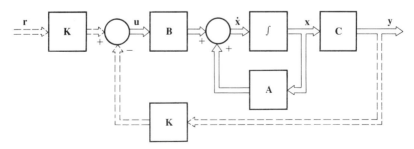

Fig. 7.7 Addition of feedback to the open loop system $\dot{x} = Ax + Bu$ via the output $y = Cx$

The closed loop system matrix is $\mathbf{A} - \mathbf{BKC}$ and the closed loop distribution matrix is $\mathbf{BK}$ for the reference input $\mathbf{r}$. The form of the relationship is similar to the open loop equation and might be written

$$\dot{\mathbf{x}} = \mathbf{A}_c \mathbf{x} + \mathbf{B}_c \mathbf{r}$$

where

$$\mathbf{A}_c = \mathbf{A} - \mathbf{BKC} \quad \text{and} \quad \mathbf{B}_c = \mathbf{BK}$$

and suffix c denoting the closed loop matrices. For the regulator case, $\mathbf{r} = \mathbf{0}$, the new state equation reduces to

$$\dot{\mathbf{x}} = \mathbf{A}_c \mathbf{x}$$

and the new state dynamics are governed by the closed loop matrix $\mathbf{A}_c$ where previously in the open loop the dynamics were governed by the matrix $\mathbf{A}$. Thus the use of proportional feedback of the state variables and a suitable choice of the feedback matrix $\mathbf{K}$ enables some control to be effected over the system dynamics, i.e. the dynamics are now governed by the new eigenvalues of the matrix $(\mathbf{A} - \mathbf{BKC})$ rather than by those of the matrix $\mathbf{A}$. The feedback thus has the same effects as previously observed, it leads to the elimination or reduction of error and it can improve system performance in terms of response times. The new dynamics are shown by the movement in the eigenvalues.

If the direct dynamics of the system are known to the extent that the matrix $\mathbf{A}_c$ may be fully specified, i.e. $\mathbf{A}_c = \mathbf{A}_{c\ desired}$, then the proportional controller $\mathbf{K}$ may be obtained since

$$\mathbf{A}_{c\ desired} = \mathbf{A} - \mathbf{BKC}$$

or

$$\mathbf{K} = \mathbf{B}^{-1}(\mathbf{A} - \mathbf{A}_{c\ desired})\mathbf{C}^{-1}.$$

Obviously the full matrix $\mathbf{K}$ can only be completely specified and calculated if both $\mathbf{B}$ and $\mathbf{C}$ have inverses, i.e. $\mathbf{B}$ and $\mathbf{C}$ must be square and non-singular. To establish a controller $\mathbf{K}$ on the basis of a specification of the closed loop eigenvalues (poles) representing system 'speed' alone rather than on all gains in the system (i.e. not a full specification of $\mathbf{A}_c$) then the condition may be relaxed with regard to $\mathbf{B}^{-1}$ and provided only that the system is controllable and $\mathbf{C}^{-1}$ exists it is possible to design the multivariable output feedback controller to give the specific closed loop eigenvalues.

Example A system is subject to two controls, u_1 and u_2 and the system equations are

$$\begin{bmatrix} \dot{x}_1 \\ \dot{x}_2 \end{bmatrix} = \begin{bmatrix} 0 & 1 \\ -1 & -1 \end{bmatrix}\begin{bmatrix} x_1 \\ x_2 \end{bmatrix} + \begin{bmatrix} 1 & 0 \\ 1 & 1 \end{bmatrix}\begin{bmatrix} u_1 \\ u_2 \end{bmatrix}$$

$$\begin{bmatrix} y_1 \\ y_2 \end{bmatrix} = \begin{bmatrix} 1 & 0 \\ 0 & 1 \end{bmatrix}\begin{bmatrix} x_1 \\ x_2 \end{bmatrix}, \quad \text{i.e.} \quad \mathbf{y} = \mathbf{I}\mathbf{x} = \mathbf{x}$$

It is proposed to introduce state variable feedback control so that the closed loop matrix $\mathbf{A}_c$ is $\begin{bmatrix} 0 & 1 \\ -1 & -1\cdot5 \end{bmatrix}$, i.e. the closed loop system equation sought for the regulator problem is

$$\begin{bmatrix} \dot{x}_1 \\ \dot{x}_2 \end{bmatrix} = \begin{bmatrix} 0 & 1 \\ -1 & -1\cdot5 \end{bmatrix} \begin{bmatrix} x_1 \\ x_2 \end{bmatrix}.$$

Determine the state feedback control law $\mathbf{K}$ to give the sought-after closed loop system matrix.

The eigenvalues for the two cases can be evaluated to show the effect required.

'External' control

$$\lambda_{1,2} = -0\cdot5 \pm j\sqrt{0\cdot75}$$

'State feedback'

$$\lambda_{1,2} = -0\cdot75 \pm j\sqrt{0\cdot4375}.$$

These values confirm stability in both cases and show the relative placing of the poles in the complex plane. The block diagram with state feedback is shown in Fig. 7.8. As $\mathbf{y} = \mathbf{x}$, the state feedback and output feedback are identical.

With $\mathbf{r} = \mathbf{0}$, $\mathbf{u} = -\mathbf{Kx}$ and for the closed loop

$$\dot{\mathbf{x}} = (\mathbf{A} - \mathbf{BK})\mathbf{x}$$

$$= \mathbf{A}_c\mathbf{x}.$$

Thus

$$\mathbf{A} - \mathbf{BK} = \mathbf{A}_c$$

i.e.

$$\mathbf{K} = \mathbf{B}^{-1}(\mathbf{A} - \mathbf{A}_c).$$

$\mathbf{B}$ can be inverted and $\mathbf{B}^{-1} = \begin{bmatrix} 1 & 0 \\ -1 & 1 \end{bmatrix}$.

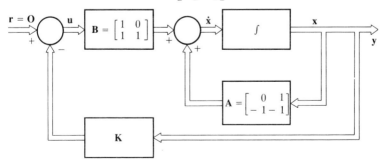

Fig. 7.8 State feedback control system

Thus

$$\mathbf{K} = \begin{bmatrix} 1 & 0 \\ -1 & 1 \end{bmatrix} \begin{bmatrix} 0 & 0 \\ 0 & 0.5 \end{bmatrix} = \begin{bmatrix} 0 & 0 \\ 0 & 0.5 \end{bmatrix}$$

so that

$$\mathbf{u} = \begin{bmatrix} 0 & 0 \\ 0 & -0.5 \end{bmatrix} \begin{bmatrix} x_1 \\ x_2 \end{bmatrix} = \begin{bmatrix} 0 \\ -0.5 \end{bmatrix} x_2$$

i.e. only the variable x_2 is required in the control law.

Substitution of this control input back into the original system equation will confirm that the desired closed loop matrix $\mathbf{A}_c$ is obtained.

$$\begin{aligned}
\dot{\mathbf{x}} &= \mathbf{A}\mathbf{x} + \mathbf{B}\mathbf{u} \\
&= \begin{bmatrix} 0 & 1 \\ -1 & -1 \end{bmatrix} \begin{bmatrix} x_1 \\ x_2 \end{bmatrix} + \begin{bmatrix} 1 & 0 \\ 1 & 1 \end{bmatrix} \begin{bmatrix} 0 & 0 \\ 0 & -0.5 \end{bmatrix} \begin{bmatrix} x_1 \\ x_2 \end{bmatrix} \\
&= \begin{bmatrix} 0+0 & 1+0 \\ -1+0 & -1-0.5 \end{bmatrix} \begin{bmatrix} x_1 \\ x_2 \end{bmatrix} = \begin{bmatrix} 0 & 1 \\ -1 & -1.5 \end{bmatrix} \begin{bmatrix} x_1 \\ x_2 \end{bmatrix}. \\
&= \mathbf{A}_c \mathbf{x}
\end{aligned}$$

If the state values $\mathbf{x}$ may be fed back directly, either from direct measurement or estimation, then the inverse $\mathbf{C}^{-1}$ does not appear explicitly in the control problem. We have now for state control the error vector $\mathbf{r} - \mathbf{x}$ and with $\mathbf{u}$ being formed on the basis of this error

$$\mathbf{u} = \mathbf{K}(\mathbf{r} - \mathbf{x})$$

$$\dot{\mathbf{x}} = (\mathbf{A} - \mathbf{B}\mathbf{K})\mathbf{x} + \mathbf{B}\mathbf{K}\mathbf{r}$$

or for the regulator case

$$\begin{aligned}
\dot{\mathbf{x}} &= (\mathbf{A} - \mathbf{B}\mathbf{K})\mathbf{x} \\
&= \mathbf{A}_c \mathbf{x}.
\end{aligned} \tag{7.13}$$

For the uncontrolled system the speed of response is given by the eigenvalues of $\mathbf{A}$, i.e. the roots of

$$|\lambda \mathbf{I} - \mathbf{A}| = 0.$$

If we specify new eigenvalues for the controlled system speed of response, say $\lambda_1, \lambda_2, \ldots, \lambda_n$ then we may equate the characteristic equation for $\mathbf{A}_c$ and for the specified closed loop dynamics,

$$|\lambda \mathbf{I} - \mathbf{A}_c| = (\lambda - \lambda_1)(\lambda - \lambda_2) \ldots (\lambda - \lambda_n) = 0$$

i.e.

$$|\lambda \mathbf{I} - (\mathbf{A} - \mathbf{B}\mathbf{K})| = (\lambda - \lambda_1)(\lambda - \lambda_2) \ldots (\lambda - \lambda_n).$$

Insertion of the $\mathbf{A}$ and $\mathbf{B}$ matrix values and the values of λ_i, $i = 1, \ldots, n$, gives linear equations in the elements of $\mathbf{K}$ enabling the k_{ij} to be determined. Note that in all the methods of this chapter our controller is chosen with respect to the new dynamics we seek for the closed loop system and no account, except possibly by interative or graphical means, is taken of the cost or 'best' choice, i.e. there is no direct measure of the optimality or best choice.

For a scalar control u, $\mathbf{B}$ becomes a vector $\mathbf{b}$ and the controller $\mathbf{K}$ (or control law) becomes a row vector $\mathbf{k}^T$ so that for the system

$$\dot{\mathbf{x}} = \mathbf{A}\mathbf{x} + \mathbf{b}u$$

$$\mathbf{y} = \mathbf{C}\mathbf{x}$$

the scalar feedback control is

$$u = \mathbf{k}^T\mathbf{y} \quad \text{for} \quad \mathbf{r} = \mathbf{0}$$

and

$$\dot{\mathbf{x}} = (\mathbf{A} - \mathbf{b}\mathbf{k}^T\mathbf{C})\mathbf{x}.$$

An illustrative example is given of a scalar controller seeking a given $\mathbf{A}_c$ using state variable feedback.

Example If the system matrix is unchanged from the previous example but the control input u is specified of scalar form, i.e.

$$\begin{bmatrix} \dot{x}_1 \\ \dot{x}_2 \end{bmatrix} = \begin{bmatrix} 0 & 1 \\ -1 & -1 \end{bmatrix} \begin{bmatrix} x_1 \\ x_2 \end{bmatrix} + \begin{bmatrix} 1 \\ 0 \end{bmatrix} u$$

$$\mathbf{y} = \mathbf{x},$$

determine the new state feedback control law $\mathbf{k}$ so that

$$u = -\mathbf{k}^T\mathbf{x}$$

and so that the closed loop eigenvalues are as in the previous example. If u is generated by state feedback then the feedback system equation is

$$\dot{\mathbf{x}} = (\mathbf{A} - \mathbf{b}\mathbf{k}^T)\mathbf{x}.$$

If we specify that the closed loop eigenvalues are the same as before then the characteristic equation of the closed loop is

$$|\lambda\mathbf{I} - \mathbf{A}_c| = 0$$

and the roots of this are specified from the example above, $\lambda_{1,2} = -0\cdot75 \pm j0\cdot66$ i.e.

$$\lambda^2 + 1\cdot5\lambda + 1 = 0. \tag{7.14}$$

We therefore require that the characteristic equation for the system

$\dot{\mathbf{x}} = (\mathbf{A} - \mathbf{b}\mathbf{k}^T)\mathbf{x}$ should be identical to this. This we can do but it should be noted that specification of the eigenvalues λ_1, λ_2 does not specify separately each element of the closed loop matrix which is what we did in the above example i.e. $\mathbf{A}_c$ is unknown at this stage.

With $\mathbf{b} = \begin{bmatrix} 1 \\ 0 \end{bmatrix}$, $\mathbf{A} = \begin{bmatrix} 0 & 1 \\ -1 & -1 \end{bmatrix}$, let $\mathbf{k} = \begin{bmatrix} k_1 \\ k_2 \end{bmatrix}$ then

$$\mathbf{A} - \mathbf{b}\mathbf{k}^T = \begin{bmatrix} 0 - k_1 & 1 - k_2 \\ -1 - 0 & -1 - 0 \end{bmatrix}$$

$$= \begin{bmatrix} -k_1 & 1 - k_2 \\ -1 & -1 \end{bmatrix}$$

and

$$|\lambda \mathbf{I} - (\mathbf{A} - \mathbf{b}\mathbf{k}^T)| = \begin{vmatrix} \lambda + k_1 & -1 + k_2 \\ 1 & \lambda + 1 \end{vmatrix}$$

$$= \lambda^2 + \lambda(1 + k_1) + k_1 + 1 - k_2. \qquad (7.15)$$

Equating coefficients in (7.14) and (7.15)

$$1 + k_1 = 1 \cdot 5, \quad \text{i.e.} \quad k_1 = 0 \cdot 5$$
$$1 + k_1 - k_2 = 1, \quad \text{i.e.} \quad k_2 = 0 \cdot 5$$

so that

$$\mathbf{k} = \begin{bmatrix} 0 \cdot 5 \\ 0 \cdot 5 \end{bmatrix}$$

and the scalar u is

$$u = -[0 \cdot 5 \quad 0 \cdot 5] \begin{bmatrix} x_1 \\ x_2 \end{bmatrix}$$

$$= -0 \cdot 5(x_1 + x_2).$$

Substitution back into the system equation yields

$$\dot{\mathbf{x}} = \mathbf{A}\mathbf{x} + \mathbf{b}u$$

$$= \begin{bmatrix} 0 & 1 \\ -1 & -1 \end{bmatrix} \begin{bmatrix} x_1 \\ x_2 \end{bmatrix} + \begin{bmatrix} 1 \\ 0 \end{bmatrix} (-0 \cdot 5(x_1 + x_2))$$

$$= \begin{bmatrix} -0 \cdot 5 & 0 \cdot 5 \\ -1 & -1 \end{bmatrix} \begin{bmatrix} x_1 \\ x_2 \end{bmatrix}$$

$$= \mathbf{A}_c'\mathbf{x}.$$

The closed loop system matrix is different from that in the previous case but we can check that the desired eigenvalues are obtained.

$$|\lambda \mathbf{I} - \mathbf{A}_c'| = \begin{vmatrix} \lambda + 0.5 & -0.5 \\ 1 & \lambda + 1 \end{vmatrix}$$

$$= \lambda^2 + 1.5\lambda + 1$$

$$= 0$$

and

$$\lambda_{1,2} = -0.75 \pm j0.66$$

as required.

7.6 Modal control

Although the output and feedback of section 7.5 leads to a change in the system dynamics it is limited in its useful applicability by the requirement that $\mathbf{C}^{-1}$ exists, i.e. that the full state vector may be measured. If $\mathbf{C}^{-1}$ does not exist then the problem may be eased by using observers or by transforming the state equation into the canonical form (section 3.4). The response of a system may be represented as the sum of its modes, equation (3.87) and the speed of the plant dynamics is governed by the slower modes, i.e. those formed in conjunction with the smallest eigenvalue (nearest to the imaginary axis) and hence with an associated long time constant. If we can concentrate on improving the dynamics associated with the slow modes – or dominant modes – then the whole performance is improved. By working in what is known as the 'modal domain' it is also easier to retain the important (slower) dynamics when reducing the complexity of a higher order model by going to a reduced lower order model, and it is also easier to deal with the problems of inexact measurements (Rosenbrock 1962; Gould 1969).

The objective of modal control is therefore to speed up the dynamics of the system by paying particular attention to the slower (smaller real part) eigenvalues, but it may form part of a larger overall control scheme. The following detail is restricted to the case where full state measurement is available. Remember however the pre-exponential coefficient of a modal term may be large even though the exponential decays rapidly. Thus the dynamics over a short time scale may have very active fast modes and if there is a large variation in the order of the pre-exponential then 'neglect' of the faster modes in favour of the slower, dominant, modes will not give good short-time accuracy. The general system equations are again

$$\dot{\mathbf{x}} = \mathbf{A}\mathbf{x} + \mathbf{B}\mathbf{u}$$

and

$$\mathbf{y} = \mathbf{Cx}$$

Replace **Bu** simply by the control vector $\mathbf{u}_c$ of the same order as $\mathbf{x}$, i.e.

$$\dot{\mathbf{x}} = \mathbf{Ax} + \mathbf{u}_c$$

as our interest is now focussed on finding $\mathbf{u}_c$ from the properties of **A**. Using the transformation

$$\mathbf{x} = \mathbf{Ez}$$

where **E** is the eigenvector matrix then,

$$\dot{\mathbf{z}} = \mathbf{E}^{-1}\mathbf{AEz} + \mathbf{E}^{-1}\mathbf{u}_c$$
$$= \mathbf{\Lambda z} + \mathbf{u}_z$$

where

$$\mathbf{u}_z = \mathbf{E}^{-1}\mathbf{u}_c$$

and $\mathbf{u}_z$ is the control vector in the transformed modal domain (cf. equation 3.93). $\mathbf{\Lambda}$ is the matrix whose diagonal is comprised of the eigenvalues of **A**. Retaining the notation of $\mathbf{z}$ for the transformed state vector repeat the procedure of section 7.4 and specify a desired closed loop matrix $\mathbf{\Lambda}_{desired}$, the elements of $\mathbf{\Lambda}_{desired}$ being the desired values of the closed loop eigenvalues of the system. These are specified in the modal domain but remember, however, that the transformation from the state $\mathbf{x}$ to the modal domain $\mathbf{z}$ means that the eigenvalues are the same in both cases. These values are to be achieved by the modal domain feedback control comprised of proportional action only, i.e. by using the reference input $\mathbf{r} = \mathbf{0}$,

$$\mathbf{u}_z = -\mathbf{K}_z\mathbf{z}.$$

The closed loop equation (7.12) is then

$$\dot{\mathbf{z}} = (\mathbf{\Lambda} - \mathbf{K}_z)\mathbf{z}.$$

The feedback controller matrix is that which is required when

$$\dot{\mathbf{z}} = \mathbf{\Lambda}_{desired}\mathbf{z}$$

i.e. the control matrix is given by

$$\mathbf{K}_z = (\mathbf{\Lambda} - \mathbf{\Lambda}_{desired}). \tag{7.16}$$

It is important to note that it is necessary to work either completely in the original system state, or as here in the modal domain, e.g. the controller, although eventually to be applied in the original, $\mathbf{x}$, system, is specified in the modal, $\mathbf{z}$, domain.

The $\mathbf{\Lambda}$, $\mathbf{\Lambda}_{desired}$ and $\mathbf{K}_z$ are all diagonal so that the matrix $\mathbf{K}_z$ is readily determined from the open loop behaviour found in the eigenvalue matrix $\mathbf{\Lambda}$ and the desired closed loop behaviour expressed in the matrix $\mathbf{\Lambda}_{desired}$.

If $\mathbf{u}_z = -\mathbf{K}_z\mathbf{z}$ then $\mathbf{u}_c$ is determined from the relations

$$\mathbf{u}_z = -\mathbf{K}_z\mathbf{z}$$
$$= -\mathbf{K}_z\mathbf{E}^{-1}\mathbf{x}$$

and

$$\mathbf{u}_z = \mathbf{E}^{-1}\mathbf{u}_c$$

so that

$$\mathbf{u}_c = -\mathbf{E}\mathbf{K}_z\mathbf{E}^{-1}\mathbf{x}.$$

Thus in terms of the original system the feedback control is given by

$$\mathbf{u}_c = -\mathbf{K}\mathbf{x}$$

where

$$\mathbf{K} = \mathbf{E}\mathbf{K}_z\mathbf{E}^{-1}$$

To sum up, the full operation of determining $\mathbf{K}$ is carried out through firstly the transformation, $\mathbf{x} = \mathbf{E}\mathbf{z}$ and secondly the specification of a control vector $\mathbf{u}_z$ in the modal domain. Only finally is the controller obtained for the original state space definition of the system. The equations representing the procedure are shown by the block diagram of Fig. 7.9. The key to the significance of the method lies in equation (7.16). From equation (7.16) it is seen that any one element of the transition matrix $\mathbf{\Lambda}$ can be changed using the matrix $\mathbf{K}_z$ without changing any of the other elements of $\mathbf{\Lambda}$. Thus completely independent control of the eigenvalues of the system is possible and in terms of the transformed variables this means also that the dynamics of any z_i can be changed without changing the dynamics of any other z_j. The slower modes may thus be speeded up without fear of adversely affecting the other modes (eigenvalues) of the system.

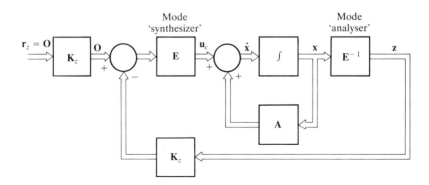

Fig. 7.9 Generation of control via the modal domain $(\mathbf{C} = \mathbf{I})$

Alternative development and extension of the cases considered may be found in the literature (Ellis and White 1965; Gould and Murray-Lasso 1966; Rosenbrock 1962). When a non-ideal situation exists in that **y** is of lower order than **x** and hence **C** is not square, interaction will occur and ideal behaviour will only be approximated to. Note that in the ideal case the eigenvalues of the closed loop system, i.e. of $\Lambda_{desired}$, can be increased in magnitude without limit using the feedback control $\mathbf{K}_z$ and hence in theory the response of the closed loop system with modal control can be made infinitely fast in any or all modes. In practice of course normal practical power limitations prevent this and 'complete' control infers all states variables can be measured and a high order controller used.

Example   For the system $\dot{\mathbf{x}} = \mathbf{Ax} + \mathbf{u}_c$ determine a state feedback controller using the modal control technique so that the greater time constant is halved. The system matrix **A** is $\begin{bmatrix} -4 & -3 \\ -1 & -2 \end{bmatrix}$. Halving the larger time constant means doubling the smaller eigenvalue. This gives faster system dynamics.

To use the canonical transformation it is necessary first to determine the eigenvalues and eigenvectors. The eigenvalues are first calculated:

$$|\lambda \mathbf{I} - \mathbf{A}| = 0$$

i.e.

$$\begin{vmatrix} \lambda + 4 & 3 \\ 1 & \lambda + 2 \end{vmatrix} = 0$$

$$\lambda^2 + 6\lambda + 5 = 0$$

so

$$\lambda_1 = -5, \qquad \lambda_2 = -1.$$

The eigenvectors $\mathbf{e}_i$ are now obtainable using the procedure of Chapter 3 (equation (3.88)):

$$\lambda_i \mathbf{e}_i = \mathbf{A} \mathbf{e}_i.$$

For the second order case

$$\mathbf{e}_1 = \begin{bmatrix} e_{11} \\ e_{21} \end{bmatrix}; \qquad \mathbf{e}_2 = \begin{bmatrix} e_{12} \\ e_{22} \end{bmatrix}.$$

For $\lambda_1 = -5$

$$-5e_{11} = -4e_{11} - 3e_{21}$$
$$-5e_{21} = -e_{11} - 2e_{21}$$

i.e.

$$-e_{11} = -3e_{21}$$

so that

$$\mathbf{e}_1 = \begin{bmatrix} 3 \\ 1 \end{bmatrix}, \quad \text{selecting} \quad e_{21} = 1.$$

For $\lambda_2 = -1$

$$-e_{12} = -4e_{12} - 3e_{22}$$
$$-e_{22} = -e_{12} - 2e_{22}$$

i.e.

$$e_{12} = -e_{22}$$

so that

$$\mathbf{e}_2 = \begin{bmatrix} -1 \\ 1 \end{bmatrix}, \quad \text{selecting} \quad e_{22} = 1.$$

There is no need to normalize the eigenvectors. The modal transition matrix $\mathbf{E} = [\mathbf{e}_1 \quad \mathbf{e}_2] = \begin{bmatrix} 3 & -1 \\ 1 & 1 \end{bmatrix}$ and the inverse $\mathbf{E}^{-1} = \dfrac{1}{4} \begin{bmatrix} 1 & 1 \\ -1 & 3 \end{bmatrix}$. Now the transformation $\mathbf{x} = \mathbf{E}\mathbf{z}$ is used. Thus the transformed equation becomes

$$\dot{\mathbf{z}} = \mathbf{E}^{-1}\mathbf{A}\mathbf{E}\mathbf{z} + \mathbf{E}^{-1}\mathbf{u}_c$$
$$= \mathbf{\Lambda}\mathbf{z} + \mathbf{u}_z$$

i.e.

$$\begin{bmatrix} \dot{z}_1 \\ \dot{z}_2 \end{bmatrix} = \begin{bmatrix} -5 & 0 \\ 0 & -1 \end{bmatrix} \begin{bmatrix} z_1 \\ z_2 \end{bmatrix} + \mathbf{u}_z \quad \text{where} \quad \mathbf{u}_z = \frac{1}{4} \begin{bmatrix} 1 & 1 \\ -1 & 3 \end{bmatrix} \mathbf{u}_c.$$

The desired closed loop system matrix $\mathbf{\Lambda}_{desired}$ is $\begin{bmatrix} -5 & 0 \\ 0 & -2 \end{bmatrix}$ and if the control

$$\mathbf{u}_z = -\mathbf{K}_z \mathbf{z}.$$

Then

$$\begin{bmatrix} \dot{z}_1 \\ \dot{z}_2 \end{bmatrix}_{desired} = \begin{bmatrix} -5 & 0 \\ 0 & -2 \end{bmatrix} \begin{bmatrix} z_1 \\ z_2 \end{bmatrix}$$
$$= \left(\begin{bmatrix} -5 & 0 \\ 0 & -1 \end{bmatrix} - \mathbf{K}_z \right) \begin{bmatrix} z_1 \\ z_2 \end{bmatrix}$$

Hence

$$\mathbf{K}_z = \Lambda - \Lambda_{desired}$$

$$= \begin{bmatrix} 0 & 0 \\ 0 & 1 \end{bmatrix} \text{ from equation (7.16).}$$

The matrix $\mathbf{K}_z$ is a diagonal matrix but as we are only interested in changing one mode, only one element is non-zero.

Transforming back to the original system we need to use $\mathbf{E}$ for the first time directly. For $\mathbf{u}_c = -\mathbf{K}\mathbf{x}$ with $\mathbf{u}_z = \mathbf{E}^{-1}\mathbf{u}_c$

$$\mathbf{u}_c = \mathbf{E}\mathbf{u}_z$$
$$= -\mathbf{E}\mathbf{K}_z \mathbf{z}$$
$$= -\mathbf{E}\mathbf{K}_z \mathbf{E}^{-1}\mathbf{x}$$

and

$$\mathbf{K} = \mathbf{E}\mathbf{K}_z \mathbf{E}^{-1}$$

$$= \begin{bmatrix} 3 & -1 \\ 1 & 1 \end{bmatrix} \begin{bmatrix} 0 & 0 \\ 0 & 1 \end{bmatrix} \begin{bmatrix} \frac{1}{4} & \frac{1}{4} \\ -\frac{1}{4} & \frac{3}{4} \end{bmatrix}$$

$$= \begin{bmatrix} \frac{1}{4} & -\frac{3}{4} \\ -\frac{1}{4} & \frac{3}{4} \end{bmatrix}.$$

The new system equation with the required control is then

$$\dot{\mathbf{x}} = \mathbf{A}\mathbf{x} + \mathbf{u}_c$$
$$= (\mathbf{A} - \mathbf{K})\mathbf{x}$$
$$= \begin{bmatrix} -\frac{17}{4} & -\frac{9}{4} \\ -\frac{3}{4} & -\frac{11}{4} \end{bmatrix} \mathbf{x}.$$

To check that this has yielded the desired effect, calculate the eigenvalues of $(\mathbf{A} - \mathbf{K})$. The characteristic equation after simplification is $\lambda^2 + 7\lambda + 10 = 0$, i.e. $\lambda_1 = -5$, $\lambda_2 = -2$, which is the result aimed at by the use of the modal control. Note that although one eigenvalue only is to be changed $\mathbf{K}$ itself is a full matrix utilizing feedback of both x_1 and x_2.

7.7 The use of observers with state feedback

The contents of sections 7.3 and 7.5 may be combined so that an observer-state feedback controller is formed. All that is required is that those states not directly available to the controller are estimated by an observer, and these estimates then used in the feedback loop.

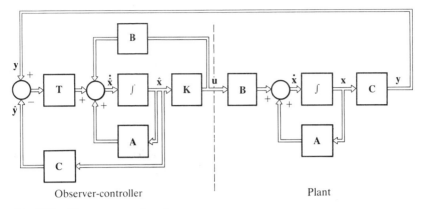

Fig. 7.10 Observer-controller-plant

We still require a model of the system for the observer to reproduce the states but given these states we can design the state feedback controller to give us, in theory, any desired closed loop dynamics.

As mentioned earlier the observer-controller combination merge in operation into a single unit and their relationship to the plant may be drawn as Fig. 7.10. The controller output **u** is now **Kx̂** instead of **Kx**. The observer equation is as before

$$\dot{\hat{\mathbf{x}}} = (\mathbf{A} - \mathbf{TC})\hat{\mathbf{x}} + \mathbf{Bu} + \mathbf{TCx}$$

and the plant equations are

$$\dot{\mathbf{x}} = \mathbf{Ax} + \mathbf{Bu}$$

$$\mathbf{y} = \mathbf{Cx}.$$

With now $\mathbf{u} = \mathbf{K\hat{x}}$
we get

$$\dot{\mathbf{x}} = \mathbf{Ax} + \mathbf{BK\hat{x}}$$

and

$$\dot{\hat{\mathbf{x}}} = (\mathbf{A} - \mathbf{TC} + \mathbf{BK})\hat{\mathbf{x}} + \mathbf{TCx}$$

or

$$\begin{bmatrix} \dot{\mathbf{x}} \\ \dot{\hat{\mathbf{x}}} \end{bmatrix} = \begin{bmatrix} \mathbf{A} & \mathbf{BK} \\ \mathbf{TC} & \mathbf{A} - \mathbf{TC} + \mathbf{BK} \end{bmatrix} \begin{bmatrix} \mathbf{x} \\ \hat{\mathbf{x}} \end{bmatrix}. \tag{7.17}$$

The estimate $\hat{\mathbf{x}}$ may be replaced by the error vector $\hat{\mathbf{x}} - \mathbf{x}$, e.g. using the relationship

$$\begin{bmatrix} \mathbf{x} \\ \mathbf{e} \end{bmatrix} = \begin{bmatrix} \mathbf{I} & \mathbf{0} \\ -\mathbf{I} & \mathbf{I} \end{bmatrix} \begin{bmatrix} \mathbf{x} \\ \hat{\mathbf{x}} \end{bmatrix}$$

so that

$$\begin{bmatrix} \dot{x} \\ \dot{e} \end{bmatrix} = \begin{bmatrix} A+BK & \vdots & BK \\ \cdots & \cdots & \cdots \\ 0 & \vdots & A-TC \end{bmatrix} \begin{bmatrix} x \\ e \end{bmatrix}. \tag{7.18}$$

As before, equation (7.18) shows that the error is decoupled from the state. The eigenvalues of equation (7.18) can be shown to be those of the observer, i.e. of $A-TC$ and of the feedback loop, i.e. of $A+BK$. Thus we get the same performance when using the estimate in the controller as we do when using the measured states directly. Indeed it may be physically more desirable to estimate a state in this way, rather than endeavour to measure it directly. Again performance falls off if the system matrices are not well known for the model.

Although fast observer dynamics and high feedback gain would seem to give us all we ask for, such a combination is sensitive to noise and power limitations and a compromise must be sought on both counts. This leads to the study of optimal observers and controllers under stochastic conditions (Kwakernaak and Sivan, 1972).

7.8 Derivative terms in the control vector

The general linear system equation has been continuously expressed as $\dot{x} = Ax + Bu$ and the relationship between the state space model and system equations in differential equation form has been shown earlier. Having now gained considerable experience in the use of this equation it is worth while to consider the case where the control input itself contains derivative terms, e.g. consider the third order equation

$$\dddot{x} + a_1\ddot{x} + a_2\dot{x} + a_3x = b_3u + b_2\dot{u} + b_1\ddot{u}$$

The general pattern to establish the state equation has been to let $x_1 = x$, $x_2 = \dot{x}_1 = \dot{x}$, $x_3 = \dot{x}_2 = \ddot{x}$ so that $\dot{x}_3 = \dddot{x}$ and we write:

$$\dot{x}_1 = x_2$$
$$\dot{x}_2 = x_3$$
$$\dot{x}_3 = -a_3x_1 - a_2x_2 - a_1x_3 + b_3u + b_2\dot{u} + b_1\ddot{u}.$$

If this form is used the state variables tend to infinity at t equals zero because of the time derivatives of u for non-zero initial conditions and a unique solution may not be reached. A set of state variables must be sought which eliminates the derivative terms on the right-hand side of the state equations, e.g. $\dot{u}$, $\ddot{u}$. This may be possible by inspection in low order cases and alternatives will exist, so that the state vector is not unique, although of course the system behaviour as described by all possible state vectors will be the same. One set of state variables may be found by using the following rule for the choice of state variables (Ogata 1970).

If

$$\overset{n}{x} + a_1 \overset{n-1}{x} + \ldots a_{n-2}\ddot{x} + a_{n-1}\dot{x} + a_n x = b_0 \overset{n}{u} + b_1 \overset{n-1}{u} + \ldots b_{n-1}\dot{u} + b_n u$$

is the differential equation of the system then *a* set of state variables will be

$$x_1 = x - \beta_0 u$$

$$x_2 = \dot{x} - \beta_0 \dot{u} - \beta_1 u = \dot{x}_1 - \beta_1 u$$

$$x_3 = \ddot{x} - \beta_0 \ddot{u} - \beta_1 \dot{u} - \beta_2 u = \dot{x}_2 - \beta_2 u$$

.

.

$$x_n = \overset{n-1}{x} - \beta_0 \overset{n-1}{u} - \beta_1 \overset{n-2}{u} - \ldots - \beta_{n-2}\dot{u} - \beta_{n-1}u = \dot{x}_{n-1} - \beta_{n-1}u$$

where $\beta_0, \beta_1, \ldots, \beta_n$ are defined by

$$\beta_0 = b_0$$

$$\beta_1 = b_1 - a_1\beta_0$$

$$\beta_2 = b_2 - a_1\beta_1 - a_2\beta_0$$

.

.

$$\beta_n = b_n - a_1\beta_{n-1} - a_2\beta_{n-2} + \ldots a_{n-1}\beta_1 - a_n\beta_0.$$

The state equation then has the form

$$\dot{\mathbf{x}} = \mathbf{A}\mathbf{x} + \mathbf{b}u$$

or in full:

$$
\begin{bmatrix} \dot{x}_1 \\ \dot{x}_2 \\ \cdot \\ \cdot \\ \cdot \\ \dot{x}_{n-1} \\ \dot{x}_n \end{bmatrix} = \begin{bmatrix} 0 & 1 & 0 & \ldots & 0 \\ 0 & 0 & 1 & \ldots & 0 \\ \cdot & & & & \\ \cdot & & & & \\ \cdot & & & & \\ 0 & 0 & 0 & & 1 \\ -a_n & -a_{n-1} & & \ldots & -a_1 \end{bmatrix} \begin{bmatrix} x_1 \\ x_2 \\ \cdot \\ \cdot \\ \cdot \\ x_{n-1} \\ x_n \end{bmatrix} - \begin{bmatrix} \beta_1 \\ \beta_2 \\ \cdot \\ \cdot \\ \cdot \\ \beta_{n-1} \\ \beta_n \end{bmatrix} u.
$$

Note the order of coefficients carefully in the basic equation for β_i.

Example Write a suitable state space system equation for the system described by $\ddot{x} + \dot{x} + x = 2\dot{u} + u$.

Direct application of the rules given may be used or inspection alone may yield the desired state variables. The difficulty arises because of the derivative term $2\dot{u}$.

Try $x_1 = x$

and, for example,

$$x_2 = \dot{x} - 2u$$
$$= \dot{x}_1 - 2u$$

i.e.

$$\dot{x}_1 = x_2 + 2u$$

then

$$\dot{x}_2 = \ddot{x} - 2\dot{u}$$
$$= -\dot{x} - x + u$$
$$= -x_2 - u - x_1,$$

using the original equation. Then the equation may be written

$$\begin{bmatrix} \dot{x}_1 \\ \dot{x}_2 \end{bmatrix} = \begin{bmatrix} 0 & 1 \\ -1 & -1 \end{bmatrix} \begin{bmatrix} x_1 \\ x_2 \end{bmatrix} + \begin{bmatrix} 2 \\ -1 \end{bmatrix} u.$$

Alternatively applying the rules above we see that

$$\beta_0 = 0$$
$$\beta_1 = 2$$
$$\beta_2 = -1$$

and a new set of state variables are:

$$x_1 = x$$
$$x_2 = \dot{x}_1 - 2u$$

and so

$$\begin{bmatrix} \dot{x}_1 \\ \dot{x}_2 \end{bmatrix} = \begin{bmatrix} 0 & 1 \\ -1 & -1 \end{bmatrix} \begin{bmatrix} x_1 \\ x_2 \end{bmatrix} + \begin{bmatrix} 2 \\ -1 \end{bmatrix} u.$$

In this case a choice by inspection to remove the derivative term from the control vector is as easy, and probably less accident prone than using the given method. Also here both forms are identical but any selection of state variables does not give a unique set of state variables. For example we could have chosen

$$x_1 = x$$
$$x_2 = \dot{x} + x - 2u$$

i.e.

$$\dot{x}_1 = x_2 - x_1 + 2u$$

so that

$$\dot{x}_2 = \ddot{x} + \dot{x} - 2\dot{u}$$
$$= -x + u, \quad \text{(using the original equation)}$$

and the state equation becomes

$$\begin{bmatrix} \dot{x}_1 \\ \dot{x}_2 \end{bmatrix} = \begin{bmatrix} -1 & 1 \\ -1 & 0 \end{bmatrix} \begin{bmatrix} x_1 \\ x_2 \end{bmatrix} + \begin{bmatrix} 2 \\ 1 \end{bmatrix} u.$$

Naturally, as this is describing the same linear system its eigenvalues will be the same as in our first choice and its behaviour in terms of the original variable x will be identical. In all cases the aim is to remove the derivative term from the right-hand side.

7.9 Further examples

To complete this chapter we consider two physical systems in which we may compare the techniques of state feedback controllers with some of the characteristics introduced in the 'classical' treatment. In the first example of an electrical servomechanism one can show how proportional plus velocity feedback is related to state feedback in a second order system. However, state feedback, like simple proportional feedback, leaves us with the probability of steady state errors. In the second of these examples the use of integral action within a state feedback system is made possible by the definition of an additional state variable whose time derivative is equal to an error signal.

Example A small position control system is illustrated by Fig. 7.11. It is possible to measure both output shaft position and output shaft velocity. Control may be effected by output feedback, or feedback with velocity feedback, in the classical sense. In addition a small analogue 'computer' controller is available which enables the two states, shaft position and shaft velocity, to be combined in a state

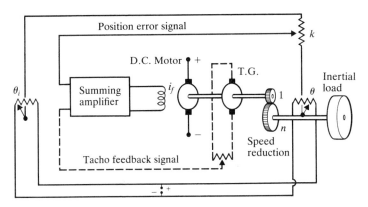

Fig. 7.11 Schematic of position control system

feedback controller. The output from this controller is a combination of weighted state values and forms the input to the D.C. motor via necessary amplification. Input and output shaft positions are measured by potentiometers. We shall see how the dynamics, the speed of settling of the feedback system, may be speeded up using 'single loop' control and state feedback control.

For the 'plant' take the servomotor, gear reduction and load. The amplifier of gain K may be included in the controller. The voltage 'output' v from each potentiometer is $K_1\theta$, where θ is the rotation from a datum and the gain of each potentiometer is assumed equal and constant at K_1. The load is a viscously damped inertia load with damping f and moment of inertia J referred to motor shaft. The inertia and damping of the motor are included within these values and the electrical (nonlinear) damping we shall assume for model simplicity to be small in comparison. If the total amplifier gain is K, and the motor field inductance is low so that there is appreciably no lag between the amplifier output and the field current i_f then

$$i_f = Ke$$

and the motor shaft torque is

$$T = K'i_f = K'Ke$$

where we assume that there is no saturation, the motor armature current is constant and e is the position error voltage. This motor torque acts upon the combined load so that

$$Jn\ddot{\theta} + fn\dot{\theta} = T$$
$$= K'Ke$$

or in transfer function form

$$(Jns^2 + fns)\theta(s) = K'Ke(s)$$

i.e.

$$\frac{\theta(s)}{e(s)} = \frac{K'K}{Jns^2 + fns}$$
$$= \frac{K'K}{s(J/f \cdot s + 1)nf}.$$

The coefficient J/f is the motor time constant, T_m. If amplifier-field current lag is taken into account then there is an additional lag term in $\theta(s)/e(s)$. If with simple feedback the amplifier input e is produced from a simple potentiometer such that

$$e = k(v_i - v)$$
$$= kK_1(\theta_i - \theta)$$

then the closed loop transfer function, on eliminating e, is

$$\frac{\theta(s)}{\theta_i(s)} = \frac{K'KK_1k}{K'KK_1k + fsn(1 + T_ms)}$$

$$= \frac{K'KK_1k/(T_mnf)}{s^2 + s/T_m + K'KK_1k/(T_mfn)}.$$

The block diagram is as in Fig. 7.12(a). With velocity feedback from the tachogenerator on the motorshaft an additional feedback loop is added, the mixer amplifier being used to sum and amplify both feedbacks (Fig. 7.12(b)). The velocity feedback signal is

$$e_v = k_v(n\dot{\theta}) = k'_v K_g(n\dot{\theta})$$

with k_v being combined tacho k'_v and potentiometer gains, K_g, and the overall transfer function is now

$$\frac{\theta(s)}{\theta_i(s)} = \frac{kKK'K_1}{s^2nT_mf + n(f + k_vKK')s + kKK'K_1}$$

$$= \frac{kKK'K_1/(nT_mf)}{s^2 + s(f + k_vKK')/(T_mf) + kKK'K_1/(nT_mf)}.$$

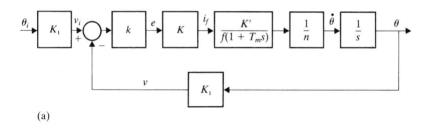

(a)

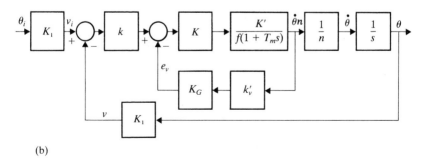

(b)

Fig. 7.12 (a) Direct and (b) velocity plus position feedback

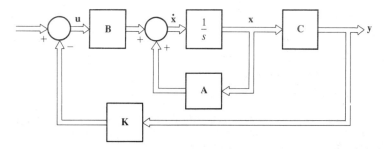

Fig. 7.13   General feedback regulator problem

This system may be described in terms of the state variables, position θ and velocity $\dot{\theta}$. If we measure tnese as indicated in Fig. 7.12 then we may describe the regulator problem by the diagram of Fig. 7.13. The gain K of the amplifier is included in the **B** matrix. The motor-load dynamics are given by the i_f, θ relation,

$$K' i_f = Jn\ddot{\theta} + fn\dot{\theta}.$$

Let the two state variables be deviations from the equilibrium state with $x_1 = \theta$, $x_2 = \dot{\theta}$ so that both position and tacho voltage are zero at **x** = **0**, then $\ddot{\theta} = \dot{x}_2$ and

$$\dot{x}_1 = x_2$$

$$\dot{x}_2 = -\frac{f}{J} x_2 + \frac{K'}{Jn} i_f.$$

Now the current i_f is the amplifier output. If it is determined by state feedback of x_1, x_2

$$i_f = -K[k_1 \quad k_2]\begin{bmatrix} x_1 \\ x_2 \end{bmatrix}$$

since i_f is scalar and **x** is a 2×1 vector. This is a 'dyadic' control form, $\mathbf{k}^T\mathbf{x}$. (Note that k_1 and k_2 effectively replace k, k_v, K_1.)
Thus

$$\dot{x}_2 = -\frac{f}{J} x_2 - \frac{K'K}{Jn}[k_1 \quad k_2]\begin{bmatrix} x_1 \\ x_2 \end{bmatrix}$$

and

$$\begin{bmatrix} \dot{x}_1 \\ \dot{x}_2 \end{bmatrix} = \begin{bmatrix} 0 & 1 \\ 0 & -f/J \end{bmatrix}\begin{bmatrix} x_1 \\ x_2 \end{bmatrix} - \frac{K'K}{Jn}\begin{bmatrix} 0 & 0 \\ k_1 & k_2 \end{bmatrix}\begin{bmatrix} x_1 \\ x_2 \end{bmatrix}$$

$$\begin{bmatrix} 0 & 1 \\ -\dfrac{k_1 K'K}{Jn} & -\dfrac{k_2 K'K}{Jn} - \dfrac{f}{J} \end{bmatrix}\begin{bmatrix} x_1 \\ x_2 \end{bmatrix}$$

$$= \begin{bmatrix} 0 & 1 \\ a_{21} & a_{22} \end{bmatrix}\begin{bmatrix} x_1 \\ x_2 \end{bmatrix}$$

Experimental determination of the system parameters shows that under load, $K_1 = 9$ V rad^{-1}, $k'_v = 0\cdot2$ Vs rad^{-1}, $K = 600$ mA V^{-1}, $T_m = 0\cdot5$ s, $J = 0\cdot01$ Nm, $n = 16$, $K' = 2\cdot7$ Nm mA^{-1}. Then $f = 0\cdot02$ Nm s rad^{-1}.

The three control system equations for the lightly damped high gain system become:

Single feedback:

$$\frac{\theta(s)}{\theta_i(s)} = \frac{9\cdot12 \times 10^4 k}{s^2 + 2s + 9\cdot12 \times 10^4 k}.$$

Position plus velocity feedback:

$$\frac{\theta(s)}{\theta_i(s)} = \frac{9\cdot12 \times 10^4 k}{s^2 + (2 + 32400 K_g)s + 9\cdot12 \times 10^4 k}.$$

State variable feedback:

$$\begin{bmatrix} \dot{\theta} \\ \ddot{\theta} \end{bmatrix} = \begin{bmatrix} \dot{x}_1 \\ \dot{x}_2 \end{bmatrix} = \begin{bmatrix} 0 & 1 \\ -1\cdot012 \times 10^4 k_1 & -1\cdot012 \times 10^4 k_2 - 2 \end{bmatrix} \begin{bmatrix} x_1 \\ x_2 \end{bmatrix}$$

Note that if k'_v and K_1 are retained, $k_1 = K_1 k'_1$, $k_2 = k'_v k'_2$ and with fixed K_1 and k'_v we are determining the gains k'_1 and k'_2. It can be seen that these are equivalent to k and k_v respectively.

Note also that in this example we can compare the control characteristics using the 'classical' approach and the state variable approach because in this particular case our state variable feedback controller is synonymous with the position plus velocity feedback controller, since each relies on measurement of the two variables output position and output speed and in each a weighted combination is used for the control signal. Such a direct equivalence is the exception rather than the rule since in higher order systems where the number of states available for measurement, or estimation, is high the state variable feedback gives potentially more control ability. However, even in such systems the practical use of single, tight, feedback loops working in a single loop independent way is still of prime importance.

Figure 7.14 shows the response of the system with simple feedback to an ideal unit impulse input in θ_i, a sudden movement and return of the input potentiometer. The solution takes the form of equation (3.21). Only the gain k can be varied to improve the response. Figure 7.14 also shows the response to the same input using velocity feedback. Again the solution takes the form of equation (3.21).

Figure 7.15 shows the response of the state variable feedback controller for given specifications of eigenvalues, showing the reduction to rest of the system from a given non-zero state. Equations (3.66) and (3.69) are used for the solution. In this example velocity feedback and state feedback can be made to act in the same way. By 'tradition'

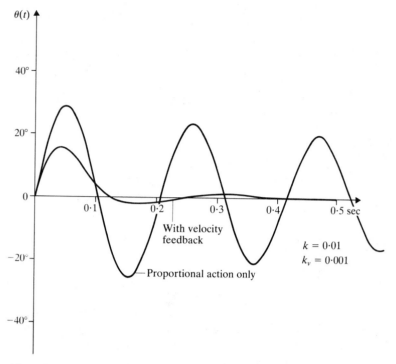

Fig. 7.14 Response using 'classical' position and velocity feedback (impulse response)

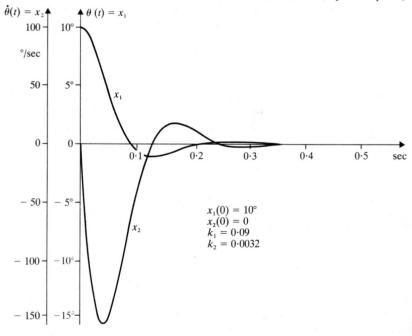

Fig. 7.15 Response expressed in state variables x_1 and x_2 using state feedback, $\mathbf{k}^T\mathbf{x}$, from initial displacement $\mathbf{x}(0)$

in the first case we could specify a required natural frequency and damping and hence evaluate the two gains k and k_v, and in the second case we would again use this specification or just specify the equivalence by asking for specific eigenvalues directly and evaluating the gains k_1 and k_2 in this manner. This is shown again in detail in the following example in which the equivalence of integral control is also specified.

Within practical control problems such as this the effort is very largely devoted to describing the system and the hardware implementation of control decisions once the basic method has been learned. In the absence of any 'optimal' control criterion included in the control law derivation, e.g. Chapter 10, the precise performance for a given set of eigenvalues, gains etc. may be assessed by simulation or experiment.

Example In the example of section 2.4 (Fig. 2.9), an hydraulic pilot valve and actuator driving a damped inertia load with Hook's law restraining force was analysed. Such hydraulic systems are used in aircraft and marine craft as well as in static high power applications. Under certain assumptions the open loop movement y of the load is related to the pilot valve spool movement x by the equation

$$\frac{Y(s)}{X(s)} = \frac{f/M}{s^2 + (f + \lambda K)/(MK) \cdot s + k/M}$$

where f, λ, K, k and M are constant parameters of the system. It has been seen that similar equations arise in electrical and mechanical dynamic systems. Because of the significance of position and speed in many systems in practice in that they govern satisfactory dynamic and steady state performance and control power requirements, the second order system illustrates the major features of state feedback systems. Only by making certain assumptions regarding second order effects, small excursions from a datum etc. is it possible to represent a real system, e.g. aircraft, ship and power components, chemical mixers, natural behaviour, in this way.

In Chapter 4 it was shown that proportional action alone can lead, in the absence of a pure integrator type of element in the plant, to an offset in the steady state response to a demand signal. The introduction of integral action effectively removes this offset. An extension of this concept to a state variable feedback control system is included in this example. A second order system is required to obtain a steady state response without offset to a step demand input at the same time as having closed loop dynamics as specified by eigenvalue selection. It will be seen that in the same way that integral action added to the order of the classical control problem, integral action now leads to the introduction of a further state variable.

Consider therefore a system represented by the general second order transfer function (Fig. 7.16(a)). If we put simple feedback about

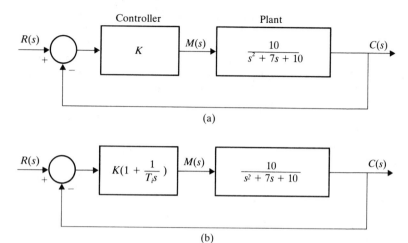

Fig. 7.16 Simple system with feedback and (a) proportional action control, (b) proportional plus integral action control

this with a proportional controller of gain K we obtain the closed loop transfer function which indicates directly a steady state error, i.e. the closed loop gain is not unity, except as $K \to \infty$, but is $K/(K+1)$.

Plant:

$$\frac{C(s)}{M(s)} = \frac{10}{s^2 + 7s + 10}$$

Open loop:

$$\frac{C(s)}{R(s)} = \frac{10K}{s^2 + 7s + 10}$$

Closed loop:

$$\frac{C(s)}{R(s)} = \frac{10K}{s^2 + 7s + 10(K+1)}$$

The 'droop', the reduction from unity gain to $K/(K+1)$, is removed by the introduction of integral action (Fig. 7.16(b)).
Closed loop with integral action:

$$\frac{C(s)}{R(s)} = \frac{10K(1 + T_i s)}{T_i s(s^2 + 7s + 10) + K(T_i s + 1)10}$$

and the steady state gain is $10K/10K$, i.e. unity.

The introduction of derivative action or velocity feedback does nothing to reduce steady state error and similarly state feedback control relying on output and time derivatives or system states will not do so either. Putting this example into state variable form let $x_1 = c$, $x_2 = \dot{x}_1 = \dot{c}$ so that

$$\dot{x}_1 = x_2$$
$$\dot{x}_2 = -10x_1 - 7x_2 + 10m$$

i.e.

$$\begin{bmatrix} \dot{x}_1 \\ \dot{x}_2 \end{bmatrix} = \begin{bmatrix} 0 & 1 \\ -10 & -7 \end{bmatrix}\begin{bmatrix} x_1 \\ x_2 \end{bmatrix} + \begin{bmatrix} 0 \\ 10 \end{bmatrix}m$$

$$y = x_1.$$

The specification to remove offset is that

$$r = y_{\text{steady state}} \quad (= y_{s.s})$$

where r is the reference input and $y_{s.s}$ is the steady state output.
Define an additional variable

$$\dot{x}_3 = r - y$$

so that in the steady state if $\dot{x}_3 = 0$, $r = y_{s.s}$ $(= x_1)$ which is our required condition. Add this to our state vector and we have

$$\begin{bmatrix} \dot{x}_1 \\ \dot{x}_2 \\ \dot{x}_3 \end{bmatrix} = \begin{bmatrix} 0 & 1 & 0 \\ -10 & -7 & 0 \\ -1 & 0 & 0 \end{bmatrix}\begin{bmatrix} x_1 \\ x_2 \\ x_3 \end{bmatrix} + \begin{bmatrix} 0 \\ 10 \\ 0 \end{bmatrix}m + \begin{bmatrix} 0 \\ 0 \\ 1 \end{bmatrix}r.$$

Now we may specify a control law of the form

$$m = [k_1 \quad k_2 \quad k_3][x_1 \quad x_2 \quad x_3]^T$$
$$= k_1x_1 + k_2x_2 + k_3x_3,$$

and proceed in the usual state feedback way. Figure 7.17 illustrates the system representation where we see that we now have the normal state feedback from x_1 and x_2 plus a feedback loop from the output y to produce an error signal, defined as $\dot{x}_3$ to which we apply integral action. Thus although fully representing the system by state feedback it retains the classical use of integral action.

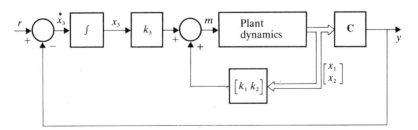

Fig. 7.17 Addition of integral action to state feedback system

Substituting for m in the state equation gives

$$\begin{bmatrix} \dot{x}_1 \\ \dot{x}_2 \\ \dot{x}_3 \end{bmatrix} = \begin{bmatrix} 0 & 1 & 0 \\ -10+10k_1 & -7+10k_2 & 10k_3 \\ -1 & 0 & 0 \end{bmatrix} \begin{bmatrix} x_1 \\ x_2 \\ x_3 \end{bmatrix} + \begin{bmatrix} 0 \\ 0 \\ 1 \end{bmatrix} r$$

$$= \mathbf{A}_{controlled}\mathbf{x} + \mathbf{Dr}.$$

We are now able to write the characteristic equation for $\mathbf{A}_{controlled}$ and evaluate k_1, k_2 and k_3 for our chosen eigenvalues. The characteristic equation is

$$|\lambda\mathbf{I}-\mathbf{A}_c| = 0$$
$$= \lambda^3 + (7-10k_2)\lambda^2 + (10-10k_1)\lambda + 10k_3.$$

If we wish the eigenvalues to be at -6, -5, -5, i.e. a non-oscillatory response, the sought-after polynomial is

$$(\lambda+6)(\lambda+5)(\lambda+5) = 0$$
$$= \lambda^2 + 16\lambda^2 + 85\lambda + 150.$$

Equating coefficients gives the feedback controller gains as

$$k_3 = 15\cdot0$$
$$k_1 = -7\cdot5$$
$$k_2 = -0\cdot9$$

Note that instead of selecting the state variables as c and $\dot{c}$ one could use the procedure

$$\frac{C(s)}{M(s)} = \frac{10}{(s+2)(s+5)}$$

and with the equivalence of Fig. 7.18, we could use

$$\dot{x}_1 = -5x_1 + x_2$$
$$\dot{x}_2 = -2x_2 + 10m.$$

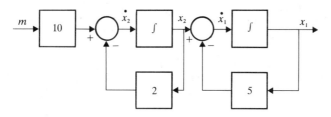

Fig. 7.18 Alternative state variable selection

7.9 Summary

For the state space forms of system equations it is necessary to be able to determine whether or not the system is controllable and observable. If the system is not fully controllable then some of the state variables will vary without us possessing the ability to control them. If some of the state variables cannot be assessed from the output measurements then the full system behaviour cannot be evaluated, although in practice some of the unobservable states may be of low significance.

The problem of stability – or lack of it – extends naturally into the study of multivariable systems and this may be assessed either by the transfer function matrix or by methods such as that of Liapunov and extensions of it, which are applicable in a wider sense as they encompass nonlinear systems also.

Control via feedback leads to the use of state variable feedback and establishment of a control law matrix in the feedback loop. By considering the response of the system as a sum of modes of response in which each mode is associated with only one eigenvalue, modal control is established. The object now is to speed up the slower modes of the open loop response when the loop is closed without affecting the other modes of the system. Although in the ideal case of full measurement this is possible, in practice there may be a lack or inaccuracy of measurements and the amount of ideal measurement and control itself may be excessive in costs and components.

Chapter 8
Nonlinear control systems

8.1 Introduction

Despite the fact that most physical systems have at best some non-linear characteristics it is possible in many situations to treat them as linear, as subject to the principle of superposition and hence we gain the advantage of being able to use a standard linear theory. When the nonlinearity is of an extent when the use of superposition cannot be justified it may be possible still to use the linear theory by restricting considerations to operation about a particular point and linearizing about that point, and this is how in fact most 'linear' systems arise. When system behaviour reaches the point where it is significantly different for different types of input or even different magnitudes of input then alternative methods of analysis must be used. An all embracing theory equivalent in its widespread use to the linear theory is not available but two techniques which are of widespread applicability in the treatment of severe nonlinearities are the phase plane and describing function, (Poincaré 1882; Kochenburger 1950). The use of Liapunov's direct method is also a possible area of benefit but as seen the difficulty arises in finding the Liapunov functions. The study of nonlinear systems is now fairly extensive and the treatment here will be restricted mainly to that of the describing function and phase plane (for single input–single output representation).

Nonlinearity of a system does not necessarily mean that all the system elements or units need be considered as nonlinear. A 'separable' system is one in which the nonlinear part can be described in a chosen manner, say by a describing function, and the remainder may be represented by its usual linear description, e.g. a transfer function. Because of the single nonlinearity though, the nature of the overall system output will be dependent on the magnitude of the signal input. In linear systems we saw that this relationship between input type and magnitude and output was simply that of superposition. The stability of nonlinear systems described by state space notation and with continuous nonlinear characteristics may be investigated by the Liapunov methods of Chapter 7.

8.2 The describing function

In describing function analysis concern is basically with the frequency response of the system. If the input x to an element, whose input–output relationship is given as

$$y = f(x)$$

where $f(x)$ is a nonlinear function of x, is sinusoidal, then the output will be comprised of a component having the same frequency as x plus components of higher harmonics. There may also be a constant component present in the output. The basis of using the describing function comes from the assumption that the higher harmonics are damped out in the remainder of the system and only the dominant fundamental frequency component effectively passes through the system. This filtering out of the high frequency components is consistent with the frequency response analysis of linear systems covered in Chapter 5 where it can be seen that attentuation increases rapidly with frequency increase in systems comprised of lags and higher order denominator-type transfer functions. The higher the system order then the better is this assumption. With this in mind the separable system having both nonlinear and linear elements may be depicted in Fig. 8.1. Since the input is sinusoidal and we consider only the fundamental, the linear transfer function is the frequency function $G(j\omega)$. It will be seen that the use of the describing function enables the extension of the linear theory to handle, with acceptable accuracy, nonlinear components.

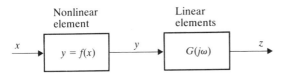

Fig. 8.1 Separable input–output system with nonlinear element

The periodic response y to sinusoidal input signal of angular frequency ω, and amplitude X, i.e.

$$x = X \sin \omega t$$

will be given by the general expression

$$y(t) = A_0 + A_1 \sin \omega t + B_1 \cos \omega t + A_2 \sin 2\omega t + B_2 \cos 2\omega t + \dots$$

$$= A_0 + \sum_{n=1}^{\infty} [A_n \sin n\omega t + B_n \cos n\omega t]$$

which is the Fourier expansion of y. Any periodic function may be expressed as a sum in this way. The major considerations are then how many terms to include and how to evaluate the A_i, B_i in numerical work. The coefficients are

$$A_0 = \text{mean value} = \frac{1}{2\pi} \int_{-\pi}^{\pi} y(\omega t) \, d(\omega t)$$

$$A_n = \frac{1}{\pi} \int_{-\pi}^{\pi} y(\omega t) \sin(n\omega t) \, d(\omega t)$$

$$B_n = \frac{1}{\pi} \int_{-\pi}^{\pi} y(\omega t) \cos(n\omega t) \, d(\omega t).$$

The limits may be taken between zero and 2π instead of $-\pi$ to π. The following alternative expression may be used,

$$y(t) = Y_0 + \sum_{n=1}^{\infty} Y_n \sin(n\omega t + \phi_n) \tag{8.1}$$

where

$$Y_n = \sqrt{(A_n^2 + B_n^2)}$$

$$\phi_n = \tan^{-1}\left(\frac{B_n}{A_n}\right).$$

Before taking these expressions further an examination of the type of nonlinear function $f(x)$ with which we are concerned will be of use. Examples are shown in Fig. 8.2 and these are classed as asymmetric, even symmetric, odd symmetric and multiple (two) value characteristics.

The type of characteristic shown in Fig. 8.2 arise largely in electrical and mechanical systems although some process characteristics may approximate to them. In fact the electrical and mechanical components are a most important part of what are generally considered as process systems, e.g. valves, electric stepper motors and controllers, linkages, amplifiers.

The class of characteristic affects the form of the Fourier series of $y(t)$ and these effects are summarized below:

Asymmetric. The series will contain all sine and cosine terms.

Even symmetric. A_0 will be zero, all A_n will be zero and only even harmonic cosine terms will appear.

Odd symmetric. A_0 will be zero, all B_n will be zero and only odd harmonic sine terms will appear.

Two valued. Both coefficients A_n and B_n now appear and a phase lag is also introduced in the sine and cosine arguments.

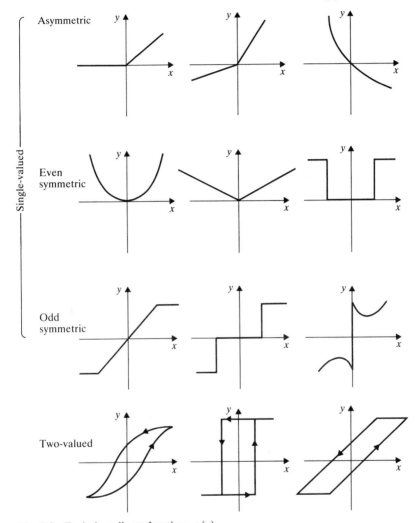

Fig. 8.2 Typical nonlinear functions, y(x)

When the output y from the nonlinear element is passed through the linear element(s) the normal transfer function and frequency response relationships hold but as the higher harmonics are considered as being filtered out the input to the linear components which is used is simply

$$y(t) \simeq Y_0 + Y_1 \sin(\omega t + \phi_1)$$

or in the case of the symmetric nonlinearity which we shall deal with

$$y(t) \simeq Y_1 \sin(\omega t + \phi_1). \tag{8.2}$$

That is, only the basic frequency response of the linear element is required. Remembering that the input to the nonlinear element is $x = X \sin \omega t$, it can be seen that the required part of the output–input relation for this element is completely defined by the magnitude ratio Y_1/X and the phase angle ϕ_1. The Y_1 and ϕ_1 in turn are defined by the Fourier series for y, equation (8.1), so that y/x is given by the describing function

$$\frac{y}{x} = N = \frac{\sqrt{(A_1^2 + B_1^2)}}{X} \; \bigg/ \tan^{-1}\left(\frac{B_1}{A_1}\right). \tag{8.3}$$

That is, both x and y are sinusoidal, with amplitude ratio $\sqrt{(A_1^2 + B_1^2)}/X$ and phase angle $\tan^{-1}(B_1/A_1)$. The output $y(t)$, as we shall see, is dependent on the magnitude of the input so that both Y_1 and ϕ_1 are themselves functions of the amplitude X, i.e. equation (8.3) may be written

$$N(X) = \frac{Y_1(X)}{X} \; \big/ \underline{\phi(X)}$$

$$= \frac{Y_1(X)}{X} \; e^{j\phi_1(X)} \tag{8.4}$$

so that N is a function of the input magnitude X. Note that $N(X)$ is also possibly a function of ω because of the nature of Y_1 and ϕ_1 and that to establish the describing function for a nonlinearity the output of the nonlinearity to the sinusoidal input must be known so that the Fourier series can be found. This is made clearer by the following examples of common nonlinearities.

Describing functions for common nonlinearities

On-off relay

Let $y = K \operatorname{sign} x$. The input, output and block diagram are shown in Fig. 8.3. The output $y(t)$ which is obtained by considering the output y to each specific value of x is odd symmetrical about $t = 0, 2\pi, \ldots$, so that we will have only odd sine terms in the Fourier expansion and all B_n will be zero. Then

$$y \approx Y_1 \sin(\omega t + \phi_1)$$

and

$$\begin{aligned} Y_1 = A_1 &= \frac{1}{\pi} \int_{-\pi}^{\pi} y(\omega t) \sin \omega t \, d(\omega t) \\ &= \frac{2}{\pi} \int_0^{\pi} y(\omega t) \sin \omega t \, d(\omega t) \\ &= \frac{2}{\pi} \int_0^{\pi} K \sin(\omega t) \, d(\omega t) \\ &= \frac{2K}{\pi} \left[-\cos \omega t \right]_0^{\pi} = \frac{4K}{\pi}. \end{aligned}$$

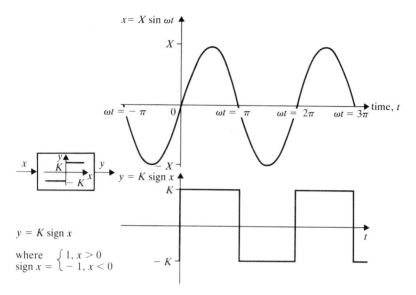

Fig. 8.3 The nonlinear function $y = K$ sign x

The value $y(\omega t) = K$ between ωt zero and $\omega t = \pi$ is taken from our knowledge of the output (Fig. 8.3). The phase angle ϕ_1, $\tan^{-1}(B_1/A_1)$, is zero and this agrees also with our knowledge from Fig. 8.3 also, i.e. y and x are in phase.

Therefore

$$\frac{y}{x} = N(X)$$

$$= \frac{4K}{\pi X}.$$

In this case X appears explicitly in the function N and because of the odd symmetry the phase angle is zero and there is no phase lag. For $x = X \sin \omega t$,

$$y = N(X) \cdot x$$

$$= \frac{4K}{\pi} \cdot \sin \omega t.$$

Dead zone

A dead zone gives a function such that the output remains constant, e.g. zero, while the input varies over a limited range (Fig. 8.4). It is commonly associated with a relay and occurs also in back lash (hysteresis) characteristics. The output for a sinusoidal input is shown in

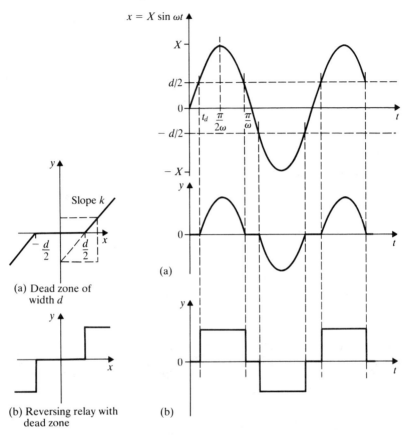

(a)

(b)

(a) Dead zone of
   width d

(b) Reversing relay with
   dead zone

Fig. 8.4   Dead zone characteristic

Fig. 8.4 also and may be plotted by taking specific pairs off the $x-y$ characteristic. Dead zone may be a deliberately imposed nonlinearity, e.g. to reduce the oscillations about a relay contact and the wear of the relay contacts. It will appear as the dead zone, say, in a thermostatically controlled system where a band of acceptable temperature exists, switching only occurring at the limits of the band. The characteristic shown is ideal in that a sudden change in slope is shown, where in practice this transition may be less well defined. If the dead zone is of amplitude $d/2$, giving a dead zone width of d, then for the amplitude X less than $d/2$ there will be no output y. If X is greater than $d/2$ the output will have the form of Fig. 8.4. From the $x-y$ diagram (Fig. 8.4(a)):

$$y = k(x - d/2)$$
$$= k(X \sin \omega t - d/2) \quad \text{for} \quad x > d/2.$$

Referring the function to the time scale it is seen that for positive time and the half cycle from $t=0$ to $\omega t=\pi$, i.e. to $t=\pi/\omega$,

$$y = \begin{cases} 0 & \text{for} \quad 0<t<t_d \\ k(X\sin\omega t - d/2) & t_d<t<\pi/\omega - t_d \\ 0 & \pi/\omega - t_d<t<\pi/\omega. \end{cases}$$

It is again noted that the function is odd symmetric and of zero mean so that only odd harmonic sines appear in its Fourier series and the first term of the series is again

$$y(t) = Y_1 \sin(\omega t + \phi_1).$$

Immediately we see that $\phi_1 = \tan^{-1}(B_1/A_1) = \text{zero}$ so that

$$y(t) = Y_1 \sin(\omega t)$$

and

$$\begin{aligned} Y_1 = A_1 &= \frac{1}{\pi}\int_{-\pi}^{\pi} y(\omega t)\sin\omega t\, d(\omega t) \\ &= \frac{4}{\pi}\int_0^{\pi/2} y(\omega t)\sin\omega t\, d(\omega t) \\ &= \frac{4}{\pi}\int_{\omega t_d}^{\pi/2} k(X\sin\omega t - d/2)\sin\omega t\, d(\omega t), \end{aligned}$$

since for ωt less than ωt_d, $x<d/2$ and $y=0$ and for ωt between ωt_d and $\pi/2$, $x>d/2$ and $y=k(X\sin\omega t - d/2)$. Also

$$d/2 = X\sin\omega t_d$$

i.e.

$$\omega t_d = \sin^{-1}(d/2X).$$

The integral Y_1 is then

$$\begin{aligned} Y_1 &= \frac{4}{\pi}\int_{\omega t_d}^{\pi/2}(kX\sin^2\omega t - kX\sin\omega t_d\sin\omega t)\, d(\omega t) \\ &= \frac{4kX}{\pi}\left[\frac{\pi}{4} - \frac{\omega t_d}{2} - \frac{\sin 2\omega t_d}{2}\right] \\ &= kX\left[1 - \frac{2}{\pi}\sin^{-1}\left(\frac{d}{2X}\right) - \frac{1}{\pi}\sin\left(2\sin^{-1}\left(\frac{d}{2X}\right)\right)\right]. \end{aligned}$$

The describing function

$$\begin{aligned} N &= \frac{Y_1}{X} \\ &= k\left[1 - \frac{2}{\pi}\sin^{-1}\left(\frac{d}{2X}\right) - \frac{1}{\pi}\sin\left(2\sin^{-1}\left(\frac{d}{2X}\right)\right)\right] \end{aligned}$$

for $X>d/2$. For $X<d/2$, $N=0$.

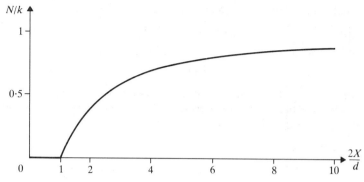

Fig. 8.5   Describing function for ideal dead zone

Again N is an explicit function of X the amplitude of the input. Hence, depending on the known amplitude X of the input x, N will have a value determined by this equation and shown in Fig. 8.5.

Saturation

A further 'piece-wise linear' nonlinearity is the saturation characteristic. This contains similar linear regions to the dead zone and the describing function may be established similarly. The ideal saturation function and the input–output curves are shown in Fig. 8.6. Real saturation will have rounded corners to the $x - y$ curve.

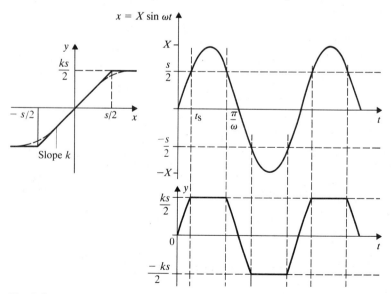

Fig. 8.6   Saturation nonlinearity

Proceeding as before, saturation occurs at time t_s where

$$x = \frac{s}{2} = X \sin \omega t_s$$

i.e.

$$\omega t_s = \sin^{-1}(s/2X)$$

and

$$y = \begin{cases} kx = kX \sin \omega t & 0 < t < t_s \\ kX \sin \omega t_s = ks/2 & t_s < t < \pi/\omega - t_s \\ kX \sin \omega t & \pi/\omega - t_s < t < \pi/\omega. \end{cases}$$

Again this is an odd symmetric function so that

$$Y_1 = A_1 = \frac{1}{\pi} \int_{-\pi}^{\pi} y(\omega t) \sin \omega t \, d(\omega t)$$

$$= \frac{4}{\pi} \int_{0}^{\pi/2} y(\omega t) \sin \omega t \, d(\omega t)$$

$$= \frac{4}{\pi} \int_{0}^{\omega t_s} kX \sin \omega t . \sin \omega t \, d(\omega t) + \frac{4}{\pi} \int_{\omega t_s}^{\pi/2} \frac{ks}{2} \sin \omega t . \, d(\omega t)$$

$$= \frac{4k}{\pi} \left[\frac{X}{2} \sin^{-1}\left(\frac{s}{2X}\right) - \frac{X}{4} \sin\left(2 \sin^{-1}\left(\frac{s}{2X}\right)\right) \right.$$
$$\left. + \frac{s}{2} \cos\left(\sin^{-1}\left(\frac{s}{2X}\right)\right) \right]$$

$$= kX \left[\frac{2}{\pi} \sin^{-1}\left(\frac{s}{2X}\right) + \frac{1}{\pi} \sin\left(2 \sin^{-1}\left(\frac{s}{2X}\right)\right) \right].$$

The describing function

$$N = \frac{Y_1}{X}$$

$$= k \left[\frac{2}{\pi} \sin^{-1}\left(\frac{s}{2X}\right) + \frac{1}{\pi} \sin\left(2 \sin^{-1}\left(\frac{s}{2X}\right)\right) \right] \quad \text{for} \quad X > s/2.$$

This is shown in Fig. 8.7 and note again that $N = N(X)$. The describing function for the full range of X in both Figs. 8.5 and 8.7 is a piece-wise function of (i) the exact relationship over a limited range plus (ii) a fundamental approximation over the remaining range of X. An analysis of the effects of neglecting the higher harmonics is given by D'Azzo and Houpis (1960).

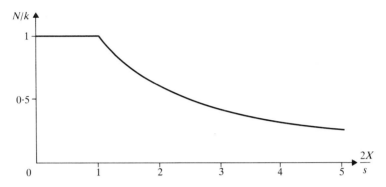

Fig. 8.7 Describing function for ideal saturation

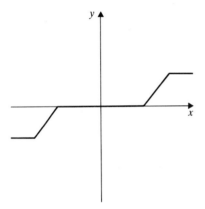

Fig. 8.8 Dead zone plus saturation

For more complicated singular nonlinearities the same method is followed, e.g. a function y may show the characteristics of dead zone and saturation combined, (Fig. 8.8). No additional difficulty arises in this case if the procedure illustrated is followed.

Backlash

A simple multivalued nonlinearity is backlash but even this has a fairly complex describing function. Backlash may be 'friction controlled' so that the driven member (y) remains stationary when not in contact with the input member (x), or 'inertia controlled' when y continues to move at constant speed after contact with the drive input ceases. Between the ideal cases the real physical system within which backlash is present will have both inertia and friction forces. Full details of these cases are available and may be treated individually, but the effect of

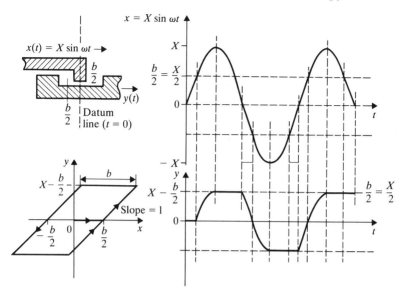

Fig. 8.9 Friction–controlled backlash ($b = X$)

friction controlled backlash is illustrated in (Fig. 8.9). For small movements of the input, $X < b/2$, about the central datum there will be no output movements. For greater values of input amplitude y will move as shown in the $x - y$ curve so one value of x will have two corresponding values of y depending on the direction of movement of x.

Both sine and cosine terms appear in the Fourier series and proceeding as before yields

$$A_1 = X\left[\frac{1}{\pi} - \frac{2}{\pi}\left(1 - \frac{b}{X}\right)\cos \alpha_b + \frac{1}{\pi}\cos 2\alpha_b\right]$$

$$B_1 = X\left[1 - \frac{\alpha}{\pi} + \frac{2}{\pi}\left(1 - \frac{b}{X}\right)\sin \alpha_b - \frac{1}{2\pi}\sin 2\alpha_b\right]$$

where

$$\alpha_b = \cos^{-1}(1 - b/X)$$

and

$$Y_1 = \sqrt{(A_1^2 + B_1^2)}$$
$$\phi_1 = \tan^{-1}(B_1/A_1).$$

Thus as well as the amplitude of the output depending on the input amplitude, there is an amplitude-dependent phase angle also.

Describing function and stability in control systems

Let us return now to the plant of Fig. 8.1 containing both nonlinear and linear elements and add unity feedback (Fig. 8.10). The nonlinear element is represented by its describing function $N(X)$ and the linear element by its frequency transfer function $G(j\omega)$. Because we are assuming any input to be sinusoidal – or nearly so – and the linear elements to have filtered out the harmonics higher than the fundamental, the feedback signal will also be essentially sinusoidal so that our basic requirements are not violated.

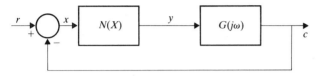

Fig. 8.10   Closed loop nonlinear control system

The input to the nonlinear element is the error signal and the closed loop transfer function is

$$\frac{C(j\omega)}{R(j\omega)} = \frac{N(X)G(j\omega)}{1 + N(X)G(j\omega)}. \tag{8.4}$$

The characteristic equation is

$$1 + N(X)G(j\omega) = 0$$

or equivalently

$$G(j\omega) = -\frac{1}{N(X)}. \tag{8.5}$$

As the linear system characteristic equation gives us the limiting conditions for stability, the $G(j\omega)$ locus passing through the $(-1, j0)$ point, so this equation gives the condition for the nonlinear system output to be a limit cycle, i.e. continuous oscillation, if $G(j\omega)$ intersects the $-1/N(X)$ plot.

The significance of equation (8.5) can be seen using the complex plane in which we plot both the $-1/N(X)$ locus and the $G(j\omega)$ locus. If no phase angle is introduced by $N(X)$ the locus of $-1/N(X)$ will be along the real axis (Fig. 8.11). Otherwise it will form a locus in the imaginary plane displaced away from the axis.

For the saturation characteristic the locus of $-1/N(X)$ runs from -1 at $X=0$ to minus infinity as $N(X)$ tends to zero at large X. The $G(j\omega)$ plot is the normal Nyquist plot for the linear element. If the curves intersect, say at X_i, ω_i on the two loci respectively, then these values satisfy equation (8.5) and correspond to limit cycle behaviour, i.e. continuous sustained oscillations, which may approach a sinusoidal nature, in the absence of an input, i.e. $r=0$. If the loci do not intersect the system is stable (curve B) and oscillations die out.

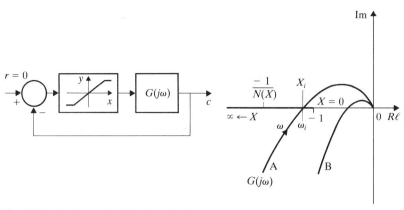

Fig. 8.11 Stability analysis for feedback system with nonlinearity

In general, if any part of the $-1/N(X)$ locus is enclosed on the right-hand side of the linear $G(j\omega)$ plot then the system is potentially 'unstable', A. If it is not enclosed at all then the system is stable, B. Obviously, since $N(X)$ is an approximation these criteria must be applied with some care and if the two curves only just intersect or are nearly tangent to each other stability – or lack of it – is not proved. However, instability might easily arise with slight changes in parameters so a useful indication is still given even for inaccurately represented systems within limits. The higher the order of the linearity and as its low pass characteristics are improved, the better the use that can be trustfully made of the describing function.

In conclusion, note that if two nonlinearities occur they should first be united as one and a combined describing function found. This is in order to maintain the requirements for a near sinusoidal signal entering each element described by a describing function.

Example A nonlinear system may be represented by a linear third order element in series with a dead zone nonlinearity as shown (Fig. 8.12). If the $G(j\omega)$ plot of the linear element cuts the real axis at -4 at a frequency of 6 rad/sec, what is the amplitude of the possible limit cycle (if any)?

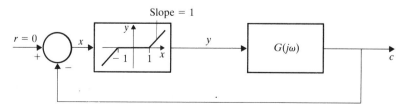

Fig. 8.12 Dead zone nonlinearity with feedback

As the system stands, the describing function of the dead zone, see text, is

$$N = 0, \qquad X < d/2$$

$$N = k\left[1 - \frac{2}{\pi}\sin^{-1}\left(\frac{d}{2X}\right) - \frac{1}{\pi}\sin\left(2\sin^{-1}\left(\frac{d}{2X}\right)\right)\right], \qquad X > d/2$$

where k is the slope of dead zone characteristic outside of the dead zone and d is the overall dead zone width. For the given system

$$N = 0, \qquad X < 1$$

$$N = 1 - \frac{2}{\pi}\sin^{-1}\left(\frac{1}{X}\right) - \frac{1}{\pi}\sin\left(2\sin^{-1}\left(\frac{1}{X}\right)\right), \qquad X > 1.$$

This has all real values and for the limit cycle criterion

$$\frac{1}{N(X)} = -G(j\omega)$$

and we see that if the two curves intersect at $G(j\omega) = -4$ then $N(X) = 0{\cdot}25$. From the plot of $N(X)$ vs X (Fig. 8.15), the corresponding value of X for $N(X) = 0{\cdot}25$ is about $1{\cdot}5$. This is the amplitude of the limit cycles as the value X is greater than unity.

Example Will the addition of an ideal relay with output $\pm K$ in series with the dead zone in the example above increase or decrease the amplitude of the limit cycle, and by how much? With the addition of an ideal relay the two nonlinearities combine to give a relay with dead zone, Fig. 8.13. The new describing function must now be derived.

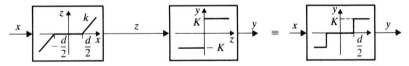

Fig. 8.13 Combination of two nonlinearities

For a sinusoidal input, the output will be as shown in Fig. 8.14. From Fig. 8.14

$$y = 0 \quad \text{for} \quad 0 < x < d/2$$

and on a time scale

$$y = 0, \qquad 0 < t < d/2$$

$$y = K, \qquad t_d < t < \pi/\omega - t_d$$

$$y = 0, \qquad \pi/\omega - t_d < t < \pi/\omega$$

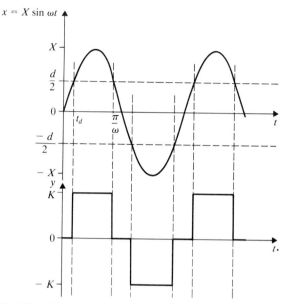

Fig. 8.14 Characteristic output for relay with dead zone

and t_d is given by

$$d/2 = X \sin \omega t_d,$$

i.e.

$$\omega t_d = \sin^{-1}\left(\frac{d/2}{X}\right).$$

The output of the nonlinearity, y, is single-valued and odd symmetric. It will contain only sine terms in its Fourier series and the first term will be

$$y(t) \simeq A_1 \sin \omega t$$

with

$$A_1 = \frac{1}{\pi} \int_{-\pi}^{\pi} y(\omega t) \sin \omega t \, d(\omega t)$$

$$= \frac{4}{\pi} \int_{0}^{\pi/2} y(\omega t) \sin \omega t \, d(\omega t)$$

$$= \frac{4}{\pi} \int_{\omega t_d}^{\pi/2} K \sin \omega t \, d(\omega t)$$

$$= \frac{4}{\pi} K \cos \omega t_d$$

$$= \frac{4}{\pi} K \cos \left(\sin^{-1}\left(\frac{d}{2X}\right)\right).$$

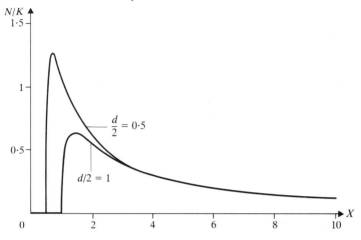

Fig. 8.15 Describing function for relay with dead zone

Then

$$N = \frac{A_1}{X}$$

$$= \frac{4K}{\pi X} \cos\left(\sin^{-1}\left(\frac{d}{2X}\right)\right).$$

Thus the addition of a second nonlinearity does not necessarily add to the complexity of the describing function. Fig. 8.15 shows N for $d/2 = 0{\cdot}5$ and 1. For $K = 1, d/2 = 1$ then for $1/N = 4, X$ now has two values, $X \approx 5$ and $X = 1{\cdot}02$. The $1/N$ locus still runs along the real axis but now from $-1{\cdot}6$ to minus infinity. Of the two values of limit cycle amplitude, $X = 5$, $X = 1{\cdot}02$ which do we require? Refer now to Fig. 8.16. We see that if we take the first value $X = 1{\cdot}02$ and give the amplitude a slight disturbance to increase it, we move to the right along the real axis and the new $-1/N(X)$ is enclosed by the $G(j\omega)$ curve. This corresponds to unstable behaviour, similar to the enclosure of the critical point $(-1, j0)$ for linear systems, and the amplitude will continue to increase until it arrives back at the $G(j\omega)$ plot at an amplitude $X = 5$. A movement now of increasing X takes the point $-1/N(X)$ further to the left, a stable system, and the amplitude thus decays until X falls back to 5. The limit cycle at $X = 1{\cdot}02$ will thus be unstable and physically unobservable and we require the value $X = 5$.

The frequency of the limit cycle is unchanged since the describing function is still real (i.e. has no phase angle) but the amplitude has been considerably, and undesirably, increased by the addition of the relay.

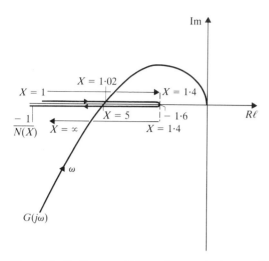

Fig. 8.16   Nonlinear and linear elements in the complex plane

8.3 Phase plane

The phase plane is used for the representation of the dynamics of a second order system. It may be seen as a simple two-dimensional case of the trajectories in n-dimensional state space for the general system dynamics,

$$\dot{\mathbf{x}} = \mathbf{A}\mathbf{x}$$

given the initial state $\mathbf{x}(0)$ at some time. The second order dynamic system may initially be represented by a single equation in which $f(\dot{x}, x)$ is a general function which may be nonlinear,

$$\ddot{x} + f(\dot{x}, x) = 0$$

or by the state variable form

$$\dot{x}_1 + f_1(x_1, x_2) = 0$$
$$\dot{x}_2 + f_2(x_1, x_2) = 0.$$

The functions f, f_1, f_2 may be linear or nonlinear and the phase plane is especially useful for dealing with the sharp nonlinearities which have been discussed above. Its main limitation lies in the low order of system which may be treated in this way, i.e. second order.

The basis of the phase plane is to remove the explicit relationships in time t and to establish an equation containing only $\dot{x}$ and x (or x_1 and x_2).

Starting with the second order equation

$$\ddot{x} + f(\dot{x}, x) = 0$$

let $x_1 = x$, $x_2 = \dot{x}$, then

$$\dot{x}_2 + f(x_2, x_1) = 0$$

or

$$\dot{x}_2 = -f(x_2, x_1).$$

Divide by $\dot{x}_1 (= x_2)$ and this gives

$$\frac{\dot{x}_2}{\dot{x}_1} = \frac{dx_2}{dx_1} \quad \left(= \frac{dx_2/dt}{dx_1/dt} \right)$$

$$= -\frac{f(x_2, x_1)}{x_2}. \tag{8.6}$$

Alternatively, if we start with the two equations where x_2 is now not necessarily $\dot{x}_1$,

$$\dot{x}_1 + f_1(x_1, x_2) = 0$$
$$\dot{x}_2 + f_2(x_1, x_2) = 0$$

then similarly,

$$\frac{\dot{x}_2}{\dot{x}_1} = \frac{dx_2}{dx_1}$$

$$= \frac{f_2(x_1, x_2)}{f_1(x_1, x_2)}. \tag{8.7}$$

Equation (8.6) is thus just a special case of the general form (equation (8.7)).

The system trajectories, the integral relationship between x_2 and x_1, are plotted in the (x_1, x_2) plane from equation (8.7) and for a given set of boundary conditions $x_1(0)$, $x_2(0)$ there is a unique trajectory. Time, although explicit in the solution, can be shown along the trajectories. Although equation (8.7) is the general form the majority of systems will be described by a state variable x and a second derived state variable, e.g. the time derivative of x, $\dot{x}$. If x is a displacement $\dot{x}$ is a velocity, if x is a velocity $\dot{x}$ is an acceleration, if x is a concentration of a component in a reactive mixture, $\dot{x}$ is the rate of generation of that component and so on. Equation (8.6) is then applied. If equations (8.6) and (8.7) can be solved explicitly for x_2 i.e. $x_2 = x_2(x_1)$, then the trajectories may be plotted directly in the phase plane. This is possible for the general linear equation. If the system is nonlinear or piece wise linear then a graphical method may either have to be used or will at least prove to be the easier method. The explicit solution method is self-explanatory and attention will be turned to the graphical construction.

The isocline method

The isocline method is based on establishing lines of constant slope, $dx_2/dx_1 = constant$, λ. The constant is thus the slope of the tangent to the x_1, x_2 curve. Equation (8.7) then gives the equation of the lines of constant slope, or isoclines of the system, since on incorporating equation (8.7),

$$\lambda f_1(x_1, x_2) = f_2(x_1, x_2).$$

For each value of λ an algebraic equation is formed between x_1 and x_2. The isoclines can thus, if required, be plotted in full. To draw the phase plane trajectories short lines are drawn across the isoclines and these are the tangents to the trajectories at that point. The general shape of the trajectories is thus built up by the 'field' of short line tangents. The method is illustrated using a simple function. The dynamics of a second order system are described by the differential equation,

$$\ddot{x} + 2\dot{x} + 1 = 0.$$

Let $x_1 = x$, $x_2 = dx/dt$, then

$$\frac{d\dot{x}}{dt} + 2\dot{x} + 1 = 0$$

and

$$\frac{d\dot{x}}{dx} = \frac{d\dot{x}/dt}{dx/dt}$$

$$= -\frac{2\dot{x}+1}{\dot{x}}$$

or

$$\frac{dx_2}{dx_1} = -\frac{2x_2+1}{x_2}.$$

A relationship of particular use in phase plane construction is

$$\frac{d^2x}{dt^2} = \frac{d}{dt}\left(\frac{dx}{dt}\right)$$

$$= \frac{d}{dx}\left(\frac{dx}{dt}\right)\frac{dx}{dt}$$

i.e.

$$\ddot{x} = \frac{d\dot{x}}{dx} \cdot \dot{x}.$$

The equation for the isoclines is given by constant values of λ i.e. of

$$\frac{d\dot{x}}{dx}, \quad \text{or of} \quad \frac{dx_2}{dx_1}.$$

Thus

$$\lambda = -\frac{2\dot{x}+1}{\dot{x}} = constant$$

i.e.

$$\dot{x} = -\frac{1}{(\lambda+2)}.$$

For all values of λ, the isoclines in this case are lines parallel to the x axis (Figure 8.17). Note the 'nonlinear' spacing of the isoclines as λ varies.

For given initial conditions $x(0)$, $\dot{x}(0)$ it is now possible to sketch in the trajectories. Say the initial conditions correspond to point P, then the dynamics are given by path A from which it is seen that the steady state (large t) is given by $x \to -\infty$, $\dot{x} \to -0 \cdot 5$. Other trajectories are also shown for other initial conditions, each set of initial conditions giving a fresh trajectory.

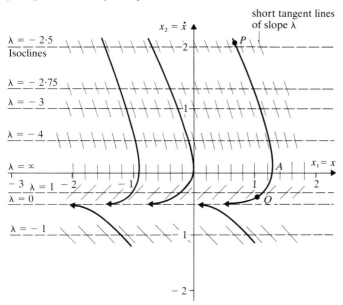

Fig. 8.17 Construction of isoclines and tangents of trajectories

The accuracy of the trajectories drawn in this way depends entirely on the number of the isoclines and of the number and accuracy of the short tangent lines since any errors continue to propagate through the trajectory.

Each set of equations or equation describing the dynamics of a second order system will have its own phase plane portrait of isoclines and trajectories and some examples will be shown. Extensive examples are shown in the literature, e.g. Naslin (1965).

An alternative graphical method is the delta method which depends on building up trajectories in a phase plane with normalized co-ordinates by a series of short arcs.

Although the time variable does not appear directly on the trajectory nor is needed for the plotting of a trajectory, it may be entered on to the trajectory and the time required to go between two specific points, e.g. P and Q on trajectory A may be evaluated. From the relationship $\dot{x} = dx/dt$, the increment

$$\Delta t \simeq \frac{\Delta x}{\dot{x}_{average}}$$

and hence from the trajectory a small interval Δx is selected, the average $\dot{x}$ for this interval is read off and Δt calculated. Proceeding by a series of such steps the time between two points may be estimated.

Alternatively,

$$\Delta t = \int_{t_1}^{t_1 + \Delta t} \frac{dx}{\dot{x}}$$

and the inverse curve $1/\dot{x}$ vs x may be plotted and Δt evaluated by graphical integration. The Δt may now be equal to the full time step P to Q since graphical integration may be over any range. However, if $\dot{x}$ takes on a zero value then this integration is not generally suitable.

Singularities in the phase plane

The principal features of the phase plane trajectories are governed by the coefficients in the system equation(s). These coefficients control also the eigenvalues of the system and hence we should see features representing stability and instability corresponding to negative and positive real parts of the system eigenvalues. For a second order system it is just as easy to work from the characteristic equation to determine system behaviour in the region of the origin. Then we may consider the equations, if not already linear, to be linearized in the vicinity and we can write

$$\ddot{x} + 2\zeta\omega_n\dot{x} + \omega_n^2 x = 0$$

for the unforced system. The eigenvalues are the roots of the equation

$$\lambda^2 + 2\zeta\omega_n\lambda + \omega_n^2 = 0$$

i.e.

$$\lambda_{1,2} = -\zeta\omega_n \pm \omega_n \sqrt{(\zeta^2 - 1)}.$$

For this linear autonomous system the only singular point, i.e. at which $d\dot{x}/dx$ is indeterminate is at the origin. According to the values of ω_n and ζ and hence of λ_1 and λ_2 the following pattern of trajectories occurs at the singular point, Table 8.1.

Thus, as might be expected stable systems have trajectories leading to the origin, unstable systems have trajectories leading away from the origin. Between the two the case of limiting stability, or of undamped behaviour, is shown by the limit cycles, about a centre. Continuous oscillation now occurs. In general if all trajectories converge to the limit cycle it is said to be stable. If the trajectories move away from the limit cycle when the system is disturbed then the limit cycle is unstable, those trajectories within it converge to a stable focus, node or other limit cycle and those outside diverge – possibly to another limit cycle.

Continuous trajectories forming limit cycles are found for linear 'conservative' systems which have no dissipative term, i.e. damping, $\zeta = 0$. The linear equation is then

$$\ddot{x} + \omega_n^2 x = 0.$$

Substituting $\dot{x}$ gives on separating the variables $\dot{x}$ and x

$$\omega_n^2 x \, dx = -\dot{x} \, d\dot{x}$$

or on integrating

$$\frac{\dot{x}^2}{\omega_n^2} + x^2 = constant.$$

The trajectories are thus ellipses. They become circles if plotted in the $\dot{x}/\omega_n$, x plane, the radius depending on the initial conditions (total 'energy' in the dynamic system). In a conservative system decrease in kinetic energy, $\dot{x}$, leads to increase in potential energy, x.

If a constant forcing function is applied to the general linear system so that the equation is

$$\ddot{x} + 2\zeta\omega_n\dot{x} + \omega_n^2 x = k\omega_n^2$$

then a shift of co-ordinates (by the substitution $x' = x - k$) gives

$$\frac{d^2}{dt^2}(x - k) + 2\zeta\omega_n \frac{d}{dt}(x - k) + \omega_n^2(x - k) = 0$$

or

$$\ddot{x}' + 2\zeta\omega_n\dot{x}' + \omega_n^2 x' = 0$$

and shows that all features of the phase plane are shifted and the singular points are moved from the origin to $x = k$, $(x' = 0)$. (For other forcing functions a further redefinition of state variables may be used.)

Table 8.1   Phase plane singularities

Value of ζ and ω	Form of eigenvalue, $\lambda_{1,2}$	Nature of phase plane trajectories in the vicinity of the origin	
$\zeta > 1$ $\omega_n^2 > 0$	Real and in left-hand plane	Stable node	
$0 < \zeta < 1$ $\omega_n^2 > 0$	Complex conjugate pair in left-hand plane	Stable focus	
$\zeta = 0$ $\omega_n^2 > 0$	Conjugate pair on imaginary axis	Centre	
$-1 < \zeta < 0$ $\omega_n^2 > 0$	Complex conjugate pair in right-hand plane	Unstable focus	
$\zeta < -1$ $\omega_n^2 > 0$	Real and in right-hand plane	Unstable node	
Any ζ $\omega^2{}_n < 0$	Real, one in left-hand plane, one in right-hand plane	Saddle point	

Note that the singular points have been described for linear systems. To describe them in these terms for nonlinear systems it must be possible to describe them by normal linearization, the Taylor series, in the vicinity of the singular point. The gross nonlinearities, such as saturation, we shall see present no difficulty if they form a piece-wise linear system and nonlinear systems generally also give rise to phase plane singularities or limit cycles.

Piece-wise linear systems

To establish the pattern of dividing the phase plane into regions for piece-wise linear systems consider a second order linear system in series with a nonlinearity, without feedback. Two straightforward nonlinearities are considered but even the most complex piece wise linear nonlinearity can be treated in a similar way.

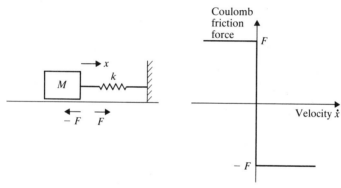

Fig. 8.18 Coulomb friction in a simple system

Coulomb friction

Coulomb (sliding) friction occurs in all physical systems, e.g. valves, motors. This friction force is constant in magnitude but always in opposition to the direction of motion. It is zero in the absence of motion. It can be seen (Figure 8.18) that it has similar characteristics, allowing for the different co-ordinates, to the ideal relay. The dynamics of a system incorporating this nonlinearity may be represented by two equations, one for positive velocities and one for negative velocities. For the simple mechanical system of Fig. 8.18, in which the mass is disturbed from its equilibrium position by a distance x, the equations are

$$M\ddot{x} + kx + F = 0 \quad \dot{x} > 0$$
$$M\ddot{x} + kx - F = 0 \quad \dot{x} < 0$$

i.e.

$$M\ddot{x} + (\text{sign } \dot{x})F + kx = 0$$

in the absence of viscous friction. In addition we see that if the inertia forces do not exceed the friction plus spring force the system comes rapidly to rest. Rewriting the equation as

$$\ddot{x} + (\text{sign } \dot{x})F/M + k/M \cdot x = 0$$

and then eliminating the time variable using $\ddot{x} = d\dot{x}/dt$ and $d\dot{x}/dx = (d\dot{x}/dt)/(dx/dt)$,

$$\frac{d\dot{x}}{dx} = -\frac{(\text{sign } \dot{x})F/M + k/M \cdot x}{\dot{x}}.$$

The phase plane is thus divided by the x axis, there being one equation for the upper half, positive $\dot{x}$ and one for the lower half, negative $\dot{x}$. In addition, if the spring force when the mass is at rest ($\dot{x} = 0$) is less than the friction force F there will be no subsequent movement, i.e. when $|x| < F/k$ and $\dot{x} = 0$. The phase portrait of possible trajectories is shown in Fig. 8.19. The heavy line indicates the region of singular points when $d\dot{x}/dx$ is indeterminate, as distinct from a single singular point. If the system is displaced from its central equilibrium point to $x = A$ and then released from rest it will come to rest at position $-B$ without oscillation. If it were released from C it would travel to $-D$, the equation of motion would change as $\dot{x}$ became zero and it would then continue and come to rest at $-E$. The phase plane curves may be

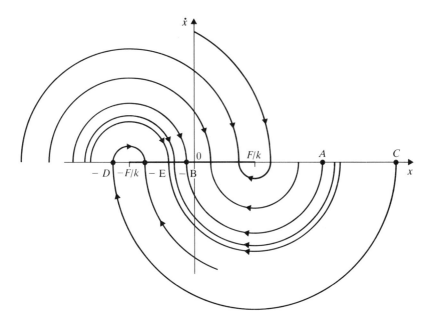

Fig. 8.19 Phase plane and Coulomb friction

drawn in this case either by isoclines for each region or by solution of the equation

$$\ddot{x} \pm F/M + k/M \, . \, x = 0$$

for each region, i.e. for each sign of $\dot{x}$ in turn.

As the system comes to rest within a band $\pm F/k$ of its central position it can be seen that if the desired position after a disturbance is the origin a large static error can occur if there is a nonlinearity having this form of characteristic in a system. The superposition of a high frequency small amplitude 'dither' input, intentionally or accidentally, will enable the system error to be removed, the effect of this being shown in Fig. 8.20.

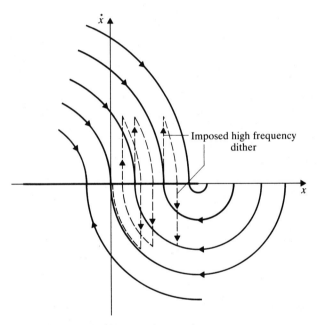

Fig. 8.20   Use of dither to reduce static error

Saturation

Saturation is a nonlinearity already discussed in the section on the describing function. The effect of this on the phase plane representation will be shown and then considered in more detail in feedback systems. If the velocity $\dot{x}$ is given as a function $f(x)$ of the displacement x but is then limited to a maximum magnitude this may be represented by Fig. 8.21. Because of the saturation the velocity is limited within the region $\pm k_s/2$. With the ideal saturation curve the cut off is sharp and trajectories B and C run along the saturation boundary until the point D is reached when they follow the trajectory AD to the origin. AD is

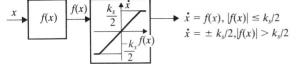

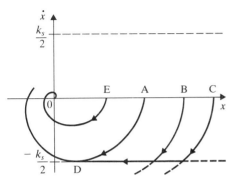

$$\dot{x} = f(x), \; |f(x)| \le k_s/2$$
$$\dot{x} = \pm k_s/2, |f(x)| > k_s/2$$

Fig. 8.21 Effect of saturation on the phase plane

tangent to the saturation boundary. Note that when the magnitude of the velocity $f(x)$ falls below the saturation values those trajectories which have reached the saturation limit do not pick up their original trajectory but follow a common one independent of their initial conditions. Trajectories starting so that saturation is not reached, e.g. at E, remain unique over their entire length. Note also that the saturation reduces overshoot and hence selection of nonlinearity may actually improve system response. However, the velocity is restricted and so we expect a longer settling time which is undesirable.

Nonlinearities in feedback systems using the phase plane

The phase plane may be used to show the behaviour of systems having both nonlinearities and feedback loops. The nonlinearities may occur both in the forward and feedback paths and more than one nonlinearity may be handled at a time. In these systems it is frequently more convenient to base the phase plane on the error and its time derivative rather than on the actual system output. The two are easily related and after all in the regulator systems it is the error which is a true measure of system performance.

The number of combinations of nonlinearities and linear systems is almost limitless and as a result a full description of even a reasonable number of the possibilities becomes copious. Some ideas of the occurrence of nonlinearities in a simple electromechanical servomechanism are shown in Fig. 8.22 and include saturation, hysteresis, backlash, and Coulomb friction. In addition there may be nonlinearities in the form of nonlinear air friction effects, gear distortion and nonlinear electrical

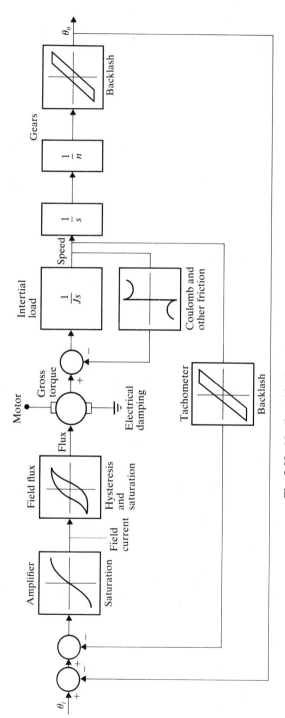

Fig. 8.22 Nonlinearities in a simple servo

damping terms in the motor. The nonlinearities such as backlash and saturation may occur in the physical feedback path also, e.g. in the tachometer. Although friction occurs in the main torque-speed relationship it appears in the block diagram as a feedback path. Two illustrative nonlinearities are saturation and Coulomb friction:

Saturation in a control system

A simple feedback control system with saturation is shown in Fig. 8.23. The output of the saturating element is restricted to lie in the range ±1.

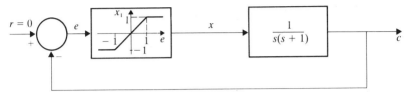

Fig. 8.23   Control system with unit-gain saturating error amplifier

The operation of the system falls into three regions within the phase plane, that before saturation is reached and the system is fully linear, and a region either side when saturation is reached when the error signal reaches or exceeds a value of either +1 or −1.

If the input r is zero, the error signal e is equal to $-c$. For linear operation we would normally use the Laplace notation

$$C(s) = \frac{X(s)}{s(s+1)}$$
$$= -E(s).$$

In phase plane analysis the time derivative is more suitable than the Laplace notation and this equation is written for operation prior to saturation (i.e. when $x = e$)

$$\ddot{e} + \dot{e} + e = 0. \qquad \text{Region I}$$

Inspection of the equation coefficients shows that the $(\dot{e}, e)$ phase plane has a stable focus at the origin (Table 8.11).

If $e > 1$ the equation becomes

$$\ddot{e} + \dot{e} + 1 = 0 \qquad \text{Region II}$$

and for $e < -1$

$$\ddot{e} + \dot{e} - 1 = 0 \qquad \text{Region III}$$

The phase plane and four trajectories for five initial conditions are shown in Fig. 8.24.

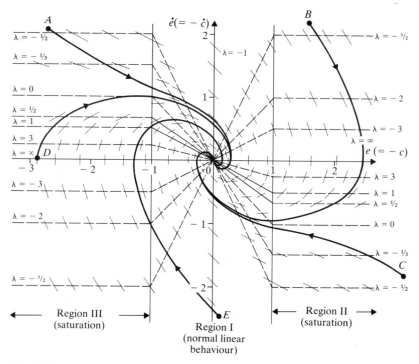

Fig. 8.24   Phase plane for system of Fig. 8.23

The $\dot{e}, e$ relationship is rewritten using the relationship

$$\frac{d^2 e}{dt^2} = \frac{d}{dt}\left(\frac{de}{dt}\right)$$

$$= \frac{d}{de}\left(\frac{de}{dt}\right)\frac{de}{dt}$$

i.e.

$$\ddot{e} = \frac{d\dot{e}}{de} \cdot \dot{e}.$$

Thus

$$\frac{d\dot{e}}{de} = -\frac{(\dot{e}+1)}{\dot{e}} \qquad \text{in region II}$$

and

$$\frac{d\dot{e}}{de} = -\frac{(\dot{e}-1)}{\dot{e}} \qquad \text{in region III.}$$

At $\dot{e} = -1$, $d\dot{e}/de$ is zero in region II, and for $\dot{e} = +1$, $d\dot{e}/de$ is zero in region III. Therefore any trajectory starting within these velocity limits will not go outside them but possibly approach them tangentially before entering region I and moving in to the origin, e.g. trajectory D.

Ideal relay and Coulomb friction

The system of Fig. 8.25 containing Coulomb friction and a relay in a mechanical system is a simplified example including feedback and forward path nonlinearities. The combination of nonlinearities leads to a greater division of the phase plane into regions. Again it is convenient to work either in the $\dot{e}$, e or $\dot{c}$, c plane, as with $r = 0$, $c = -e$.

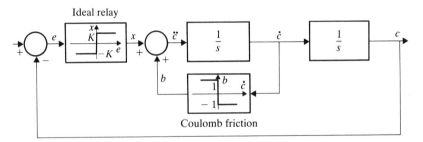

Fig. 8.25 System with relay and Coulomb friction

The system equations are

$$x = K \, \text{sign} \, e$$

for the relay and for the Coulomb friction feedback effect

$$\ddot{c} = x - \text{sign} \, \dot{c}$$

i.e.

$$\ddot{e} = -K \, \text{sign} \, e - \text{sign} \, \dot{e}$$

and

$$\frac{d\dot{e}}{de} = -\frac{(K \, \text{sign} \, e + \text{sign} \, \dot{e})}{\dot{e}}.$$

Solving this equation for $\dot{e}$ as a function of e

$$\dot{e}^2 = -2e(K \, \text{sign} \, e + \text{sign} \, \dot{e}) + constant.$$

This is the equation of the family of trajectories in the phase plane and is very dependent on the relative signs of e and $\dot{e}$ and on the value of

K, the relative 'gain' of the two nonlinearities:

Error and velocity ranges	$K = 1$	$K = 2$
$e > 0,\ \dot{e} > 0$	$\dot{e}^2 = -4e + const.$	$\dot{e}^2 = -6e + const.$
$e > 0,\ \dot{e} < 0$	$\dot{e}^2 = 0 + const.$	$\dot{e}^2 = -2e + const.$
$e < 0,\ \dot{e} > 0$	$\dot{e}^2 = 0 + const.$	$\dot{e}^2 = 2e + const.$
$e < 0,\ \dot{e} < 0$	$\dot{e}^2 = 4e + const.$	$\dot{e}^2 = 6e + const.$

The phase plane trajectories for $K = 1$, $K = 2$ are shown in Fig. 8.26(a) and (b) respectively.

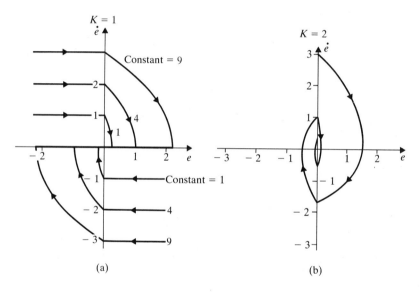

(a)   (b)

Fig. 8.26   Phase planes for system of Fig. 8.25

For $K = 2$, the system follows an oscillatory path with increasing time, finishing at the origin. With $K = 1$ the pattern is less well defined. A slight excursion into the second clockwise quadrant would indicate subsequent movement towards the origin, but at infinitely slow velocity, i.e. the system comes to rest with a finite error. The whole e axis is in fact a region of singularities and the system comes to rest. Referring to Fig. 8.25 we see that at the second summation point $x = -1$ as $\dot{c}$ 'goes positive', the summation is zero and $\ddot{c}$ is zero, and the output c remains unchanging.

For the more realistic case where $K \neq 1$ there are four different equations controlling the system behaviour, one in each quadrant. As the error changes sign each time so the relay switches. This is a typical feature of relay phase plane diagrams, the axis being a switching line. As the derivative of error, i.e. system velocity, changes sign the friction force switches in sign also.

If a trajectory on passing into one region of the phase plane is directed back to the region from which it has come then sliding action results if the switching line is not on the zero velocity axis. The trajectory moves along the switching line to the origin (Fig. 8.27). The actual passage into each region occurs in practice because of the real time switching time which is required and the 'chatter' hence occurs in real systems. In the phase plane it may be shown by the trajectory passing straight along the switching line or by the zig-zag shown in Fig. 8.27.

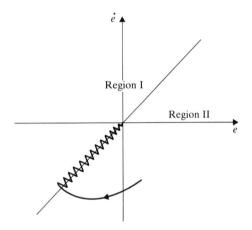

Fig. 8.27 Sliding action along a switching line between regions

For numerous other examples the reader is referred to the literature. We shall return to the phase plane in our consideration of optimal control.

Example For the system with velocity feedback shown in Fig. 8.28 construct the isoclines and show typical trajectories in the absence of an input r but with different initial states, i.e. values of the

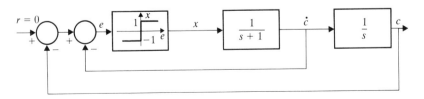

Fig. 8.28 Velocity feedback about an ideal relay

displacement c and velocity $\dot{c}$. For the nonlinear element of the system

$$x = -\text{sign}\,(\dot{c} + c)$$

and the first linear stage of the model gives

$$\ddot{c} + \dot{c} = x$$
$$= -\text{sign}\,(\dot{c} + c).$$

Thus

$$\frac{d\dot{c}}{dc} = \frac{d\dot{c}/dt}{dc/dt}$$

$$= \frac{\ddot{c}}{\dot{c}}$$

$$= -\frac{\dot{c} + \text{sign}\,(\dot{c} + c)}{\dot{c}}.$$

The switching line is given by the condition that $(\dot{c} + c)$ changes sign, i.e. by the line $\dot{c} = -c$ and this may be entered on the phase plane (Fig. 8.29).

The equation of the *isoclines* is

$$\lambda = -\frac{\dot{c} + \text{sign}\,(\dot{c} + c)}{\dot{c}}$$

$$= -1 - \frac{\text{sign}\,(\dot{c} + c)}{\dot{c}}.$$

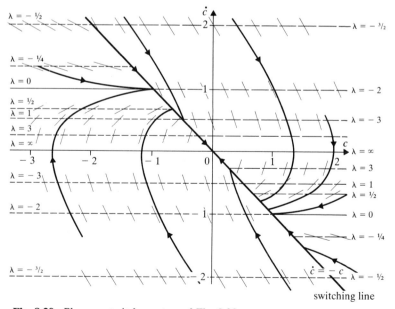

Fig. 8.29 Phase portrait for system of Fig. 8.28

For $\dot{c} = -c$, λ is indeterminate, for $\lambda = -1$, $\dot{c} = \pm\infty$. For $\dot{c} + c > 0$, above switching line, $\dot{c}\lambda = -\dot{c} - 1$, i.e. $\dot{c} = -1/(1+\lambda)$. For $\dot{c} + c < 0$, below switching line, $\dot{c}\lambda = -\dot{c} + 1$, i.e. $\dot{c} = 1/(1+\lambda)$ and the isoclines may be drawn in. The phase plane is complex but shows features of a stable system, and has 'sliding action' along the $\dot{c} + c = 0$ line, the switching line. Notice that in the absence of the relay, the second order feedback system is naturally stable also. The direction of the trajectories (or elemental section) with increasing time may be deduced at any point (or region) $\dot{c}$, c since we know that $\Delta c \simeq \dot{c}\,\Delta t$.

Example Draw the phase plane portrait for the system with hysteresis as shown in Fig. 8.30. The phase plane trajectories are governed not only by the magnitude of the error e (or output c) but also by the sign of $\dot{e}$ within the central band $-1 < e < 1$. For

$e < -1$, $x = -1$ and $\ddot{c} + \dot{c} = -1$

$\left.\begin{array}{l} -1 < e < 1 \\ \dot{e} < 0 \end{array}\right\}$, $x = -1$ and $C + \dot{c} = -1$

$\left.\begin{array}{l} -1 < e < 1 \\ \dot{e} < 0 \end{array}\right\}$, $x = 1$ and $\ddot{c} + \dot{c} = 1$

$e > 1$, $x = 1$ and $\ddot{c} + \dot{c} = 1$.

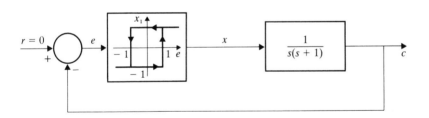

Fig. 8.30 Idealized hysteresis in a system (e.g. relay with backlash)

Noting that if $r = 0$, $e = -c$ and $\dot{e} = -\dot{c}$ let us use the $\dot{e}$, e phase plane. The phase plane is divided into two composite regions, shown by the heavy lines (Fig. 8.31). In region I

$\ddot{e} + \dot{e} = 1$

In region II

$\ddot{e} + \dot{e} = -1$

i.e.

$$\frac{d\dot{e}}{de} = \frac{-\dot{e} \pm 1}{\dot{e}}$$

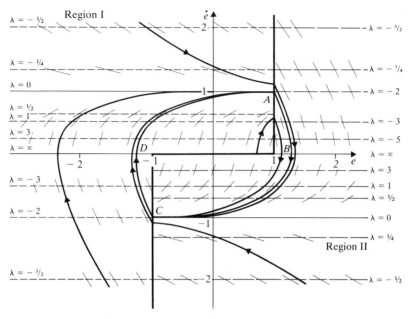

Fig. 8.31   Phase plane for system of Fig. 8.30

and for regions I and II the isocline equations are

$$\lambda = \frac{-\dot{e} \pm 1}{\dot{e}}$$

respectively.

These are now drawn in and show that whatever the starting point for the trajectories a limit cycle is eventually established *ABCDA* with the extreme velocities ±1. Any initial condition with a velocity outside the value ±1 comes within the area of the phase plane sketched within one oscillation of the system. The limit cycle is stable, and is approached either from outside or inside by the trajectories.

8.4 Introduction to combinational and sequential control systems

A brief introduction is given here to basic logic and combinational control and the extension of this to sequential control. These topics are included in this chapter as they depend on elements within the control system having only two states, on and off or zero and unity. The basic concepts are used to ensure that all conditions in a plant are correct before certain actions can be taken, and for enabling a set sequence of processes to be passed through e.g. in a batch reactor. These operations may be carried out by electromagnetic, electronic, fluidic, or hydraulic constructions.

Basic definitions and algebra

Logic algebra, or Boolean algebra after the name of its developer, G. Boole, is concerned with elements which take on only two alternative values, normally designated 0 and 1. An element which may be seen as fitting into this category is the ideal relay. If a variable may be either 0 or 1 we can write

$$x = 0 \quad \text{if} \quad x \neq 1$$
$$x = 1 \quad \text{if} \quad x \neq 0$$

and the complement of the variable, e.g. 0 is the complement of 1, is indicated by x' so that for

$$x = 0, \; x' = 1$$

and in some cases $(x')' = x$.

The basic operations of logic, switching, algebra are the complement NOT, the sum OR and the product AND. These operations are shown by truth tables or Venn diagrams and a series of such operations is shown by a signal chart or network.

For two variables x_1 and x_2 the OR and AND operations are shown in Fig. 8.32. The Boolean algebra equations are:

AND

$$f = x_1 . x_2$$

OR

$$f = x_1 + x_2 \hspace{3cm} (8.8)$$

NOR

$$f = x'$$

For example for $x_1 = 1$, $x_2 = 0$,

AND

$$f = 1 . 0 = 0$$

OR

$$f = 1 + 0 = 1$$

NOR

$$f_{x_1} = x_1' = 0.$$

AND	OR

Network symbols

x_1 ———
x_2 ——— f

x_1 ———
x_2 ———

Truth tables

x_1	x_2	f
0	0	0
1	0	0
0	1	0
1	1	1

x_1	x_2	f
0	0	0
1	0	1
0	1	1
1	1	1

Venn diagrams

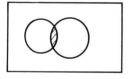

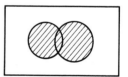

Fig. 8.32 The AND and OR logic operations

If $x_1 = 1$, $x_2 = 1$

AND

$$f = 1 \cdot 1 = 1$$

OR

$$f = 1 + 1 = 1.$$

The unity symbol may represent the presence of a signal and a zero the absence of any signal. Thus the AND element, or 'gate', requires both inputs x_1 and x_2 to give an output but there will be an output from an OR gate if either or both inputs are present. The complement NOT of the AND operation is written NAND and the complement of the OR operation is NOR.

NAND

$(x_1 . x_2)'$

NOR

$(x_1 + x_2)'$

The De Morgan relationships relating the sum and product can be seen from the corresponding Venn diagram and network, and use of the NAND and NOR concept so that (Fig. 8.33),

$$\left.\begin{array}{ll} (x_1 . x_2)' = x_1' + x_2' & \text{(NAND)} \\ (x_1 + x_2)' = x_1' . x_2' & \text{(NOR)} \end{array}\right\} \qquad (8.9)$$

Because of these last two relationships, equation (8.9), a system need only include NOT and OR or NOT and AND elements.

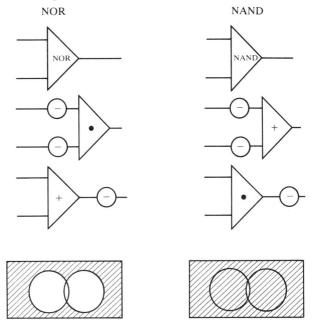

Fig. 8.33 NOR and NAND on the Venn diagram. (The symbol ⊖ represents the complement)

When more than two variables are involved then the commutative, associative and distributive laws may be used:

Commutative law.

$x_1 . x_2 . x_3 = x_3 . x_2 . x_1$

Associative law.

$$x_1 + x_2 + x_3 = x_1 + (x_2 + x_3) = (x_1 + x_2) + x_3$$
$$x_1 . x_2 . x_3 = x_1 . (x_2 . x_3) = (x_1 . x_2) . x_3$$

Distributive law.

$$x_1 . x_2 + x_1 . x_3 = x_1 . (x_2 + x_3)$$
$$(x_1 + x_2) . (x_1 + x_3) = x_1 + x_2 . x_3.$$

These are also illustrated by the Venn diagrams (Fig. 8.34).

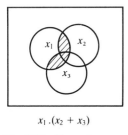   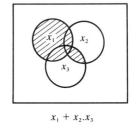

$x_1 . (x_2 + x_3)$ $x_1 + x_2 . x_3$

Fig. 8.34   Venn diagrams for distributive law

Further extensions to the more general case of n variables are possible and with more than two or three variables the Karnaugh map replaces the simple truth tables and Venn diagrams which are otherwise satisfactory (Karnaugh 1953). Even then these Karnaugh maps are really only suitable for up to about five variables. They are used for reducing or simplifying the logic relationships between the greater number of variables. (Quine 1959; McCluskey 1956).

Combinational systems

For a combinational system the output at some instance of time is determined entirely by the inputs at that same instance of time. There is no memory ability within the system. Some of these inputs may be signals indicating that interlocks are in place, levels are at a particular value and so on. It may, for example, be essential that at the time a positive displacement pump is started a particular valve is open. Inspection of equipment may not be possible until a motor is switched off and access is prevented by using a special interlock key which is required for both access and for the motor power so that both are not available simultaneously. The possible examples are thus widely spread. For simple systems logical analysis may be carried out intuitively or for more complex systems the Karnaugh maps mentioned above may be used. However, even with complex systems involving many components in the network it is possible to check a proposed network of OR and AND gates by a systematic following of the paths in a step by step manner.

Sequential systems

Within a sequential system the output depends not only on current input but also on the past history of the system and past inputs. Time, which is not a factor of Boolean algebra, becomes important and switching may be activated by an external clock, synchronous switching, or by the system itself, without signal pulses from a clock, in asynchronous systems.

To retain the memory of past inputs an additional element is required which, once activated by an input to give an output retains that output even when the input has ceased. The output is then cancelled, or transferred to another outlet of the element, only by a second input signal, which may be a pulse from a clock. By designing a sequential switching system a process may be taken through a number of logical steps. These steps may be timed from without or by the state of the system and they may include safeguards of a purely combinational nature also. An example of such a requirement is the control system for a pulp baling press. The control must allow for the filling and emptying of a hopper, movement of a rotating press table, the action of possibly two compression rams, the emptying of the baled pulp etc. These actions must be allied to interlocks assuring the correct location of each piece of the machinery and the successful completion of the previous operation in the sequence. Such systems involve tens or hundreds of logic elements and although flow charts, tables, Karnaugh maps etc. aid in checking the control system much still depends on the judgement and careful piece-wise analysis of the system.

Sequential control may also take a fairly simple form, e.g. in the operation of batch processes where process conditions form the input signals. The switching is then based on a time elapse or the reaching of a particular process variable value or a combination of these. Implementation may be achieved by quite simple mechanical and electromechanical units in many cases.

Example Check the safety of the press controlled in part by the network shown (Fig. 8.35). The signal r controls a main ram and requires that both the interlock on the table is in place t, that the hopper is full, h, and that the hopper chute is clear, c. These last two positions may be overridden for manual testing but the table must always be correctly aligned. Automatic or manual operation is controlled, under safeguards, from another network. By inspection alone we see that in the manual mode the override inputs both h' and c'. This is sufficient to give an output from both OR gates and hence the upper AND gate. However the ram will not be activated (r) unless the second AND gate receives both the table position signal t and the output from the first AND gate. In automatic mode correct chute position c and completion of hopper sequence h is required to trigger the OR gates and the upper AND gates. The OR gates are only required to enable

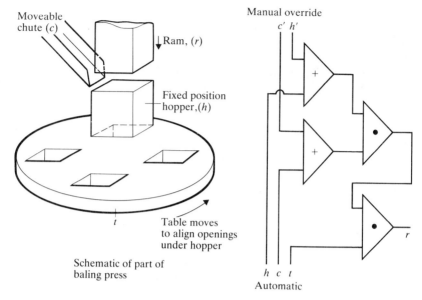

Fig. 8.35 Simplified section of press operation

interchange between manual mode and automatic mode. The table position must still be correct before ram operation can proceed.

8.5 Summary

A nonlinear system is one which does not satisfy the superposition criterion for a linear system. The nonlinearity may be such that by treating perturbations about a fixed point the system equations can be linearized and the performance assessed by the theory of linear systems for small perturbations. Because of the many forms that nonlinearities take a unified theory covering all possibilities and applications is not available but for acute nonlinearities, such as are produced by ideal relays and on-off mechanisms, two widely applicable methods are the describing function and phase (-state) plane methods. In the describing function method we rely on separating the system into nonlinear and linear elements and determining by a truncated Fourier series an equivalent amplitude-dependent gain and phase angle for the non-linearity when the system input is a sine wave or approximately so. When use is made of the phase plane the equation of the system dynamics has a form which depends on that part of the nonlinear characteristic being used. With such a piece-wise linear nonlinearity the phase plane, the plot of the time derivative (or some other derivative possibly) of a variable against the variable – normally the output or error – may be divided into regions. Passage from one region to another is across a switching line and the system equation then changes. Both describing function and phase plane methods can be used in feedback systems.

An entirely different form of control is one which relies on logical switching to take a process through its operation and to monitor and ensure safe system behaviour. The analysis of such systems is by Boolean algebra which may be used, by careful division of a large system network, for systems with extensive sequential and combinational operations.

Chapter 9
Sampled-data systems

9.1 Continuous and sampled systems

So far system representation has been by means of continuous (smooth) functions of time. This seems to be a reasonable approach in that the system variables change continuously with time and differential equations arise naturally from the model formation. Measurement feedback and controller action have also been seen as continuous signals.

In contrast to the continuous system the need arises to look at discrete time and sampled data systems. These arise where information about a system may only be obtainable by sampling, e.g. gas chromotography, where information is delivered in a discrete form, or where an information channel is shared to transmit signals from more than one source, e.g. if a digital computer which relies on taking signals at discrete time intervals forms part of the system. If the original signal is continuous then the sampling of the signal at discrete times is a form of signal modulation. Although sampling may be at varying rates or random we shall consider it to be at a fixed rate so that in a given interval of time a whole known number of measurements, or signals generally, will be conveyed.

As when the continuous system dynamic equations were developed in Chapter 3, single input–single output systems will be dealt with initially followed by the multivariable case. Criteria for sampling intervals and for the stability of sampled data systems will be covered and it will be seen how similarities and equivalents to continuous systems arise and are used. First, let us look at the relation between the continuous and sampled signals.

With a uniform sampling period the signal $u(t)$ of Fig. 9.1 will be sampled at the instances of time, 0, T, $2T$, $3T$ etc. and the sampled values constitute the basis of the system analysis and are the sampled-data or discrete time function $u(nT)$. In addition to using a discrete time interval it may be necessary to approximate the fully continuous

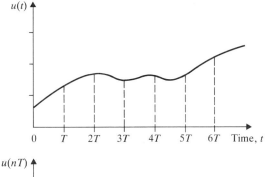

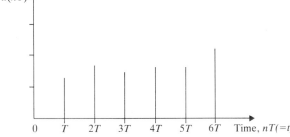

Fig. 9.1 Continuous and sampled-data functions

(analogue) value of $u(t)$ by quantized values which can be expressed by a limited set of digits, e.g. a value 6·54 may have to be read as 6·5 only. In digital systems both will be necessary but concentration is focussed here on the discretization of the time base alone.

9.2 Single input–single output systems

The z-transform

The z-transform plays a similar role in discrete time systems to that occupied by the Laplace transform in continuous system analysis.

Referring to Fig. 9.1 one might ask what happens in the system between the sampling intervals? The ideal sampler may be regarded as a switch which closes for an infinitely short time at which time u is measured. In practice, this assumption is justified when the duration of sampling is a short time compared with the time constant (dynamics) of the rest of the system. The sampled signal w, a set of spikes, is transformed to a continuous signal x by passing it through a hold device. If the signal is held constant this is a zero hold device. This sampler and hold combination as part of the overall process or system thus has a continuous input and a continuous output, with the hold itself having a constant output over a sampling period of time T duration, after a pulse input from the sample. For a single signal the input–output relationship of the zero-hold is as shown in Fig. 9.2, the

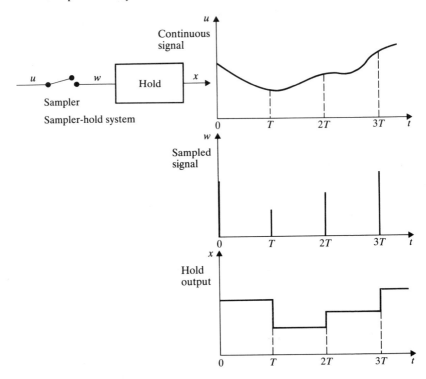

Fig. 9.2 Input–output of sampler and hold

output x holding a constant value for time duration T. The hold is a linear system with input w and output x, the Laplace transforms of the input and output being, for a unit input pulse,

$$W(s) = 1$$

and

$$X(s) = \frac{1}{s} - \frac{1}{s} \cdot e^{-sT}$$

$$= \frac{1}{s}(1 - e^{-sT})$$

as $X(s)$ has the form of a unit step followed by a reverse unit step delayed by a time T (Fig. 9.2). The transfer function of the hold is thus

$$G_h(s) = \frac{X(s)}{W(s)}$$

$$= \frac{1 - e^{-sT}}{s}. \tag{9.1}$$

If the sampler produces a pulse w at intervals of time T then the hold output x will be a function made up of constant value segments as shown in Fig. 9.2. During the interval of time from kT to $(k+1)T$ the output at time $(kT+t')$, $0 \leq t' < T$ will be

$$x(kT+t') = x(kT).$$

The function $G_h(s)$, equation (9.1), was obtained for a unit impulse input to the hold. This actual hold input will vary in magnitude according to the magnitude of the continuous function being sampled so that each impulse will be a unit impulse $\delta(t)$ times the magnitude of the variable $u(t)$ at the instant of sampling. The hold input, a function of time, is made up of the series of values

$$w(t) = \delta_s(t)u(t)$$

where $\delta_s(t)$ is a series of the unit impulse at the particular times t. If $\delta_s(t)$ is a series of unit impulses at regular intervals of time, 0, T, $2T$, ... then $w(t)$ will be a series of impulses at the same time intervals, but weighted by $u(t)$. The unit impulse at time kT may be expressed by the expression incorporating a pure delay, i.e.

$$\delta(t = kT) = \delta(t - kT),$$

Fig. 9.3. Since this has a finite value (i.e. other than zero) of unity only at the times kT where $k = 0, 1, \ldots$ the output $w(t)$ may be expressed

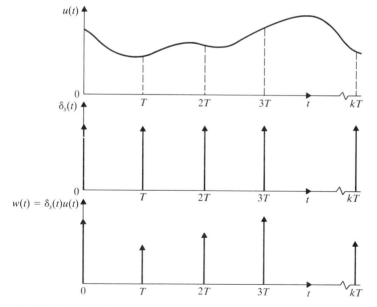

Fig. 9.3 Relationship between the continuous signal $u(t)$, series of unit delta functions $\delta_s(t)$, and the sampled signal $w(t)$

as the summation,

$$w(t) = \delta(t)u(t=0) + \delta(t-T)u(T) + \delta(t-2T)u(2T) + \ldots$$

or

$$w(t) = \sum_{k=0}^{k=\infty} \delta(t-kT)u(kT) \tag{9.2}$$

for a function $u(t)$ for $t \geq 0$.

Although the summation is taken over all kT and $w(t)$ is a continuous function of time at any time, which is a multiple of T, the value of $w(t)$ is given only by the single term

$$w(kT) = \delta(t-kT)u(kT) \tag{9.3}$$

and at times other than the exact times kT, $w(t)$ is zero. The term $\delta(t-jT)$ is zero for all values of k not equal to j. Thus equation (9.2) reduces to equation (9.3). Note that as with the continuous systems we are restricting the consideration to variables for which $u(t)$ is zero for $t < 0$. (The variable $w(t)$ is frequently given the notation $u^*(t)$ to denote its relationship to $u(t)$.)

With this introduction to the structural representation of a sampled signal we can define the z-transformation which deals with the relationship between such discrete signals.

Define first the complex variable z,

$$z = e^{Ts}$$

where s is the Laplace operator, then the z-transform of a variable is defined by

$$W(z) = \mathscr{Z}[w(t)]$$

$$= \sum_{k=0}^{\infty} w(kT)z^{-k}. \tag{9.4}$$

Using the definition of z explicitly in equation (9.4) gives

$$W(z) = \sum_{k=0}^{\infty} w(kT)e^{-kTs}. \tag{9.5}$$

Compare this with the Laplace transform for a variable $w(t)$

$$W(s) = \mathscr{L}[w(t)]$$

$$= \int_0^\infty w(t)e^{-ts}\,dt \tag{1.1}$$

and observe the similarity of form. Returning to equation (9.2) and taking Laplace transforms term by term of both sides of this gives

$$W(s) = \sum_{k=0}^{\infty} u(kT)e^{-kTs}$$

since $u(kT)$ is solely a weighting on each term of the impulse series. But by the definition of equation (9.5)

$$\sum_{k=0}^{\infty} u(kT)e^{-kTs} = U(z)$$

Thus

$$W(s) = U(z). \tag{9.6}$$

Thus we see that the z-transform of the continuous signal $u(t)$ is the Laplace transform of the sampled signal $w(kT)$. Although we take the z-transform of a continuous function the transform only depends on the value of $u(t)$ at the specific times $0, T, 2T, \ldots$. Thus it contains only the incomplete information given by the function $u(t)$ at these times and hence an inverse will not lead to the unique continuous function $u(t)$ itself but only to specific values of a (continuous) function at $0, T, 2T$, etc.

Because $U(z)$ depends only on the specific values $u(0), u(T), \ldots$, the function $U(z)$ will equal the z-transform of the sampled data function $w(t)$ also, i.e. $W(z)$. Thus

$$W(z) = U(z)$$

$$= \sum_{k=0}^{\infty} u(kT)z^{-k}. \tag{9.7}$$

A table of z-transforms may be established in a similar way to that used for building up the table of Laplace transforms (section 1.4).

Example For example for

$$u(t) = e^{-at}, \qquad u(t) = 0 \quad \text{for} \quad t < 0$$
$$u(kT) = e^{-akT}$$

and

$$U(z) = \sum_{k=0}^{\infty} u(kT)z^{-k}$$
$$= 1 + e^{-aT}z^{-1} + e^{-2aT}z^{-2} + \ldots e^{-akT}z^{-k} + \ldots .$$

This is a standard geometric progression with the sum

$$U(z) = \frac{z}{z - e^{-aT}}.$$

In the same way that $U(s)$ is expressed as a function of s, so is $U(z)$ expressed at a function of the complex variable z. Some of the more common z-transforms are shown in Table 9.1 (cf. Table 1.1).

Table 9.1   Some z-transform pairs (T is the sampling interval)

$f(t)$ or $f(kT)$	$F(z)$
Unit impulse $\delta(t)$	1
Delayed impulse $\delta(t-kT)$	z^{-k}
Unit step $U(t)=1$	$\dfrac{z}{z-1}$
t	$\dfrac{Tz}{(z-1)^2}$
t^2	$\dfrac{T^2z(z+1)}{(z-1)^3}$
e^{-at}	$\dfrac{z}{z-e^{-aT}}$
te^{-at}	$\dfrac{Tze^{-aT}}{(z-e^{-aT})^2}$
t^2e^{-at}	$\dfrac{T^2(z^2+1)e^{-aT}}{(z-e^{-aT})^3}$
$\sin \omega t$	$\dfrac{z \sin \omega T}{z^2-2z \cos \omega T+1}$
$\cos \omega t$	$\dfrac{z(z-\cos \omega T)}{z^2-2z \cos \omega T+1}$
$\sinh at$	$\dfrac{z \sinh aT}{z^2-2z \cosh aT+1}$
$\cosh at$	$\dfrac{z^2-z \cosh aT}{z^2-2z \cosh aT+1}$
$e^{-at} \sin \omega t$	$\dfrac{e^{-aT}z \sin \omega T}{z^2-2e^{-aT} \cos \omega T+e^{-2aT}}$
$e^{-at} \cos \omega t$	$\dfrac{z^2-e^{-aT} \cos \omega T}{z^2-2e^{-aT} \cos \omega T+e^{-2aT}}$

Example   Determine the z-transform for the continuous functions, a step input $f_1(t)=U(t)$ and a ramp input $f_2(t)=at$.

(i) Step: If

$$f_1(t)=U(t)=\begin{cases}1, & t\geqslant 0 \\ 0, & t<0\end{cases}$$

$$F_1(z)=\sum_{k=0}^{\infty} f_1(kT)z^{-k}$$

$$=\sum_{k=0}^{\infty} 1 . z^{-k}$$

$$=1+z^{-1}+z^{-2}+\ldots z^{-k}+\ldots$$

$$=z/(z-1)$$

(ii) Ramp:

$$F_2(z) = \sum_{k=0}^{\infty} f_2(kT)z^{-k}$$

$$= \sum_{k=0}^{\infty} akT \cdot z^{-k}$$

$$= a(0 + Tz^{-1} + 2Tz^{-2} + 3Tz^{-3} + \ldots kTz^{-k} + \ldots)$$

Thus

$$(z-1)F_2(z) = a(T + Tz^{-1} + Tz^{-2} + \ldots)$$

$$= aT(1 + z^{-1} + z^{-2} + \ldots)$$

$$= \frac{aTz}{z-1}$$

Therefore

$$F_2(z) = \frac{aTz}{(z-1)^2}$$

Analogous to the useful theorems of linearity, differentiation etc. which were quoted for the Laplace transforms, similar theorems may be derived for z-transforms.

Linearity.

$$\mathscr{Z}[f_1(t) + f_2(t)] = \mathscr{Z}[f_1(t)] + \mathscr{Z}[f_2(t)]$$

Constant multiplication.

$$\mathscr{Z}[af(t)] = a\mathscr{Z}[f(t)]$$

Shift theorem.

$$\mathscr{Z}[f(k+1)T] = z\mathscr{Z}[f(kT)] - zf(0)$$

and in general (equivalent to the nth derivative and Laplace transform):

$$\mathscr{Z}[f(k+n)T] = z^n\mathscr{Z}[f(kT)] - (z^nf(0) + \ldots zf((n-1)T))$$

Translation.

$$\mathscr{Z}[f(k-n)T] = z^{-n}\mathscr{Z}[f(kT)]$$

Final value theorem.

$$f(\infty) = \lim_{z \to 1}[(1 - z^{-1})F(z)], \quad F(z) = \mathscr{Z}[f(t)]$$

Initial value theorem.

$$f(0) = \lim_{z \to \infty} F(z).$$

The z-transform has been introduced to enable the dynamics of a system with discrete data to be studied (although it may also be used in the direct solution of difference equations). Its usefulness can only be fully realized if, like in the case of the Laplace transform, it is possible to revert to the time domain from the system equation involving the z-transform. That is, we require an inverse z-transform method.

Table 9.1 itself provides a look-up table for some inverse z-transforms and the use of such a table is the most convenient method of obtaining the inverse where readily available. This table may be expanded or individual inversions obtained by utilizing partial fraction and power series methods of expansion.

The partial fraction expansion is similar to that used for the Laplace transform except that it can be seen in Table 9.1 that nearly all z-transforms contain the factor z in the numerator. The general transform is then expanded, in the case of distinct roots, to

$$F(z) = \frac{b_1 z}{z - z_1} + \frac{b_2 z}{z - z_2} + \dots$$

and

$$f(t) = f(kT) = b_1 z_1^k + b_2 z_2^k + \dots b_n z_n^k + \dots$$

where

$$z_1 = e^{-a_1 T}, \; z_2 = e^{-a_2 T} \text{ etc. as in Table 9.1.}$$

i.e.

$$\begin{aligned} f(t) &= f(kT) \\ &= b_1 e^{-a_1 kT} + b_2 e^{-a_2 kT} + \dots . \end{aligned}$$

Example   Suppose

$$U(z) = \frac{z}{(z-2)(z-4)}.$$

Expand this into partial fractions

$$U(z) = \frac{1}{2} \left\{ \frac{-z}{z-2} + \frac{z}{z-4} \right\}.$$

Notice that in Table 9.1 the equivalent term is $z/(z - e^{-aT})$ so that for each fraction respectively $e^{-aT} = 2$, $e^{-aT} = 4$. Inversion of $z/(z - e^{-aT})$ is the term e^{-at} so that as

$$\begin{aligned} e^{-at} &= e^{-a(kT)} \\ &= (e^{-aT})^k \end{aligned}$$

so

$$u(t) = u(kT) = \tfrac{1}{2}(-2^k + 4^k).$$

Thus for

$$
\begin{array}{ll}
k = 0, & u(0) = 0 \\
k = 1, & u(T) = 1 \\
k = 2, & u(2T) = 6 \\
k = 3, & u(3T) = 28 \text{ etc.}
\end{array}
$$

Alternatively the z-transform may be expanded in a convergent power series in z^{-1}. This effectively is an expansion of the definition, equation (9.4),

$$
\begin{aligned}
F(z) &= \sum_{k=0}^{\infty} f(kT)z^{-k} \\
&= f(0) + f(T)z^{-1} + f(2T)z^{-2} + \ldots f(kT)z^{-k} + \ldots .
\end{aligned}
$$

The expansion of a given $U(z)$ is obtained by the long division of the numerator by the denominator, each being expressed as a polynomial in ascending powers of z^{-1}.

Example Consider the above example again,

$$
\begin{aligned}
U(z) &= \frac{z}{(z-2)(z-4)} \\
&= \frac{z}{z^2 - 6z + 8}.
\end{aligned}
$$

To obtain this in powers of z^{-1} rewrite $U(z)$ as

$$U(z) = \frac{z^{-1}}{1 - 6z^{-1} + 8z^{-2}}$$

by dividing numerator and denominator by z^2. Long division yields the series

$$U(z) = z^{-1} + 6z^{-2} + 28z^{-3} + 120z^{-4} + \ldots .$$

From Table 9.1 this corresponds immediately to the series of delayed pulses

$$
\begin{aligned}
u(kT) &= u(kT)\delta(t - kT) \\
&= u(0)\delta(t) + u(T)\delta(t - T) + u(2T)\delta(t - 2T) + \ldots
\end{aligned}
$$

so by inspection

$$u(0) = 0$$
$$u(T) = \mathscr{Z}^{-1}[z^{-1}] = 1$$
$$u(2T) = \mathscr{Z}^{-1}[6z^{-2}] = 6$$
$$u(3T) = \mathscr{Z}^{-1}[28z^{-3}] = 28$$
$$u(4T) = \mathscr{Z}^{-1}[120z^{-4}] = 120 \text{ etc.}$$

which agree with the partial fraction solution.

A further method of inversion is by the direct use of the inversion integral $\left(x(kT) = \dfrac{1}{2\pi j} \oint X(z) z^{k-1}\, dz \right)$ but we may also proceed by more direct comparison between the continuous and sampled systems as below.

The pulse transfer function

The z-transform itself is utilized in systems where discrete data is a feature and the pulse transfer function has an equivalent role to play in such systems as the transfer function based on the Laplace transform plays in continuous linear systems. However, generally systems which contain discrete data signals, samplers, and hold devices also have the usual linear elements and it is these mixed systems which are therefore of prime interest. In fact the zero-order hold is itself a linear element, equation (9.1), so that the discrete time linear system may be considered as being comprised of one or more samplers producing the pulses, plus continuous linear elements.

Consider the system of Fig. 9.4. In the absence of feedback this represents the system, for which the pulse transfer function is required. The pulse transfer function is by definition

$$G(z) = \frac{X(z)}{U(z)}. \tag{9.8}$$

It is convenient now to use the * notation to denote sampled signals to prevent a growth in the number of variables and to indicate clearly the sampled signal.

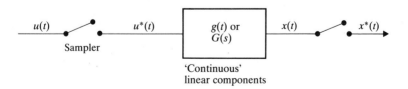

Fig. 9.4   General discrete time system

Interest for a particular system might be centred on the continuous output $x(t)$ or on the sampled signal $x^*(t)$. Where the output is also sampled the two samplers will be considered as being simultaneous in their action with sampling period T and both of small sampling duration compared with the remaining system dynamics. For time $t < 0$ both $u(t)$ and $x(t)$ are zero. It was noted, equation (9.7), that the z-transforms of $u^*(t)$ and $x^*(t)$ are equal to the z-transforms of $u(t)$ and $x(t)$ respectively so that for the system of Fig. 9.4 without feedback the pulse transfer function of the overall system is also that for the continuous elements alone.

In Chapter 2, section 2.7, the convolution integral was introduced. In the terms of that section, $x(t)$ is the output and $u(t)$ the input and

$$x(t) = \int_0^t g(t-\tau)u(\tau)\, d\tau$$
$$= g(t) * u(t).$$

For the plant with input $u^*(t)$ we may use this to establish the output $x(t)$ and we may develop the input–output relationship also on a step by step basis. For an input impulse at $t = 0$ of magnitude $u(0)$ the output $x(t)$ is given by the convolution integral

$$x(t) = \int_0^t g(t-\tau)u(0)\delta(t)\, d\tau$$
$$= u(0)g(t)$$

since the response to a unit impulse is equal to the system weighting function $g(t)$. The input is $u(0)\delta(t)$. For an impulse at $t = T$ of magnitude $u(T)$ the output $x(t)$ is given by the convolution integral again, this time

$$x(t) = \int_0^t g(t-\tau)u(T)\delta(t-T)\, d\tau$$
$$= u(T)g(t-T),$$

the response to a delayed unit impulse being the weighting function, also delayed by the same duration T and so on. Thus for the sequence of inputs given by the sampled data input up to $u(kT)$ the output $x(t)$ is given by

$$x(t) = u(0)g(t) + u(T)g(t-T) + u(2T)g(t-2T)$$
$$+ \dots u(kT)g(t-kT),$$

where $t \leqslant kT$. The response at time t is unaffected by inputs at values of $kT > t$ and so $g(t-nT)$ where $nT > t$ is zero. The sum $x(t)$ is written at time $t = kT$ as,

$$x(t) = x(kT) = \sum_{i=0}^{k} u(iT)g(kT - iT)$$

$$= u(kT) * g(kT) \qquad (9.9)$$

where $u(kT) * g(kT)$ indicates the 'convolution summation'. We can see that taking a limit on the sampling time as $T \to$ zero leads to the convolution integral. The $x(t)$ is still a continuous function but its values are evaluated only at the intervals of time 0, $T, \dots kT$ by the convolution summation. Hence its values are also the values given by the sampled signal $x^*(t)$ at these instances at intervals of T.

The z-transform of $x(kT)$ may be obtained by the double summation

$$X(z) = \sum_{k=0}^{\infty} x(kT)z^{-k}$$

$$= \sum_{k=0}^{\infty} \left[\sum_{i=0}^{\infty} u(iT)g(kT - iT) \right] z^{-k}$$

using the substitution of equation (9.9).

Introducing the variable m where $k = m + i$

$$X(z) = \sum_{m=0}^{\infty} \sum_{i=0}^{\infty} u(iT)g(mT)z^{-(m+i)}$$

$$= \sum_{m=0}^{\infty} \left[g(mT)z^{-m} \sum_{i=0}^{\infty} u(iT)z^{-i} \right]$$

$$= \sum_{m=0}^{\infty} g(mT)z^{-m}U(z).$$

Thus

$$\frac{X(z)}{U(z)} = \sum_{m=0}^{\infty} g(mT)z^{-m}$$

$$= g(0) + g(T)z^{-1} + \dots g_j(jT)z^{-j} + \dots$$

$$= G(z), \text{ the pulse transfer function.} \qquad (9.10)$$

since the pulse transfer function is defined by equation (9.8). Thus Fig. 9.4 may be reduced to Fig. 9.5. The general similarity with the transfer function is now apparent. Now, starting with a given transfer function $G(s)$, the inverse may be used to obtain $g(t)$ and in turn $G(z)$ is obtained from the summation of equation (9.10).

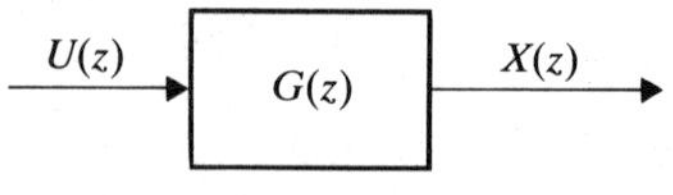

Fig. 9.5 The pulse transfer function $G(z)$

Example   For the second order system of Fig. 9.6, obtain the pulse transfer function $G(z)$.

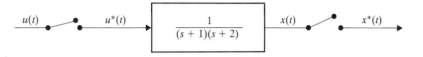

Fig. 9.6   Second order sampled-data system

Now

$$G(s) = \frac{1}{(s+1)(s+2)}$$

so that

$$g(t) = \mathscr{L}^{-1}\left[\frac{1}{(s+1)(s+2)}\right]$$
$$= e^{-t} - e^{-2t}$$

and

$$G(z) = \sum_{m=0}^{\infty} g(mT)z^{-m}$$
$$= (1-1) + (e^{-T} - e^{-2T})z^{-1} + (e^{-2T} - e^{-4T})z^{-2} + \dots$$
$$= \{1 + e^{-T}z^{-1} + e^{-2T}z^{-2} + \dots\} - \{1 + e^{-2T}z^{-1} + e^{-4T}z^{-2} + \dots\}$$
$$= \frac{1}{1 - z^{-1}e^{-T}} - \frac{1}{1 - z^{-1}e^{-2T}}$$
$$= \frac{z(e^{-T} - e^{-2T})}{(z - e^{-T})(z - e^{-2T})}.$$

The presence of a sampler between two continuous elements needs treating with care. If there is no sampler, then $g(t)$ is obtained from $\mathscr{L}^{-1}[G_1(s)G_2(s)] = \mathscr{L}^{-1}[G(s)]$, Fig. 9.7(a) and

$$\frac{X(z)}{U(z)} = G_1G_2(z)$$
$$= G(z). \tag{9.11}$$

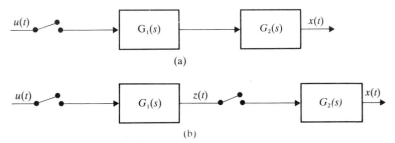

Fig. 9.7   Additional sampler between continuous elements

If there is a second sampler, synchronized with the first, then, Fig. 9.7(b),

$$\frac{Z(z)}{U(z)} = G_1(z)$$

$$\frac{X(z)}{Z(z)} = G_2(z)$$

$$\frac{X(z)}{U(z)} = G_1(z)G_2(z)$$

$$\neq G(z). \tag{9.12}$$

Example   Obtain the pulse transfer function for the system shown in Fig. 9.8 for a sampling interval T. The pulse transfer function is

$$G(z) = \frac{X(z)}{U(z)}$$

$$= \sum_{k=0}^{\infty} g(kT)z^{-k}.$$

The transfer function of the zero order hold is $(1 - e^{-sT})/s$ so that

$$G(s) = \frac{1 - e^{-sT}}{s(s+1)(s+2)}.$$

Proceeding by way of the time function $g(t)$, $G(s)$ is inverted:

$$G(s) = \left\{ \frac{0.5}{s} - \frac{1}{s+1} + \frac{0.5}{s+2} \right\} - e^{-sT} \left\{ \frac{0.5}{s} - \frac{1}{s+1} + \frac{0.5}{s+2} \right\}$$

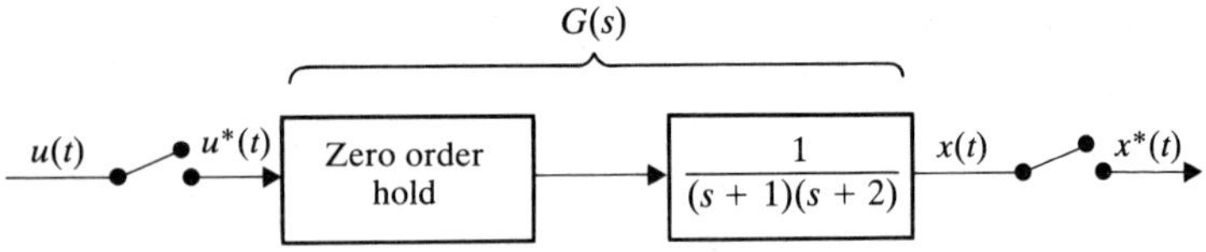

Fig. 9.8   Second order system with sampler and hold

and

$$g(t) = (0 \cdot 5 - e^{-t} + 0 \cdot 5 e^{-2t}) - (0 \cdot 5 - e^{-(t-T)} + 0 \cdot 5 e^{-2(t-T)}) U(t - T)$$

so that $g(kT) = 0$ for $k = 0$ and for $k = 1, 2 \ldots ,$

$$g(kT) = (0 \cdot 5 - e^{-kT} + 0 \cdot 5 e^{-2kT}) - (0 \cdot 5 - e^{-(k-1)T} + 0 \cdot 5 e^{-2(k-1)T}).$$

Hence

$$G(z) = \sum_{k=0}^{\infty} \{ -e^{-kT} + 0 \cdot 5 e^{-2kT} + e^{-(k-1)T} - 0 \cdot 5 e^{-2(k-1)T} \} z^{-k}$$
$$+ (0 \cdot 5 - e^T + 0 \cdot 5 e^{2T})$$

$$= \frac{-z}{z - e^{-T}} + \frac{0 \cdot 5 z}{z - e^{-2T}} + \frac{e^T z}{z - e^{-T}} - \frac{0 \cdot 5 e^{2T} z}{z - e^{-2T}}$$
$$+ 0 \cdot 5 - e^T + 0 \cdot 5 e^{2T}$$

$$= \frac{(0 \cdot 5 e^{-2T} - e^{-T} + 0 \cdot 5) z + 0 \cdot 5 e^{-3T} - e^{-2T} + 0 \cdot 5 e^{-T}}{(z - e^{-T})(z - e^{-2T})}$$

and

$$x(kT) = \mathcal{Z}^{-1} [G(z) U(z)].$$

To evaluate $x(kT)$ it is possible to substitute the numerical value for T, i.e. the given sampling interval, multiply the subsequent $G(z)$ by $U(z)$, e.g. $z/(z - 1)$ for a unit step input, or $aTz/(z - 1)^2$ for a ramp input, and then express the $G(z)U(z)$ as a series in z^{-1} by long division. The long division will yield a series of terms $x(kT) = \mathcal{Z}^{-1}[a_1 z^{-1} + a_2 z^{-2} + \ldots]$ so that

$$x(kT) = a_1 \delta(t - T) + a_2 \delta(t - 2T) + \ldots$$

and

$$x(0) = 0$$
$$x(T) = a_1$$
$$x(2T) = a_2$$
$$x(3T) = a_3 \text{ etc. (or the term by term form of } G(z) \text{ may be used).}$$

Example What is the pulse transfer function for the system shown in Fig. 9.9?

Taking the inner loop first,

$$C(s) = G_2(s) Y^*(s)$$
$$= G_2(s) \{ G_1^*(s) E^*(s) - G_1 H_1^*(s) C^*(s) \}.$$

Now $E(s) = R(s) - C(s)$ therefore

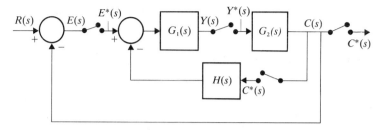

Fig. 9.9   Feedback system with a number of samplers

$$C(s) = G_2(z)\{G_1^*(s)(R^*(s) - C^*(s)) - G_1 H_1^*(s)C^*(s)\}$$

and

$$C(z) = G_2(z)\{G_1(z)(R(z) - C(z)) - G_1 H_1(z)C(z)\}$$

or

$$\frac{C(z)}{R(z)} = \frac{G_2(z)G_1(z)}{1 + G_1(z)G_2(z) + G_1 H_1(z)G_2(z)}.$$

Sampled-data feedback systems

Bearing in mind the influence of samplers between the continuous elements the addition of feedback makes careful procedure even more important. Also of course the position of the samplers may be variable, e.g. they may be either in the feedback loop or in the error signal, and the output from the plant under consideration may be required either in a sampled or continuous form. However, provided we stick to the use of synchronized samples some simplifying block diagram algebra may be used, e.g. Fig. 9.10.

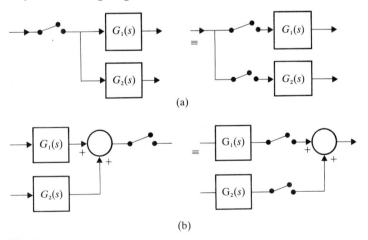

Fig. 9.10   Equivalent sampler arrangements

Sampler arrangement   Form of C(z)

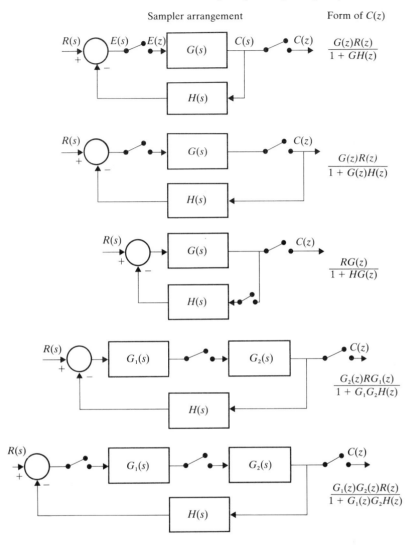

Fig. 9.11   Effect of samplers on z-transform of system output $C(z)$ (Ogata 1970)

A change in the position of the sampler results in a change in the overall pulse transfer function of the closed loop. Figure 9.11 (taken from Ogata) shows just how the pulse transfer function and output $C(z)$ of a feedback system is affected by the sampling arrangements (using zero hold). The notation $GH(z)$ indicated the z-transform derived from the combined Laplace transform $G(s)H(s)$ and is not equal to $G(z)H(z)$ which is derived from the individual transforms $G(s)$, $H(s)$. To see how these expressions are established consider the third

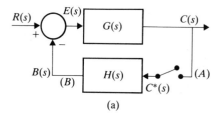

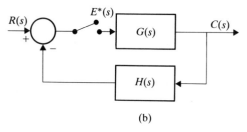

Fig. 9.12 Sampling for (a) analogue controller, (b) digital controller

example shown with the sampler at *A*, Fig. 9.12(a). This is equivalent to a controller-plant system, *G*, whose output is sampled, e.g.
by an analogue to digital converter. As shown the feedback element *H*
may incorporate a data processing or simple transducer system eventually giving an analogue, smoothed, output for comparison with the reference input *R*(*s*). If the controller in the system is a digital device, e.g. a
computer then this may be represented by Fig. 9.12(b) where in this
case analogue feedback and reference are compared before sampling.
This system is extended in section 9.5. The continuous variables are
shown as *C*(*s*) etc. and the sampled function by *C**(*s*). Analysing the
system of Fig. 9.12(a) in detail:

$$C(s) = G(s)E(s)$$
$$E(s) = R(s) - H(s)C^*(s)$$

so that

$$C(s) = G(s)[R(s) - H(s)C^*(s)]$$
$$= G(s)R(s) - G(s)H(s)C^*(s).$$

When *C*(*s*) is sampled each term is in effect sampled and also

$$C^*(s) = GR^*(s) - GH^*(s)C^*(s)$$

i.e.

$$C^*(s) = \frac{GR^*(s)}{1 + GH^*(s)}.$$

Equation (9.6) showed that the Laplace transform of a sampled function or variable is the z-transform of the continuous function so that the z-transformed output is

$$C(z) = \frac{GR(z)}{1 + GH(z)}. \tag{9.13}$$

The overall pulse transfer function $G_o(z)$ is given by equation (9.10).

$$G_o(z) = \frac{C(z)}{R(z)}$$

$$= \frac{C^*(s)}{R^*(s)}$$

$$= \frac{GR(z)}{R(z)(1 + GH(z))}.$$

Note that the overall pulse transfer function form unlike the transfer function for continuous systems, may be 'input dependent', as in this case. Also, although in Fig. 9.12(a) the output $C(s)$ is continuous the sampled output $C^*(s)$ $(= C(z))$ is only able to give the precise values of $c(t)$ at the times $0, T, 2T, \ldots$.

Stability in sampled-data feedback systems

The stability of the continuous system is determined by the denominator of the overall transfer function. Each closed loop form of the sampled-data systems shown in Fig. 9.11 is subject to a similar stability analysis, the stability again being determined by the characteristic equation. Taking the straightforward case with the error signal sampled (Fig. 9.12(b)).

$$\frac{C(z)}{R(z)} = \frac{G(z)}{1 + GH(z)}$$

$$= G_o(z)$$

and the characteristic equation is

$$1 + GH(z) = 0.$$

It can be seen that this factor, or a factor similar to it, is present in all the closed loop forms illustrated. In the earlier examples on inversion of the z-transform it was seen that either partial fractions or power series expansion led to terms suitable for inversion. Looking at the partial fractioning of $G_o(z)$, the denominator of the partial fractions, i.e. $(z - z_1)$, $(z - z_2)$ etc. are the factors of the characteristic equation. The inversion of the terms $b_1 z/(z - z_1)$, $b_2 z/(z - z_2)$ etc. give terms in the (kT) time solution of $b_1 z_1^k$, $b_2 z_2^k$ etc. Hence if $|z_1|$, $|z_2| >$ unity then with increasing time, i.e. increasing k, the output will continue to increase in size with either positive or negative sign. For stability therefore the roots z_i of $1 + GH(z) = 0$ must have modulus less than unity, i.e.

$$|z_i| < 1 \qquad \text{for all } i \tag{9.14}$$

Since z_i may be real or complex this means that the values of z_i plotted in the complex plane must lie within a circle of unit radius, referred to as the unit circle. The same conclusion follows if the closed loop pulse

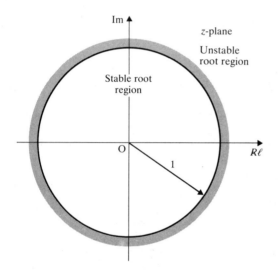

Fig. 9.13   Stability in the z-plane – the unit circle

transfer function is expanded as a series in z^{-1}, z^{-2} etc. If there is difficulty in determining z_i because of the order of $1 + GH(z)$ in z then we can use the Routh–Hurwitz method which was used for the continuous systems so that z_i need not be explicitly determined. If z is replaced by a variable, say w, by the transformation

$$z = \frac{w+1}{w-1}, \qquad z \text{ real or complex,}$$

then the unit circle in the z plane is mapped into the imaginary axis of the w plane, the inside of the unit circle mapping into the left hand of the w plane and the outside mapping into the right hand of the w plane. Hence application of the Routh criterion to the characteristic equation expressed in terms of w will determine the stability of the sampled system.

Example   The feedback system in Fig. 9.14 has a variable gain K. Determine the range of K if the system is to remain stable with a sampling interval of unity. What is the effect on the suitable range of K if the sampling interval is reduced to give very rapid sampling? The linear element transfer function is $G(s) = K/s(s+4)$ and there is no

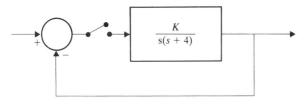

Fig. 9.14   Sampled system with feedback

hold this time. The feedback is unity so the characteristic equation, for the sampler in this position, is

$$1 + GH(z) = 1 + G(z) = 0.$$

$G(z)$ is obtained from $G(s)$ either by direct equivalence tables or through the inversion of $G(s)$

$$g(t) = \mathcal{L}^{-1}[G(s)] = \mathcal{L}^{-1} \frac{1}{4}\left[\frac{K}{s} - \frac{K}{s+4}\right]$$

$$= \frac{K}{4}(1 - e^{-4T})$$

and

$$G(z) = \frac{K}{4}\left\{\frac{z}{z-1} - \frac{z}{z - e^{-4T}}\right\}$$

$$= \frac{K}{4} \cdot \frac{z(1 - e^{-4T})}{(z-1)(z - e^{-4T})}.$$

For $T = 1$ the characteristic equation gives

$$4(z-1)(z - e^{-4}) + Kz(1 - e^{-4}) = 0 \tag{9.15}$$

i.e.

$$4(z-1)(z - 0 \cdot 0183) + Kz(0 \cdot 9817) = 0$$

or

$$4z^2 + (0 \cdot 9817K - 4 \cdot 0732)z + 0 \cdot 0732 = 0,$$

and the values of the roots of this equation z_1, z_2 gives us the information about the stability of the system. For a given K the roots could be evaluated in the usual way and $|z_1|$, $|z_2|$ checked to see if the roots fall within the unit circle or not. To determine the range of K giving stability the substitution $z = (w+1)/(w-1)$ and the subsequent use of the Routh criterion is more satisfactory as it is easily, if laboriously, applied to any order system. The equation becomes

$$4\left(\frac{w+1}{w-1}\right)^2 + (0 \cdot 9817K - 4 \cdot 0732)\left(\frac{w+1}{w-1}\right) + 0 \cdot 0732 = 0$$

i.e.

$$0 \cdot 9817Kw^2 + 7 \cdot 8536w + (8 \cdot 1464 - 0 \cdot 9817K) = 0$$

The Routh–Hurwitz array for this equation is

w^2	$0 \cdot 9817K$	$8 \cdot 1464 - 0 \cdot 9817K$
w^1	$7 \cdot 8536$	0
w^0	$8 \cdot 1464 - 0 \cdot 9817K$	0

or more simply we know that for negative real-part roots of the quadratic equation $aw^2 + bw + c = 0$ to exist only, then $ac > 0, (b > 0)$ i.e. for a positive K, $8 \cdot 1464 - 0 \cdot 9817K > 0$, the last line of the array. Thus to avoid a change in sign in the first column of the array

$$K > 0$$

and

$$8 \cdot 1464 - 0 \cdot 9817K > 0$$

so that

$$0 < K < 8 \cdot 30.$$

The Routh array becomes trivial for the second order case but it illustrates the general case, and its systematic use can prevent errors. The general form for a variable sampling interval T of this characteristic equation is from equation (9.15) above

$$4(z-1)(z - e^{-4T}) + Kz(1 - e^{-4T}) = 0.$$

As $T \rightarrow$ zero, so $e^{-4T} \rightarrow$ unity and the value of K can be progressively increased without loss of stability until in the limit the system is continuous and for the second order system stability is assured. This can be rigorously shown by developing the array with T as a variable, finding the stability-limiting range of K as a function of T and then letting T tend to zero.

The root-locus and sampled-data systems

As a plot of the roots of the characteristic equation of the continuous system enable us to see the effect of the changing the gain, or other parameters, in a system and to assess the 'nearness' to stability and the effect of extra elements, e.g. controller modes, so also does the root locus for the discrete system. It has been seen above that the unit circle in the z-plane has the same role as the imaginary axis in the s-plane for the continuous system. As parameters of the discrete system change so we might also expect the location of the roots of the equation $1 + GH(z) = 0$ to change and give us a series of loci, one for each root, in the complex plane.

Example Consider again the system of Fig. 9.14. The second order transfer function may be taken for example as a second order inertial system with a proportional controller K or a first order system with integral control action. In either case the continuous system is stable and we may plot the roots loci for the closed loop system which we know (Fig. 9.15(a)) to fall fully in the left half plane. The system is stable for all K. In this example we have already established a pulse transfer function $G(z)$ for the system and by the circle criterion a range of gain K which retains system stability. The system may be represented by Fig. 9.15(b). The closed loop characteristic equation, $1 + G(z) = 0$, has been evaluated as

$$4(z - 1)(z - e^{-4}) + Kz(1 - e^{-4}) = 0.$$

Following our root locus rules we see that the open loop poles (from $G(z)$) are at $z = 1$ and $z = e^{-4}$ and the single open loop zero is at $z = 0$. We thus expect two branches of the loci, one terminating on the diagram at $z = 0$ and the other extending off the diagram with increasing gain K. As K varies the closed loop characteristic equation roots follow the paths of Fig. 9.15(b). Stability is limiting when one of the loci cuts the unit circle, at $K = 8.3$.

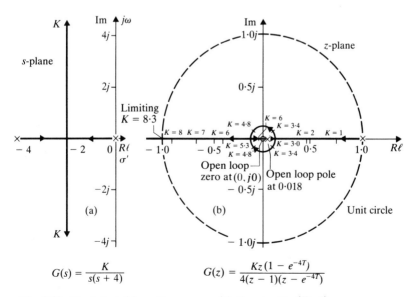

$$G(s) = \frac{K}{s(s+4)} \qquad G(z) = \frac{Kz(1 - e^{-4T})}{4(z-1)(z-e^{-4T})}$$

Fig. 9.15 Roots loci, (a) continuous case, (b) discrete case $(T = 1)$

9.3 Multivariable sampled-data systems

System equations

In Chapter 3, section 3.4, and subsequently in Chapter 7, the multivariable linear continuous time-invariant system was discussed. The system equations were

$$\dot{x} = Ax + Bu$$

$$y = Cx + Du$$

where A, B, C, D are constant matrices, x is the state vector, y the output vector and u the input or control vector. For the discrete time system the system equations are difference equations

$$x((k+1)T) = Px(kT) + Qu(kT)$$

$$y(kT) = Cx(kT) + Du(kT),$$

where we assume that the state at time $(k+1)T$ is dependent only on the state at time kT and the control at that time kT. If other control terms are present then a redefinition of variables reduces the equation back to this form. The output equation is directly equivalent to the continuous case. To simplify the notation the sampling interval T is normally understood and suffices $k+1$, k etc. used alone so that the equations may be written

$$x(k+1) = Px(k) + Qu(k) \tag{9.16}$$

$$y(k) = Cx(k) + Du(k). \tag{9.17}$$

In the general case x is an n-dimensional vector, u is an m-dimensional vector and y a p-vector. The P, Q are then $n \times n$, $n \times m$ matrices respectively and C, D are $p \times n$ and $p \times m$ respectively. We shall consider only the time-invariant case and shall see similarities with continuous system treatment, e.g. as in stability determination.

Example A sampled-data system is represented by the state equation $x(k+1) = Px(k) + Qu(k)$. If x is third order and P is the matrix

$$\begin{bmatrix} 0 & -1 & -1 \\ 2 & 1{\cdot}5 & 1 \\ 0 & 0 & 1 \end{bmatrix}$$

is the system stable or unstable in behaviour for a bounded input $u(k)$?

To check for instability in the multivariable case examine the characteristic equation which is established below, equation 9.33,

$$|zI - P| = 0$$

i.e.

$$\begin{vmatrix} z & 1 & 1 \\ -2 & z-1{\cdot}5 & -1 \\ 0 & 0 & z-1 \end{vmatrix} = z^3 - 2{\cdot}5z^2 + 3{\cdot}5z - 2$$

$$= 0.$$

This factorizes to

$$(z-1)(z^2 - 1{\cdot}5z + 2) = 0$$

so the roots are $z_1 = 1$, $z_{2,3} = 0 \cdot 75 \pm j\sqrt{1 \cdot 4375}$. The moduli of these roots are 1, $\sqrt{2}$, $\sqrt{2}$ and hence the sampled system will be unstable. Note that in the sampled system the real part of the roots z_i may be negative and even of magnitude less than one but the modulus of z_i may still be greater than unity and hence yield an unstable system.

The discrete time equations, equations (9.16) and (9.17) may arise in two ways. Firstly, there may come from the direct description of the system behaviour which is immediately obtainable in the form of difference equations. Secondly, they may be describing a continuous system which is either sampled at intervals of time or which is 'discretized' for the benefit of computation or control. So that the multivariable discrete time form of system equations do not seem remote from continuous systems consider how the sampled-data equations are arrived at for the multivariable continuous case.

The sampled-data system equations for a continuous system

The continuous system equation is

$$\dot{\mathbf{x}}(t) = \mathbf{A}\mathbf{x}(t) + \mathbf{B}\mathbf{u}(t)$$

and the solution of this equation for the linear time-invariant case is

$$\mathbf{x}(t) = e^{\mathbf{A}t}\mathbf{x}(0) + \int_0^t e^{\mathbf{A}(t-\tau)}\mathbf{B}\mathbf{u}(\tau)\, d\tau \tag{3.66}$$

where $e^{\mathbf{A}t}$ is the transition matrix.

Consider now that each element of the control vector $\mathbf{u}(t)$ only changes its value at discrete sampling times $0, T, 2T, \ldots$ and it remains constant during each interval of duration T. We wish to know the values of the state variables at these instances of time, i.e. $\mathbf{x}(0)$, $\mathbf{x}(T)$ etc. From some initial state $\mathbf{x}(0)$ the system equation (3.66) gives the values at time $t = kT$ and $t = kT + T$,

$$\mathbf{x}(kT) = e^{\mathbf{A}kT}\mathbf{x}(0) + e^{\mathbf{A}kT}\int_0^{kT} e^{-\mathbf{A}\tau}\,\mathbf{B}\mathbf{u}(\tau)\, d\tau$$

$$\mathbf{x}((k+1)T) = e^{\mathbf{A}(k+1)T}\mathbf{x}(0) + e^{\mathbf{A}(k+1)T}\int_0^{(k+1)T} e^{-\mathbf{A}\tau}\mathbf{B}\mathbf{u}(\tau)\, d\tau$$

Elimination of $\mathbf{x}(0)$ by rearranging the first equation to give $\mathbf{x}(0)$ and substitution of this in the expression for $\mathbf{x}((k+1)T)$ gives

$$\mathbf{x}((k+1)T) = e^{\mathbf{A}T}\mathbf{x}(kT) + \int_{kT}^{(k+1)T} e^{\mathbf{A}((k+1)T-\tau)}\mathbf{B}\mathbf{u}(\tau)\, d\tau$$

or by substituting

$$\lambda = (k+1)T - \tau, \qquad d\lambda = -d\tau$$

and

$$\mathbf{x}((k+1)T) = e^{\mathbf{A}T}\mathbf{x}(kT) + \int_0^T e^{\mathbf{A}\lambda}\mathbf{B}\, d\lambda \cdot \mathbf{u}(kT) \qquad (9.18)$$

as $\mathbf{u}(kT)$ is constant over the period of duration T from kT to $(k+1)T$.

Comparison of equation (9.18) with (9.16) shows that the sampled-data system equation for the given continuous system is given by equation (9.16) with

$$\mathbf{P} = e^{\mathbf{A}T} \qquad (9.19)$$

$$\mathbf{Q} = \int_0^T e^{\mathbf{A}\lambda}\mathbf{B}\, d\lambda \qquad (9.20)$$

Thus for a given T and system matrices $\mathbf{A}$ and $\mathbf{B}$, the matrices $\mathbf{P}$ and $\mathbf{Q}$ are constant but they are dependent on the sampling interval T.

Although equation (9.18) gives only the state vector at the sampling intervals, $\mathbf{x}$ may be found at intermediate times t between kT and $(k+1)T$ by using equation (3.66) in the form

$$\mathbf{x}(t) = e^{\mathbf{A}(t-kT)}\mathbf{x}(kT) + \int_{kT}^t e^{\mathbf{A}(t-\tau)}\mathbf{B}\mathbf{u}(kT)\, d\tau, \qquad kT < t < (k+1)T.$$

The $\mathbf{C}$ and $\mathbf{D}$ matrices in equation (9.17) are unchanged from those of the continuous system.

Use of the dummy variable $\lambda = \tau - kT$ prior to equation (9.18) or using a new substitution $\lambda = T - \gamma$ directly in equation (9.18) gives the alternative form of these equations

$$\mathbf{x}(k+1) = e^{\mathbf{A}T}\mathbf{x}(kT) + \int_0^T e^{\mathbf{A}(T-\lambda)}\mathbf{B}\, d\lambda \cdot \mathbf{u}(kT) \qquad (9.21)$$

and

$$\mathbf{Q} = \int_0^T e^{\mathbf{A}(T-\lambda)}\mathbf{B}\, d\lambda. \qquad (9.22)$$

Note that the integral is being evaluated over the interval $t = T$. This pair of equations are equivalent to equations (9.18) and (9.20) but this form enables the $\mathbf{P}$ and $\mathbf{Q}$ matrices to be evaluated using techniques we have already established. In Chapter 3 it was seen that the transition matrix $e^{\mathbf{A}t}$ is most conveniently evaluated using the inverse transform, so that over the interval of duration T,

$$\mathbf{P} = e^{\mathbf{A}t}$$
$$= \mathcal{L}^{-1}[(s\mathbf{I} - \mathbf{A})^{-1}]|_{t=T}. \qquad (3.69)(9.23)$$

The convolution integral forming the last part of equation (9.21) may similarly be evaluated. Now

$$\int_0^T e^{\mathbf{A}(T-\lambda)}\mathbf{B}u(kT)\, d\lambda = \int_0^T e^{\mathbf{A}(T-\lambda)}\mathbf{B}U(\lambda)\, d\lambda\, \mathbf{u}(kT)$$

where U is the unit step function, constant over the interval 0 to T and $\mathbf{u}(kT)$ is a constant vector over the period. Equation (3.70) for the general linear system gives us

$$\int_0^T e^{\mathbf{A}(T-\lambda)}\mathbf{B}U(\lambda)\, d\lambda = \mathcal{L}^{-1}[(s\mathbf{I}-\mathbf{A})^{-1}\mathbf{B}U(s)]|_{t=T}.$$

But for a unit step function $U(s)$ is $1/s$ so that evaluation over the interval T gives

$$\mathbf{Q} = \mathcal{L}^{-1}\left[\frac{(s\mathbf{I}-\mathbf{A})^{-1}\mathbf{B}}{s}\right]\Bigg|_{t=T} \tag{9.24}$$

Equations (9.23) and (9.24) enable the $\mathbf{P}$ and $\mathbf{Q}$ matrices in the discrete form

$$\mathbf{x}(k+1) = \mathbf{P}\mathbf{x}(k) + \mathbf{Q}\mathbf{u}(k) \tag{9.16}$$

to be readily evaluated from the continuous system and distribution matrices.

As in the continuous case, Chapter 7, the state variable values may be used in a feedback law, i.e.

$$\mathbf{u}(kT) = \mathbf{K}\mathbf{x}(kT)$$

By restricting the control to a scalar the matrix $\mathbf{K}$ becomes a row vector of the same order as $\mathbf{x}$. This is illustrated in the following example.

The eigenvalues of the continuous time system are the roots of

$$|s\mathbf{I}-\mathbf{A}| = 0$$

and the corresponding values for the discrete time system are the roots of

$$|z\mathbf{I}-\mathbf{P}| = 0$$

equation (9.33) below. Now it is known that if the eigenvalues of $\mathbf{A}$ are λ_i then those of $e^{\mathbf{A}T}$ are $e^{\lambda_i T}$. Since $\mathbf{P}$ has been shown, equation (9.19), to be $e^{\mathbf{A}T}$ the roots λ_i (or s_i) of the continuous system characteristic equation and z_i of the discrete system are related by

$$z_i = e^{\lambda_i T} \tag{9.25}$$

Example In this example the progressive stages of (i) modelling in continuous time a system from a physical description, (ii) the formation of the state equations for the sampled system and (iii) the evaluation of a controller using state feedback in the sampled system to give specified dynamic behaviour are all illustrated.

A large second order undamped mechanical system has a position input u and its output position movement is x. The (undamped) natural frequency is $0 \cdot 1$ rad sec^{-1}. Taking output position and velocity as state variables and using a sampling time of 10 seconds establish the discrete form state equations. The 'steady state' gain is 100.

(i) First we note that the natural dynamics are slow, corresponding to a full oscillation about every 60 seconds. The sampling interval of 10 seconds is thus sufficiently rapid for good output representation.

The general second order system differential equation is

$$\ddot{x} + 2\zeta\omega_n\dot{x} + \omega_n^2 x = k\omega_n^2 u$$

and in the absence of damping, $\zeta = 0$ and with $k = 100$,

$$\ddot{x} + \omega_n^2 x = 100\omega_n^2 u$$

If $\omega_n = 0 \cdot 1$ then

$$\ddot{x} + 0 \cdot 01x = u.$$

(The steady state gain at $\dot{x} - \ddot{x} = 0$ is thus 100, a small input movement being greatly magnified.)

With the selected state variables let

$$x_1 = x$$
$$x_2 = \dot{x}_1 = \dot{x}$$

and the state equations are

$$\dot{x}_1 = x_2$$
$$\dot{x}_2 = -0 \cdot 01x_1 + u$$

with the output $y = x_1$.

In composite notation

$$\dot{\mathbf{x}} = \mathbf{A}\mathbf{x} + \mathbf{B}u$$
$$\mathbf{y} = \mathbf{C}\mathbf{x}$$

and we have

$$\begin{bmatrix} \dot{x}_1 \\ \dot{x}_2 \end{bmatrix} = \begin{bmatrix} 0 & 1 \\ -0 \cdot 01 & 0 \end{bmatrix} \begin{bmatrix} x_1 \\ x_2 \end{bmatrix} + \begin{bmatrix} 0 \\ 1 \end{bmatrix} u$$

and

$$y = \begin{bmatrix} 1 & 0 \end{bmatrix} \begin{bmatrix} x_1 \\ x_2 \end{bmatrix}.$$

(ii) The sampled data equivalent system representation is

$$\mathbf{x}(k+1) = \mathbf{P}\mathbf{x}(k) + \mathbf{Q}\mathbf{u}(k)$$
$$\mathbf{y}(k) = \mathbf{C}\mathbf{x}(k).$$

Immediately we see that the measurement matrix $\mathbf{C}$ is the same in each instance. From equation (9.23) and (9.24) the $\mathbf{P}$ and $\mathbf{Q}$ matrices are determined using the values of the sampling interval T and the $\mathbf{A}$ and $\mathbf{B}$ matrices.

$$\mathbf{P} = \mathscr{L}^{-1}[(s\mathbf{I} - \mathbf{A})^{-1}]|_{t=T}$$

$$(s\mathbf{I} - \mathbf{A})^{-1} = \begin{bmatrix} s & -1 \\ 0\cdot01 & s \end{bmatrix}^{-1}$$

$$= \frac{1}{s^2 + 0\cdot01} \begin{bmatrix} s & 1 \\ -0\cdot01 & s \end{bmatrix}$$

$$= \begin{bmatrix} \dfrac{s}{s^2+0\cdot01} & \dfrac{1}{s^2+0\cdot01} \\ \dfrac{-0\cdot01}{s^2+0\cdot01} & \dfrac{s}{s^2+0\cdot01} \end{bmatrix}$$

so that

$$\mathbf{P} = \mathscr{L}^{-1}[(s\mathbf{I} - \mathbf{A})^{-1}]|_{t=T}$$

$$= \begin{bmatrix} \cos 0\cdot1t & 10 \sin 0\cdot1t \\ -0\cdot1 \sin 0\cdot1t & \cos 0\cdot1t \end{bmatrix}_{t=T=10}$$

$$= \begin{bmatrix} 0\cdot5402 & 8\cdot415 \\ -0\cdot08415 & 0\cdot5402 \end{bmatrix}.$$

Similarly

$$\mathbf{Q} = \mathscr{L}^{-1}\left[\frac{1}{s}(s\mathbf{I} - \mathbf{A})^{-1}\mathbf{B}\right]$$

$$= \mathscr{L}^{-1}\left[\frac{1}{s^2+0\cdot01}\begin{bmatrix} 1 & \dfrac{1}{s} \\ \dfrac{-0\cdot01}{s} & 1 \end{bmatrix}\begin{bmatrix} 0 \\ 1 \end{bmatrix}\right]$$

$$= \mathscr{L}^{-1}\begin{bmatrix} \dfrac{1}{0\cdot01}\left(\dfrac{1}{s} - \dfrac{s}{s^2+0\cdot01}\right) \\ \dfrac{1}{s^2+0\cdot01} \end{bmatrix}$$

$$= \begin{bmatrix} 100 - 100 \cos 0\cdot1t \\ 10 \sin 0\cdot1t \end{bmatrix}_{t=T=10}$$

$$= \begin{bmatrix} 45\cdot98 \\ 8\cdot415 \end{bmatrix}.$$

The discrete state equations are therefore in full

$$\begin{bmatrix} x_1(k+1) \\ x_2(k+1) \end{bmatrix} = \begin{bmatrix} 0.5402 & 8.415 \\ -0.0841 & 0.5402 \end{bmatrix} \begin{bmatrix} x_1(k) \\ x_2(k) \end{bmatrix} + \begin{bmatrix} 45.98 \\ 8.415 \end{bmatrix} u$$

$$y(k+1) = x(k+1).$$

(iii) To both improve the damping of the system and speed up its dynamics, feedback is to be added to obtain a closed loop system with dynamics equivalent to those of a continuous system with a damping factor ζ of 1.2 and an undamped natural frequency of 0.2 rad sec^{-1}. For this improvement a digital state feedback controller (using the same sampling time) is to be used. What is the closed loop characteristic polynomial for the discrete system and what controller gains **K** are required if the feedback law is

$$u(k) = \mathbf{K}\mathbf{x}(k)$$

$$= [k_1 k_2] \begin{bmatrix} x_1 \\ x_2 \end{bmatrix}?$$

If the control input is given by this relation the closed loop discrete state equation becomes

$$\mathbf{x}(k+1) = \mathbf{P}\mathbf{x}(k) + \mathbf{Q}\mathbf{K}\mathbf{x}(k)$$

$$= (\mathbf{P} + \mathbf{Q}\mathbf{K})\mathbf{x}(k).$$

From the given damping and natural frequency we know that the continuous equivalent system equation is

$$\ddot{x} + 2 \times 1.2 \times 0.2\dot{x} + 0.04x = 0$$

i.e.

$$\ddot{x} + 0.48\dot{x} + 0.04x = 0$$

The roots of the characteristic equation are given by

$$\lambda^2 + 0.48\lambda + 0.04 = 0$$

i.e.

$$\lambda_{1,2} = -0.3727, \ -0.1073.$$

These are of course the eigenvalues of the new state space system matrix $\mathbf{A}_c$.

From equation (9.25) the equivalent discrete time values are

$$z_1 = e^{\lambda_1 T}$$

$$= 0.0241$$

$$z_2 = e^{\lambda_2 T}$$

$$= 0.3419$$

and the closed loop polynomial is

$$(z - 0.0241)(z - 0.3419) = 0$$
$$z^2 - 0.366z + 0.00824 = 0.$$

With the proposed state feedback this is equal to

$$|z\mathbf{I} - (\mathbf{P} + \mathbf{QK})| = 0.$$

Substituting for $\mathbf{P}$, $\mathbf{Q}$, and $\mathbf{K}$ gives

$$\begin{vmatrix} z - 0.5402 - 45.98k_1 & -8.415 - 45.98k_2 \\ 0.08415 - 8.415k_1 & z - 0.5402 - 8.415k_2 \end{vmatrix} = 0$$

i.e.

$$z^2 - (1.0804 - 8.415k_2 - 45.98k_1)z + (1 + 8.415k_2 - 45.98k_1) = 0.$$

Equating coefficients of z in these two equivalent equations leads to

$$k_1 = 0.018$$
$$k_2 = -0.016$$

for our discrete feedback gains. These low gains are in keeping with the comparatively high gain in the rest of the system.

Solution of the sampled-data state equation using the z-transform

Equation (9.16) is the state equation for the sampled data system and to determine $\mathbf{x}(kT)$ and/or $\mathbf{y}(kT)$ the solution of this is required. Using the z-transform $\mathscr{Z}$ with equation (9.16) yields,

$$z\mathbf{X}(z) - z\mathbf{x}(0) = \mathbf{PX}(z) + \mathbf{QU}(z) \tag{9.26}$$

where

$$\mathscr{Z}[\mathbf{x}(k+1)] = z \cdot \mathscr{Z}[\mathbf{x}(k)] - z \cdot \mathbf{x}(0)$$

by the shift theorem. The matrices $\mathbf{P}$ and $\mathbf{Q}$ are constants for a fixed T and $\mathbf{X}(z)$ is $\mathscr{Z}[\mathbf{x}(k)]$ and $\mathbf{U}(z)$ is $\mathscr{Z}[\mathbf{u}(k)]$. Rearranging equation (9.26) yields

$$(z\mathbf{I} - \mathbf{P})\mathbf{X}(z) = z\mathbf{x}(0) + \mathbf{QU}(z)$$

or

$$\mathbf{X}(z) = z(z\mathbf{I} - \mathbf{P})^{-1}\mathbf{x}(0) + (z\mathbf{I} - \mathbf{P})^{-1}\mathbf{QU}(z). \tag{9.27}$$

To obtain a pulse transfer function matrix relating the input and output vectors assume zero initial conditions so that $\mathbf{x}(0) = \mathbf{0}$ then

$$\mathbf{X}(z) = (z\mathbf{I} - \mathbf{P})^{-1}\mathbf{QU}(z) \tag{9.28}$$

and

$$\mathbf{Y}(z) = \mathbf{C}\mathbf{X}(z) + \mathbf{D}\mathbf{U}(z)$$
$$= [\mathbf{C}(z\mathbf{I} - \mathbf{P})^{-1}\mathbf{Q} + \mathbf{D}]\mathbf{U}(z). \tag{9.29}$$

The pulse transfer function matrix for the multi input-multi output system in the expression $\mathbf{Y}(z) = \mathbf{G}(z)\mathbf{U}(z)$ is thus

$$\mathbf{G}(z) = \mathbf{C}(z\mathbf{I} - \mathbf{P})^{-1}\mathbf{Q} + \mathbf{D}.$$

If $\mathbf{C} = \mathbf{I}$, $\mathbf{D} = \mathbf{0}$ then $\mathbf{Y}(z) = \mathbf{X}(z)$ and the pulse transfer function matrix is

$$\mathbf{G}(z) = (z\mathbf{I} - \mathbf{P})^{-1}\mathbf{Q}. \tag{9.30}$$

The full time solution with non-zero initial condition $\mathbf{x}(0)$ is obtained by inversion of equation (9.27) to yield

$$\mathbf{x}(k) = \mathscr{Z}^{-1}[z(z\mathbf{I} - \mathbf{P})^{-1}]\mathbf{x}(0) + \mathscr{Z}^{-1}[(z\mathbf{I} - \mathbf{P})^{-1}\mathbf{Q}\mathbf{U}(z)] \tag{9.31}$$

so for the unforced system $\mathbf{u} = \mathbf{0}$ and an initial state $\mathbf{x}(0)$

$$\mathbf{x}(k) = \mathscr{Z}^{-1}[z(z\mathbf{I} - \mathbf{P})^{-1}] \, . \, \mathbf{x}(0). \tag{9.32}$$

The characteristic equation is

$$|z\mathbf{I} - \mathbf{P}| = 0 \tag{9.33}$$

and again for stability all roots of this polynomial in z must lie within the unit circle of the z-plane and the values of the roots express the rate of response, i.e. the dynamics, of the system.

An advantage of difference and discrete time equations is that recursive methods may be used for their solution. This is particularly applicable to digital computer calculation. The refinements leading to rapid computation cover a wide field in themselves but the principle is that given $\mathbf{x}(0)$ and the control input values $\mathbf{u}(0)$, $\mathbf{u}(T), \ldots$ etc. any $\mathbf{x}(kT)$ may be evaluated recursively by

$$\mathbf{x}(1) = \mathbf{P}\mathbf{x}(0) + \mathbf{Q}\mathbf{u}(0)$$
$$\mathbf{x}(2) = \mathbf{P}\mathbf{x}(1) + \mathbf{Q}\mathbf{u}(1)$$
$$= \mathbf{P}\mathbf{P}\mathbf{x}(0) + \mathbf{P}\mathbf{Q}\mathbf{u}(0) + \mathbf{Q}\mathbf{u}(1).$$

Similarly

$$\mathbf{x}(3) = \mathbf{P}^3\mathbf{x}(0) + \mathbf{P}^2\mathbf{Q}\mathbf{u}(0) + \mathbf{P}\mathbf{Q}\mathbf{u}(1) + \mathbf{Q}\mathbf{u}(2)$$

so that the general term is

$$\mathbf{x}(k) = \mathbf{P}^k\mathbf{x}(0) + \sum_{j=0}^{j=k-1} \mathbf{P}^{k-i-1}\mathbf{Q}\mathbf{u}(j). \tag{9.34}$$

These terms are directly equivalent to the two terms of equation (9.31) so that given $\mathbf{P}$ and $\mathbf{Q}$ (possibly from equations (9.23) and (9.24)) $\mathbf{x}(kT)$ is determined without evaluation of inverse z-transforms.

9.4 Sampling intervals

During the previous parts of this chapter we have assumed a sampling interval T without comment as to how this interval is chosen. If T is very small then although close representation of the continuous signal is obtained the actual implementation may be more costly, e.g. channel sharing facilities by which more than one variable is transmitted along a single communicative channel may decrease and refined duplicated analytical instruments may be required. If T is too great then gross inaccuracies may occur and too much information about the true nature of the signal being sampled will be lost. It is the later point which is the main factor in the selection of the theoretical sampling interval T. For a full consideration of sampling the literature should be consulted (Shannon 1949).

9.5 Conventional control laws and digital control

To illustrate in an introductory manner the use of sampled systems with a digital controller the use of the P.I.D. algorithm is shown. This constitutes only the briefest of introductions to a wide field with its own range of literature and digital algorithms are not restricted to those like P.I.D., which have been used with analogue controllers. The use of the digital system in optimization, static and dynamic, and in general system management also goes far beyond the use of a computer to replace or supplement analogue controllers in direct digital control schemes.

Consider the single loop control system of Fig. 9.16 where the digital controller D receives a sampled signal and the digital output, also discrete, is transferred by a hold device (zero order) to the final control element which is incorporated as part of the general plant G_p. The two samplers are taken as synchronized with a time interval of sampling T. These, with the ideal zero-hold device, are idealizations which are not fully met in practice.

If the digital controller is to be equivalent to an analogue P.I.D.

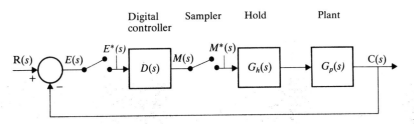

Fig. 9.16 Elements of a digital control system

controller the continuous differential equation is replaced by a difference equation, i.e. it is necessary to replace

$$m = k_p e + k_i \int e\, dt + k_d \frac{de}{dt}. \tag{4.17}$$

Writing this expression as a finite difference expression then after intervals of time T (from zero conditions)

$$m(k) = k_p e(k) + k_i \sum_{i=0}^{k} e_i T + k_d \frac{(e(k) - e(k-1))}{T}. \tag{9.35}$$

Also at the earlier time

$$m(k-1) = k_p e(k-1) + k_i \sum_{i=0}^{k-1} e_i T + k_d \frac{(e(k-1) - e(k-2))}{T}.$$

If the change in controller output m at time kT is $\Delta m(k)$ then
remembering that m is held constant over each interval of time T,

$$\Delta m(k) = m(k) - m(k-1)$$
$$= k_p(e(k) - e(k-1)) + k_i e(k) T$$
$$+ \frac{k_d}{T}(e(k) - 2e(k-1) + e(k-2)).$$

The error signal $e(k) = r(k) - c(k)$ so that if the input $r(k)$ is constant

$$\Delta m(k) = k_p(c(k-1) - c(k)) + k_i(r(k) - c(k)) T$$
$$+ \frac{k_d}{T}(-c(k) - c(k-2) + 2c(k-1)).$$

If the input $r(k)$ is constant it only appears in the integral action term.
The $\Delta m(k)$ will be the change made at kT at entry to the plant to effect
control action and will be calculated by the expression in the somputer.

Consider now the sampled-data system of Fig. 9.16. Then establishing the z-transform via the Laplace transforms

$$C(s) = G_p(s)G_h(s)M^*(s)$$
$$M(s) = D(s)E^*(s)$$
$$E(s) = R(s) - C(s).$$

Therefore

$$C(s) = G_p(s)G_h(s)D^*(s)(R^*(s) - C^*(s))$$

and

$$C(s) = \frac{G_p(s)G_h(s)D^*(s)R^*(s)}{1 + G_p(s)G_h(s)D^*(s)}.$$

The z-transform is then

$$C(z) = \frac{G_p G_h(z) D(z) R(z)}{1 + G_p G_h(z) D(z)}. \tag{9.36}$$

Note, that in comparison to the example of Fig. 9.14, the z-transform of the process is not now tied to the input form r and $C(z)/R(z)$ can be written independent of $r(t)$. To obtain $D(z)$, the digital controller z-transform, revert to equation (4.17) above with $P + I$ terms only (to avoid difficulties here with the derivative term). Then

$$D(s) = k_p + \frac{k_i}{s}$$
$$\mathcal{L}^{-1}[D(s)] = k_p \delta(t) + k_i$$

and

$$D(z) = k_p + k_i \frac{z}{z-1}.$$

This expression is used in the determination of $C(z)$ which is then inverted to give $c(t)$. If a required $C(z)/R(z)$ is known then the required $D(z)$ may be found by rearranging equation (9.36).

For further details of digital control systems one is referred to the literature (e.g. Savas 1965, Tou 1959; see also reference under Shannon).

9.6 Summary

In a sampled-data system at least one sampler is used. This produces a signal of discrete values of the sampled variable by modulation of an original continuous signal with a series of impulses at intervals of time, T, the sampling period. To describe the system behaviour the z-transform is introduced which allows manipulation of the sampled-data system is a manner similar to the use of the Laplace transform with continuous systems. A table of z-transforms and functions of time is established and use of this table and the Laplace transform table would allow a direct transformation to be made between the two transforms. This would eliminate (in some standard cases) the need to produce the z-transform via the inversion of the Laplace transform. The pulse transfer function plays an equivalent role to the continuous system transfer function and is expressed in terms of the complex variable z, instead of the s of the transfer function. Within feedback systems in particular the positioning of the sampler significantly affects the system output and the form of the overall pulse transfer function of the system.

The concept of stability becomes even more important in sampled-data systems than in continuous systems, e.g. a second order sampled data feedback system may exhibit instability which we know does not occur for continuous minimum phase second order systems with negative feedback. As well as single input–single output systems higher order systems may be of sampled-data form and the state equations then yield solutions giving the system state and output at only the discrete times 0, T, $2T, \ldots$. It is important that the sampling interval T satisfies the requirements for retaining all required information about the sampled signal. Shannon's theorem is basic to this requirement and this may be established via the Fourier transform treatment of the original and modulated signals.

Within digital control systems sampling plays an important part as the digital controllers require discrete time inputs and produce discrete time outputs. The control laws may be based either on the analogue P.I.D. controllers or on additional control modes.

Part 4

Control – C – Introduction to optimal and stochastic control systems

Chapter 10
Optimal systems

10.1 Introduction

At the level at which this book is aimed only an introduction to optimal systems is intended. Unfortunately the meaningful introduction of the subject may in itself occupy more space than one would wish for in a generally introductory text covering the field of modelling general dynamics and control. Within the study of optimal control it is necessary to specify a performance criterion, to consider continuous and sampled systems and to use the various methods for optimization etc. This chapter will tend to state conclusions, give brief outlines, and illustrate results with simple examples. It is hoped that in this way the theoretical bases may first be compared and then the methods in applications.

The aim of an optimal policy of operation may be expressed in a 'performance criterion' which is a scalar measure of performance but which may be a function of all the state variables and control variables of the system. The 'optimal policy' will be that which causes the function to be either a maximum or a minimum, normally under conditions which enforce constraints on the control, and state, variables. The selection of the performance criterion, or index, is thus basic to the concept of optimal control. The problem of synthesizing controls or of improving design on this basic index may then be undertaken in various ways which will be outlined.

10.2 The performance index

Within the control problem we are concerned mainly with reducing any error which exists between a desired state, which may either be constant or changing as a function of time, and the actual state of the system. Our 'state' which is being looked at may be incomplete in that it is solely the output of the system or process, or it may be more fully defined in terms of a fuller state vector. Some discussion of the choice of performance index for single input–single output systems has been given in section 6.2, where various integrals of the error were given as being the basis of determining the quality of control, e.g. the integral of the absolute error $\int |e| \, dt$, the integral of the square of the error $\int e^2 \, dt$, and there are various others. These performance indices are all similar in that no penalty has been included for the cost of the control itself. Is it really worth higher control costs to yield only a small annual increase in income as a result of error reduction? Also, they are all expressed in terms of an error so that looking at the more general case the important state variable quantity will be $\mathbf{x}(t)_{desired} - \mathbf{x}(t)$. By a linear transformation this deviation can itself be expressed as the state variable and this simplifies the subsequent problems (see below).

For a given system, i.e. one which we can describe by input–output or state equations, and one on which certain constraints are placed the performance index J can be given the general form

$$J = \int_0^t F(\mathbf{x}, \mathbf{u}) \, dt \tag{10.1}$$

where $F(\mathbf{x}, \mathbf{u})$ is the 'cost function'. This function has to be predetermined for a given optimization problem. The control vector $\mathbf{u}$ may depend on the initial state $\mathbf{x}(0)$ and the state $\mathbf{x}(t)$ but what we are really seeking is a $\mathbf{u}(t)$ which minimizes (or maximizes) the scalar J. Throughout we shall consider J to be a cost function for which we seek to obtain a minimum. If a problem is formulated in such a way that a maximum is sought then a J may easily be determined (e.g. by a change of sign) so that our minimization arguments still hold since

$$\max J = \min (-J).$$

The principal form of the performance index is the quadratic performance index. In this the cost function is a sum of weightings of the squares of the individual state variables and of the control variables. This leads to particularly suitable mathematical treatment as well as satisfying the qualitative consideration of penalizing both large deviations from the desired state and also increased control actions more severely, i.e.

$$J = \int_0^t (\mathbf{x}^T \mathbf{Q} \mathbf{x} + \mathbf{u}^T \mathbf{R} \mathbf{u}) \, dt. \tag{10.2}$$

If the weighting matrices $\mathbf{Q}$ and $\mathbf{R}$ are diagonal

$$J = q_{11} x_1^2 + q_{22} x_2^2 + \ldots q_{nn} x_n^2 + r_{11} u_1^2 + \ldots r_{pp} u_p^2.$$

Since we are 'penalizing' the values of the state variables $\mathbf{x}$ it is important that they are correctly defined as we are really seeking to reduce the errors by applying a penalty to the errors. As before we see that the state variables defined in terms of deviations from a desired state satisfy this requirement. For example, if the system equation is the linear expression.

$$\dot{\mathbf{x}} = \mathbf{A}\mathbf{x} + \mathbf{B}\mathbf{u}$$

and $\mathbf{x}_{desired}$ is the desired constant state and a control $\mathbf{u}$ is required to keep the system at that state, the error vector $\mathbf{e}$ is $\mathbf{x} - \mathbf{x}_{desired}$ (or we could use $\mathbf{e} = \mathbf{x}_{desired} - \mathbf{x}$) and for invariant $\mathbf{A}$ and $\mathbf{B}$,

$$\dot{\mathbf{e}} = \dot{\mathbf{x}} - \dot{\mathbf{x}}_{desired}$$
$$= \mathbf{A}(\mathbf{x} - \mathbf{x}_{desired}) + \mathbf{B}(\mathbf{u} - \mathbf{u}_{desired}).$$

In the desired state is $\dot{\mathbf{x}}_{desired}$ is $\mathbf{0}$ (regulator problem)

$$\mathbf{A}\mathbf{x}_{desired} + \mathbf{B}\mathbf{u}_{desired} = \mathbf{0}.$$

The matrix $\mathbf{B}$ is not necessarily square so premultiply this last equation by $\mathbf{B}^T$ and then

$$\mathbf{u}_{desired} = -(\mathbf{B}^T\mathbf{B})^{-1}\mathbf{B}^T\mathbf{A}\mathbf{x}_{desired}$$

in this case, provided $\mathbf{B}$ is of maximal rank.

Define the new $\mathbf{u}^*$ control vector $\mathbf{u}^* = \mathbf{u} - \mathbf{u}_{desired}$ then

$$\dot{\mathbf{e}} = \mathbf{A}\mathbf{e} + \mathbf{B}\mathbf{u}^*.$$

Now define our new state and control values by letting $\mathbf{x} = \mathbf{e}$, $\mathbf{u} = \mathbf{u}^*$ in our formulation of the problem, i.e. in

$$\dot{\mathbf{x}} = \mathbf{A}\mathbf{x} + \mathbf{B}\mathbf{u}$$

then J (equation 10.2) carries out the task of penalizing error and control movements and $\mathbf{u}$ (i.e. $\mathbf{u}^*$) represents the least squares solution.

Note that the quadratic performance index is not by any means the only index we could choose and a linear weighting, for example, could also be chosen, but this is not necessarily so convenient. Another performance index which we shall treat in some detail, is the time-optimal criterion where the time to return the system to its desired state following a deviation is minimized. Also, although in setting up the performance criteria we hope to have used the correct weightings, these can be changed with experience of the results. With the pre-scribed function J the control system, although optimal in the analytical sense, may be such that its implementation cannot in fact be justified or even physically carried out if the resulting $\mathbf{u}(t)$ is particularly complicated. A system less than 'optimal' may then in fact be used but the optimal control law and the calculation or simulation of the system under the action of such a law sets a standard for comparison of other 'sub-optimal' controllers.

10.3 Methods for determining the optimal-control law

Within this section we shall look at the major methods for determining the $u(t)$ which satisfies the performance index, assuming of course that the conditions of observability and controllability (Chapter 7) are satisfied. These methods are all limited in practice by the order and complexity of the system in which they can be successfully used and the selection of method will be influenced by each particular problem.

Calculus of variations

Although the calculus of variations has given way largely to dynamic programming and the maximum principle we shall deal with it briefly, (see e.g. Athans and Falb 1966; Hesters 1966). The calculus of variations enables one to determine the minimum (or maximum) value of a functional and the values of the function which minimize it, e.g. if the functional is J in equation (10.1) we can determine the effect that variations in $u(t)$ have on it, determine the minimum of J as $u(t)$ is varied, and also determine the time dependent control $u_0(t)$ causing this minimum J_0. The J is a functional in that it is a function having a numerical value and whose independent variable is a function which is itself variable.

The optimizations which may be dealt with by the calculus of variations may conveniently be divided into those where (i) the end-point, i.e. $x(t_f)$ is fixed at some specified time t_f and (ii) those for which $x(t_f)$ is not specified, i.e. the end-point is free. In both cases we require, for a given state equation and initial conditions $x(0)$, to minimize J over the time 0 to t_f (or in general terms possibly t_1 to t_2). The free end point case is more complex and we shall restrict ourselves to finding the optimal control $u_0(t)$ to take the system from the state $x(0)$ at t is zero to the final specified state $x(t_f)$ so as to minimize the functional J where

$$\dot{\mathbf{x}} = \mathbf{f}(\mathbf{x}, \mathbf{u})$$

and

$$J = \int_0^{t_f} F(\mathbf{x}, \mathbf{u})\, dt$$

We shall also limit our considerations to the scalar case within the fixed end-point classification, i.e.

$$\dot{x} = f(x, u)$$

$$J = \int_0^{t_f} F(x, u)\, dt.$$

The control $u_0(t)$ denotes the control function $u(t)$ giving the minimum J, i.e. it is the optimal $u(t)$.

In addition to the problems of fixed- and free-end states there may be constraints placed on the magnitude of the control. We shall consider the controls unconstrained but if large control values are undesired then they may be weighted heavily and a squared power of $u(t)$ used in the cost function F, for example. Bearing these restrictions in mind it is possible to proceed to a solution of the problem.

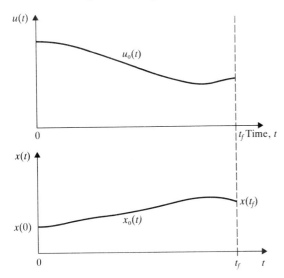

Fig. 10.1 Optimal control action $u_0(t)$ and the change of state $x_0(t)$ between the initial state $x(0)$ and final set value $x(t_f)$

Assume that the optimal control $u_0(t)$ does exist and has the form shown in Fig. 10.1. This $u_0(t)$ minimizes the integral J and the controlled path between the two end points $x(0)$ and $x(t_f)$ is that labelled $x_0(t)$. If $u_0(t)$ is the optimal path then

$$J_0 = \int_0^{t_f} F(x_0(t), u_0(t)) \, dt$$

$$= \min_u J$$

i.e. substitution of $x_0(t)$ and $u_0(t)$ into F gives the minimum of J compared with that resulting from all other possible controls $u(t)$. If $x(t)$ does not follow the optimal trajectory but deviates from it by arbitrarily small qualities $\delta x(t)$, where the deviation varies with time, and the control path varies from $u_0(t)$ also by, say, $\delta u(t)$ (where $\delta u(t)$ is not equal to $\delta x(t)$), then this situation may be shown by Fig. 10.2.

The new control and state trajectories are

$$u(t) = u_0(t) + \delta u(t)$$

$$x(t) = x_0(t) + \delta x(t)$$

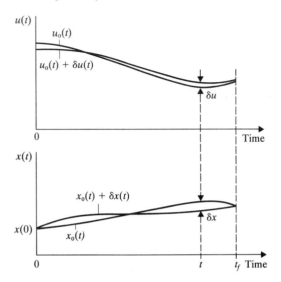

Fig. 10.2 Effect of deviations from the optimal control path

but $x(t)$ is still subject to the initial and final conditions $x(0) = x_0(0)$ and $x(t_f) = x_0(t_f)$ so that $\delta x(0) = \delta x(t_f) = 0$. For a given state equation, $\dot{x} = f(x, u)$, $\delta u(t)$ and $\delta x(t)$ are not independent and $\delta u(t)$ cannot follow just any path but must vary with variations $\delta x(t)$ so that the state equation is still satisfied.

The variation in the performance index J is given by the difference in integrals over the time zero to t_f, i.e.

$$\Delta J = J - J_0$$
$$= \int_0^{t_f} F(x_0 + \delta x, u_0 + \delta u) \, dt - \int_0^{t_f} F(x_0, u_0) \, dt$$
$$= \int_0^{t_f} [F(x_0 + \delta x, u_0 + \delta u) - F(x_0, u_0)] \, dt.$$

The increment ΔJ will be greater than (or equal to) zero. For small deviations δx, δu the Taylor expansion may be used for $F(x_0 + \delta x, u_0 + \delta u)$, i.e.

$$F(x_0 + \delta x, u_0 + \delta u) = F(x_0, u_0) + \frac{\partial F(x, u)}{\partial x}\bigg|_{x_0, u_0} . \, \delta x$$

$$+ \frac{\partial F(x, u)}{\partial u}\bigg|_{x_0, u_0} . \, \delta u + \frac{1}{2!} \frac{\partial^2 F(x, u)}{\partial x^2}\bigg|_{x_0, u_0} . \, (\delta x)^2 + \dots .$$

Taking the first order terms, ΔJ becomes, on making the substraction,

$$\Delta J = \int_0^{t_f} \left[\frac{\partial F(x, u)}{\partial x} \bigg|_{x_0, u_0} \cdot \delta x + \frac{\partial F(x, u)}{\partial u} \bigg|_{x_0, u_0} \cdot \delta u \right] dt.$$

To progress beyond this point it is necessary to be more specific either over the form of F or in the relationship f between x and u which has so far not been used. If we take the relationship

$$\dot{x} = f(x, u)$$

as the state equation then we may proceed in two ways. The control u may be eliminated using this expression from the cost function F, so that F is replaced by $F'(x, \dot{x})$ in place of $F(x, u)$ and in place of deviations in δu and δx the argument is developed in terms of $\delta \dot{x}$ and δx. Alternatively we may use the state equation and retain u explicitly so that for a non-optimal path,

$$\dot{x} = \dot{x}_0 + \delta \dot{x}$$

$$= f(x_0, u_0) + \frac{\partial f(x, u)}{\partial x} \bigg|_{x_0, u_0} \cdot \delta x + \frac{\partial f(x, u)}{\partial u} \bigg|_{x_0, u_0} \cdot \delta u$$

or in simplifying the notation

$$\delta \dot{x} = \frac{\partial f}{\partial x} \bigg|_0 \delta x + \frac{\partial f}{\partial u} \bigg|_0 \delta u.$$

Writing $F(x, u)$ as F similarly and using this expression to substitute for δu,

$$\Delta J = \int_0^{t_f} \left[\frac{\partial F}{\partial x} \bigg|_0 \cdot \delta x + \frac{\partial F}{\partial u} \bigg|_0 \cdot \delta u \right] dt$$

$$= \int_0^{t_f} \left[\left\{ \frac{\partial F}{\partial x} \bigg|_0 - \frac{\partial F}{\partial u} \bigg|_0 \frac{(\partial f/\partial x)_0}{(\partial f/\partial u)_0} \right\} \delta x + \frac{(\partial F/\partial u)_0}{(\partial f/\partial u)_0} \cdot \delta \dot{x} \right] dt. \qquad (10.3)$$

The last term can be integrated by parts since

$$\int_0^{t_f} \frac{(\partial F/\partial u)_0}{(\partial f/\partial u)_0} \delta \dot{x} \, dt = \left[\frac{(\partial F/\partial u)_0}{(\partial f/\partial u)_0} \delta x \right]_{t=0}^{t_f} - \int_0^{t_f} \delta x \cdot \frac{d}{dt} \left[\frac{(\partial F/\partial u)_0}{(\partial f/\partial u)_0} \right] dt.$$

At $t = 0$ and $t = t_f$, $\delta x = 0$ (by problem definition) and the first term of this integral is zero. ΔJ may then be written

$$\Delta J = \int_0^{t_f} \left[\frac{\partial F}{\partial x} \bigg|_0 - \frac{\partial F}{\partial u} \bigg|_0 \cdot \frac{(\partial f/\partial x)_0}{(\partial f/\partial u)_0} - \frac{d}{dt} \left\{ \frac{(\partial F/\partial u)_0}{(\partial f/\partial u)_0} \right\} \right] \delta x \, dt.$$

If J has a true minimum then if we have *small* changes δx from the optimal trajectory ΔJ will be zero, $(\partial J/\partial x)_0 = 0$. For ΔJ to be zero for any arbitrary small $\delta x(t)$ the factor multiplying δx in the last integral must be zero so that

$$\frac{\partial F}{\partial x} - \frac{\partial F}{\partial u} \cdot \frac{\partial f/\partial x}{\partial f/\partial u} - \frac{d}{dt}\left[\frac{\partial F/\partial u}{\partial f/\partial u}\right] = 0 \qquad (10.4)$$

where all integrals in its solution are evaluated on the optimal control and state trajectories. That is, for a given $F(x, u)$ and $f(x, u)$, solution of this equation gives the optimal $u(t)$ for the prescribed cost function F. Substitution of this optimal $u(t)$ into F will give the minimum performance index values J_0. Equation (10.4) is one form of an Euler or Euler–Lagrange equation. One might wonder on inspection of equation (10.4) if we have actually made much progress towards obtaining a solution. In fact we can only hope for analytic solutions (as compared to numerical integration and solution) when the state function f is linear and the cost function F is quadratic in x and u. For example, if

$$\dot{x} = ax + bu, \quad \text{i.e.} \quad f(x, u) = ax + bu$$

and

$$F = \alpha x^2 + \beta u^2$$

then

$$\frac{\partial F}{\partial x} = 2\alpha x$$

$$\frac{\partial F}{\partial u} = 2\beta u$$

$$\frac{\partial f}{\partial x} = a$$

$$\frac{\partial f}{\partial u} = b$$

and equation (10.4) is

$$2\alpha x - 2\frac{a}{b}\beta u - \frac{d}{dt}\left(2\frac{\beta}{b}u\right) = 0.$$

This is a linear differential equation in the two variables x and u and with the two boundary conditions $x(0)$ and $x(t_f)$ and the state equation it is capable of analytic solution.

Example Using the calculus of variations determine the optimal control $u_0(t)$ to take a process between two known states $x(0) = 0$ and $x(t_f)$ at time t_f. The state equation is

$$\dot{x} = f(x, u) = ax + u,$$

the cost function is

$$F(x, u) = x^2 + u^2$$

and we wish to minimize $J = \int_0^{t_f} F(x, u) \, dt$.

Starting from the Euler–Lagrange equation, (10.4), we have

$$\frac{\partial F}{\partial x} - \frac{\partial F}{\partial u} \cdot \frac{\partial f / \partial x}{\partial f / \partial u} - \frac{d}{dt}\left[\frac{\partial F / \partial u}{\partial f / \partial u}\right] = 0.$$

For the given f, F,

$$\frac{\partial F}{\partial x} = 2x \qquad \frac{\partial f}{\partial x} = a$$

$$\frac{\partial F}{\partial u} = 2u \qquad \frac{\partial f}{\partial u} = 1$$

so that equation (10.4) becomes

$$2x - 2u \cdot a - 2\dot{u} = 0. \tag{10.5}$$

The state equation is

$$\dot{x} - ax - u = 0 \tag{10.6}$$

and elimination of x or u yields a second-order differential equation in the remaining variable. Solution may be effected in two ways:

(i) elimination of u using the state equation gives

$$\ddot{x} - (1 + a^2)x = 0$$

with the solution

$$x = C_1 \exp(\sqrt{1 + a^2} \cdot t) + C_2 \exp(-\sqrt{1 + a^2} \cdot t).$$

At $t = 0$, $x = x(0) = 0$ and at $t = t_f$, $x = x(t_f)$. Substitution of these values in the general solution for x gives the values of C_1 and C_2 so that

$$C_1 = \frac{x(t_f)}{2 \sinh(\sqrt{(1 + a^2)} \cdot t_f)}$$

$$C_2 = \frac{-x(t_f)}{2 \sinh(\sqrt{(1 + a^2)} \cdot t_f)}$$

and

$$x = \frac{x(t_f)}{2 \sinh(\sqrt{(1 + a^2)} \cdot t_f)}\{\exp(\sqrt{(1 + a^2)} \cdot t) - \exp(-\sqrt{(1 + a^2)} \cdot t)\}$$

$$= \frac{x(t_f)}{\sinh(\sqrt{(1 + a^2)} \cdot t_f)} \sinh(\sqrt{(1 + a^2)} \cdot t).$$

The optimal control law is then given by

$$u_0 = \dot{x} - ax$$

$$= \frac{x(t_f)}{\sinh(\sqrt{(1+a^2)} \cdot t_f)} \{\sqrt{(1+a^2)} \cdot \cosh(\sqrt{(1+a^2)} \cdot t)$$

$$- a \sinh(\sqrt{(1+a^2)} \cdot t)\}.$$

(ii) Alternatively we may work using Laplace transforms so that on taking transforms of equation 10.5 and 10.6

$$X(s) - aU(s) - [sU(s) - u(0)] = 0$$

and

$$[sX(s) - x(0)] - aX(s) - U(s) = 0$$

In this case $x(0) = 0$ but $u(0)$ is unknown at present. Eliminating $U(s)$ gives

$$sX(s) - aX(s) - \left[\frac{X(s) + u(0)}{s + a}\right] = 0$$

$$[s(s+a) - (s+a)a - 1]X(s) = u(0)$$

i.e.

$$[s^2 - (1 + a^2)]X(s) = u(0)$$

$$X(s) = \frac{u(0)}{s^2 - (1 + a^2)}$$

so

$$x(t) = u(0)\frac{\sinh(\sqrt{(1+a^2)} \cdot t)}{\sqrt{(1+a^2)}}. \tag{10.7}$$

We have already used the boundary condition $x(0) = 0$ so use the boundary condition $x = x(t_f)$ at $t = t_f$ to get $u(0)$, i.e.

$$x(t_f) = u(0)\frac{\sinh(\sqrt{(1+a^2)} \cdot t_f)}{\sqrt{(1+a^2)}}.$$

Use $u(0)$ from this equation in (10.7) and hence

$$x(t) = \frac{x(t_f)\sinh(\sqrt{(1+a^2)} \cdot t)}{\sinh(\sqrt{(1+a^2)} \cdot t_f)}$$

agreeing as expected with $x(t)$ in (i). The u_0, the optimal control, is then found as above once more or possibly through the transform equations.

Dynamic programming

We shall continue by consideration of a single variable system in introducing Bellman's dynamic programming method. The actual working techniques will vary with the complexity of the problem, available computing facilities, and the form of the cost function. In this section consideration is restricted to a representation which illustrates the basis for the use of dynamic programming in determining an optimal control.

The fundamental principle of the dynamic programming method is the principle of optimality (Bellman 1961, 1962; Lapidus and Luus 1967).

'An optimal policy (of control) has the property that whatever the initial state and the initial decision the remaining decisions must form an optimal policy with regard to the state resulting from the first decision.' As the principle is stated in terms of sequential decisions it is convenient to consider the time and state as being discretized and especially as computation will normally require a digital computer we lose nothing in generality from the continuous case. For the single independent variable x, the system state, a grid is constructed (Fig. 10.3), comprised of discrete values of x against discrete equal intervals of time. At zero time some initial state $x(0)$ is specified. To pass from the initial state to the final state a series of decisions must be taken to determine the control to be used over each interval of time. The basis for the decision taken is again a performance index. For a continuous process this is as before

$$J = \int_0^{t_f} F(x, u) \, dt$$

and the discrete time equivalent is

$$J = \sum_{k=1}^{N} F(x(k), u(k-1))$$

the sum of incremental cost functions, when N is the number of time intervals. Each time interval is t_f/N. Consider also that the control is unconstrained, i.e. there is no limit on the value which it may take in enabling us to go from one value of x at one time instant to another value after one time increment. (In fact the presence of constraints may considerably reduce the computation.) If the range of values of x is discretized so that x may take any one of M discrete values then after one interval of time the x value may be changed by our control $u(0)$ to any one of M values, and for each new value of x and of u the 'incremental cost function' $F(x(1), u(0))$ will have a value, dependent both on the control $u(0)$ and the new state $x(1)$. In the second interval of time x may be moved to any M values again, using a control $u(1)$ and once again there are M possible values of $x(2)$ and of $F(x(2),$

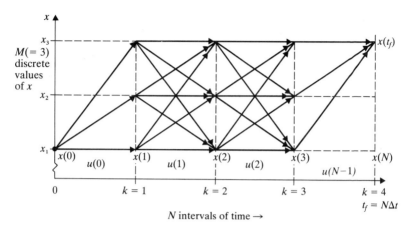

Fig. 10.3   Alternative 'trajectories' for moving in four stages from $x(0) = x_1$ to $x(t_f) = x_3$ using three discrete values of x

$u(1)$). When the beginning of the last time interval is reached, $k = N - 1$, and provided the end state is specified, $x(t_f)$, there is just the one choice of end-state over that increment of time. If the end-state is not specified then there are again M choices. The step-wise process is illustrated in Fig. 10.3, where x can take any one of three values at each interval of time and the total time is divided into four intervals. The initial state $x(0)$ is x_1, and the final state $x(t_f)$ is x_3 and these are specified. To determine which of the routes gives the best (minimum) performance index J all alternative sums

$$\sum_{k=1}^{N} F(x(k), \, u(k-1))$$

could be evaluated and the best then selected. [Note that with specified $x(0)$, $x(t_f)$ the alternative summation $\sum_{k=0}^{N-1} F(x(k), u(k))$ could be used.]

From the argument above it can be seen that for the example the total number of different J would be $3.3.3 = 27$, this being the number of different ways of proceeding from $x(0)$ to $x(4)$, e.g. by the path $x_1 \rightarrow x_3 \rightarrow x_1 \rightarrow x_2 \rightarrow x_3$. For N time intervals and M discrete values of x the general expression for the number of paths, and hence the number of values of the performance index, is M^{N-1}. For n instead of one state variable the value rises to $M^{n(N-1)}$. For two state variables, with $M = 10$, $N = 10$ the number of alternatives is 10^{18}. Even then this is for a fixed end state, $x(t_f)$. Thus, except for very simple cases as depicted, the evaluation and selection from all overall alternatives in order to select the optimal trajectory is just not a feasible proposition. Relief comes in the application of the principle of optimality and the use of dynamic programming.

The use of dynamic programming may be possible in two ways. Having discretized the time and state variable scales we can apply the principle of optimality in working back from the final state to the initial state. This may be termed the normal dynamic programming algorithm. The overall optimal forward path through the stages then leads to stipulation of the control sequence to give the minimum performance index. The alternative approach is to make just one pass through the stages in the forward direction to achieve the same result. This has been termed the dual of the normal dynamic programming method (Dreyfus 1965; Aris, Nemhauser and Wilde 1964). Depending on the specific case in question there are advantages and disadvantages to each in comparison with the other, and in simple cases of the order which can be illustrated by hand calculation there is probably little to choose between them. The two pass, backward moving first, method directly yields the continuous form of the optimal dynamic programming method invoking the principle of optimality so this will be illustrated in more detail.

(i) 'Backwards' method ('normal' method)

Let us base the derivation of the normal method on the scalar system whose discrete state equation (see Ch. 9) is

$$x(k+1) = f(x(k), u(k)) \qquad k = 0, 1, \ldots N$$

with the initial condition $x(0)$. At present the final state $x(N)$ need not be specified but the performance index is

$$J(x(0), N) = \sum_{k=1}^{N} F(x(k), u(k-1))$$

where the notation $J(x(0), N)$ signifies the performance index evaluated over the whole N time states of the process starting at $x(0)$. For any specified starting condition, $x(0)$ need not be included in the summation. As distinct from the introductory example above all possible end states are now to be considered, i.e. this is the general free end-state problem. The state $x(k)$ arises from the effect of control $u(k-1)$ on the state $x(k-1)$. Denote the optimal of the performance index by J_0 then

$$J_0(x(0), N) = \min_{u(k)} \sum_{k=1}^{N} F(x(k), u(k-1))$$

where min means that the sum is minimized with respect to the full
$u(k)$
sequence of the control function $u(k)$. (As before $u(k)$ is used for $u(kT)$ etc. where T is the duration of a time interval.) Although the process starts at $x(0)$ the final state in dynamic programming is referred to as the 'origin' as computation starts here. Consider the final $n(\leq N)$ stages of the process, i.e. those stages from $k = N-n$ to

$k = N$, then if there is optimal control over these last n stages,

$$J_0(x(N-n), n) = \min_{u(k)} \sum_{k=N-n+1}^{N} F(x(k), u(k-1)).$$

(Note that the n in $J_0(x, n)$ refers to the number of stages over which J_0 is evaluated.) This summation may be written in two parts:

$$J_0(x(N-n), n) = \lim_{u(k)} \left\{ F(x(N-n+1), u(N-n)) \right.$$

$$\left. + \sum_{k=N-n+2}^{N} F(x(k), u(k-1)) \right\},$$

i.e. one term is taken separately from the summation. By direct invoking of the principle of optimality we know that for $J_0(x(N-n), n)$ to be optimal then whatever the first decision, $u(N-n)$, the remaining policy must be optimal with regard to the resulting state $x(N-n+1)$. Thus, looking at the last equation,

$$J_0(x(N-n), n) = \min_{u(N-n)} \{F(x(N-n+1), u(N-n))\}$$

$$+ J_0(x(N-n+1), (n-1)). \quad (10.8)$$

Thus to ensure optimal operation over the last n stages the sequence of the last $(n-1)$ stages must be operated optimally. Operation of the first of the last n stages in an optimal manner is ensured by making $F(x(N-n+1), u(N-n))$ a minimum by minimizing this with respect to the control $u(N-n)$. This control is then applied to the state $x(N-n)$. Then, from the state equation.

$$x(N-n+1) = f(x(N-n), u(N-n)) \quad (10.9)$$

and for the final single state

$$J_0(x(N-1), 1) = \min_{u(N-1)} \{F(x(N), u(N-1))\} \quad (10.10)$$

Equations (10.10), (10.9) and (10.8) provide a set of recursive relationships which enable the sequence of $J_0(x(N-1), 1)$, $J_0(x(N-2), 2)$, ... $J_0(x(0), N)$ and the optimizing control sequence $u(N-1), \ldots, u(0)$ to be established. Thus, instead of the simultaneous determination of the N-stage control to minimize $J_0(x(0), N)$, each of the N stages is looked at in turn, starting at the last (the 'origin'). Following this we may move forward through the process applying $u(0)$ to $x(0)$, then $u(1)$ to $x(1)$ etc.

Assuming again that the discretized values of x only are given by the available control consider the situation shown in Fig. 10.4 where x_1, etc. are the possible discrete values of x and u_1 etc. are the possible P values of u. If $x(N)$ is 'free' then consider first that x has the value

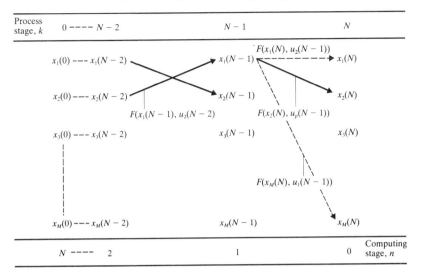

Fig. 10.4 Optimal control determination, showing stage optimal paths

$x_1(N-1)$ at the stage $(N-1)$. Apply to this in turn all the possible control values $u_1(N-1)$, $u_2(N-1)$ etc. so that, for example, application of u_2 gives $x_1(N)$, u_P gives $x_2(N)$ and so on to u_1 giving $x_M(N)$. (Not all $x_i(N)$ may in fact be given as results of this action.) In each case the performance index $J(x(N), 1)$ is calculated for this step and then the control giving the minimum performance index for this step is selected out, e.g. it may be that $[J_0(x(N-1), 1)]_{x_1} = F(x_2(N), u_P(N-1))$. The same procedure is repeated for all $x(N-1)$, i.e. $x_2(N-1), \ldots x_M(N-1)$ and in each case the minimum performance index for the last step is selected, $[J_0(x(N-1), 1)]_{x_i}$. We now have all the optimal paths leaving $x_i(N-1)$, $i = 1, \ldots M$.

Step backwards in the process now to stage $N-2$ and for all discrete possible values of $x(N-2)$, i.e. $x_i(N-2)$, repeat this procedure but going now from $x(N-2)$ to values of $x(N-1)$. From each $x_i(N-2)$ there will be a single optimal path characterized by its cost function $F(x_i(N-1), u_j(N-2))$. We forget all other paths now, i.e. except the stage optimal paths, and hence determine $J(x(N-2), 2)$ for each of the retained paths starting at stage $N-2$ and finishing at stage N. In Fig. 10.4 for example, for a stage optimal path originating at $x_2(N-2)$, the J for this path finishing at stage N would be

$$[J(x(N-2), 2)]_{x_2} = F(x_1(N-1), u_2(N-2)) + F(x_2(N), u_P(N-1))$$

and we could, by comparing all M such paths, determine the $J_0(x(N-2), 2)$ and the required two-stage control for the optimal path from stage $N-2$ to N alone. However, we now continue further back into the process in stages until stage 0 and the value $x_i(0)$ are reached. The full

$J_0(x(0), N)$ can then be determined. If an $x_i(0)$ is specified as the initial condition then the optimal path to $x(1)$ and the control $u_0(0)$ are selected, giving $x_i(1)$ and hence in turn $u_i(1)$, $x_i(2)$ etc. and the whole trajectory which is optimal may be obtained. If the final state had also been specified then only those paths leading to that state, e.g. $x_2(N)$ would be considered as possible in the last stage and the rest of the procedure would be the same. Note that at each stage, one retains the knowledge of the backward evaluated cumulative stage optima (see example below).

In practice the available discrete control actions may give rise to values of x which fall between the tabulated discrete values of Fig. 10.4. Interpolation between these values may then be necessary. If the state x is not scalar but is comprised of a number of variables, i.e. there is a state vector $\mathbf{x}$, then interpolation becomes very demanding. With an increase in state variable number the basic selection of the optimal paths is also greatly complicated, the computational requirements increasing as the power of the state vector order and a third order problem may not even be soluble by this method because of the computational demands made. By hand the problems are quite intractable.

Constraints on the control and state variables reduce the size of the problem since they reduce the number of computations necessary at each stage, or severe penalties may be imposed on constraints within the cost function. In this sense an apparent complication may lead to a reduction in computational difficulties.

The continuous form of the problem is that

$$\dot{x} = f(x, u) \tag{10.11}$$

and

$$J(x(0), t_f) = \int_0^{t_f} F(x, u, t) \, dt. \tag{10.12}$$

The time interval 0 to t_f may be divided into the two 'stages', $t = 0$ to $t = \Delta t$ and $t = \Delta t$ to $t = t_f$ so that

$$J = \int_0^{\Delta t} F(x, u, t) \, dt + \int_{\Delta t}^{t_f} F(x, u, t) \, dt.$$

By use of the principle of optimality a continuous form of the dynamic programming problem may be expressed (see Douglas 1972; Koppel 1968). A major feature of course is that the continuous system equations expressed here may be expressed in discrete terms and the discrete dynamic programming used. However, let us look at this continuous formulation. If J is the optimal J_0 then

$$J_0(x(0), t_f) = \min_{u(t)} \left[\int_0^{\Delta t} F(x, u, t) \, dt + \int_{\Delta t}^{t_f} F(x, u, t) \, dt \right].$$

For the overall control policy to be optimal the second integral must also constitute an optimal policy from $t = \Delta t$ to $t = t_f$ so that

$$J_0(x(0), t_f) = \min_{u(t)} \left[\int_0^{\Delta t} F(x, u, t) \, dt + J_0(x(\Delta t), t_f - \Delta t) \right]. \qquad (10.13)$$

If Δt is a very small increase from the initial time then the Taylor expansion may be used so that for J_0 a function of x and t, and x a function of t,

$$J_0(x(\Delta t), t_f - \Delta t) = J_0(x(0), t_f) + \frac{\partial J_0}{\partial x} \cdot \Delta x + \frac{\partial J_0}{\partial t} \cdot \Delta t + \ldots$$

$$= J_0(x(0), t_f) + \frac{\partial J_0}{\partial x} \frac{dx}{dt} \Delta t + \frac{\partial J_0}{\partial t} \Delta t + \ldots .$$

But if $\dfrac{dx}{dt} = f(x, u)$,

$$J_0(x(\Delta t), t_f - \Delta t) = J_0(x(0), t_f) + \frac{\partial J_0}{\partial x} f(x, u) \, \Delta t + \frac{\partial J_0}{\partial t} \Delta t$$

and hence from substituting this in equation (10.13)

$$J_0(x(0), t_f)$$
$$= \min_{u(t)} \left[\int_0^{\Delta t} F(x, u, t) \, dt + J_0(x(0), t_f) + \frac{\partial J_0}{\partial x} f(x, u) \, \Delta t + \frac{\partial J_0}{\partial t} \Delta t \right].$$

As J_0 is not explicitly dependent on $u(t)$, being a minimum value and $\partial J_0 / \partial t$ is similarly not dependent on $u(t)$ then we can write this as

$$0 = \min_{u(t)} \left[\int_0^{\Delta t} F(x, u, t) \, dt + \frac{\partial J_0}{\partial x} f(x, u) \, \Delta t \right] + \frac{\partial J_0}{\partial t} \Delta t.$$

Expanding $\int_0^{\Delta t} F(x, u, t) \, dt$ also by a Taylor expansion, so that

$$\int_0^{\Delta t} F(x, u, t) \, dt = F(x, u, t)|_0 \, \Delta t$$

we obtain

$$\min_{u(t)} \left[F(x, u, t) \Big|_0 \Delta t + \frac{\partial J_0}{\partial x} f(x, u) \, \Delta t \right] = -\frac{\partial J_0}{\partial t} \Delta t.$$

That is the optimal control must minimize the quantity in brackets. Although the step has been taken from $t = 0$ to $t = \Delta t$ it could have been taken just as easily over the time period of any time t to $t + \Delta t$ with $0 < t < t_f$. The equation

$$F(x, u, t) + \frac{\partial J_0}{\partial x} f(x, u) = -\frac{\partial J_0}{\partial t} \qquad (10.14)$$

must therefore be satisfied for all t for $u(t)$ to be the optimal control law. For J to be the optimal value J_0 its derivative with respect to u, $\partial J/\partial u$, must be zero so that also for unconstrained u, differentiation of equation (10.14) with respect to u yields the optimal condition

$$\frac{\partial F(x, u, t)}{\partial u} + \frac{\partial J_0}{\partial x}\frac{\partial f}{\partial u} = 0 \tag{10.15}$$

(ii) 'Forward' method

Because the method of dynamic programming utilizes stage wise optimization each optimal stage may be determined in a forward pass through the process and similar reasoning then applied to establish the optimal control sequence and optimal process trajectory. This has advantages in particular where the final time is not fixed (Dreyfus 1965; Bhavnani and Chen 1966).

Example   Using the principles of optimality and of dynamic programming determine the optimum (minimum cost) path to be followed in the discrete process shown in Fig. 10.5. The initial value $x(0)$ and the final value $x(t_f)$ are fixed and the cost incurred by each path at each stage is written on the diagram. What is the total minimum cost and the control sequence if

$$u(k) = x(k+1) - x(k),$$

i.e.

$$x(k+1) = x(k) + u(k)?$$

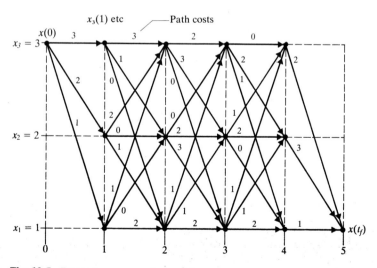

Fig. 10.5 Dynamic programming problem

As the cost function and performance index have been given for all alternative stage paths we may work backwards through the figure, using these numerical values instead of having to calculate (if we were given the form of J) the cumulative minimum cost functions according to the above and equations (10.8), (10.9) and (10.10). As $x(t_f) = x_1$ is fixed, we initially retain all three of the paths between the discrete states x_1, x_2, x_3 at stage 4 and state x_1 at stage 5, i.e. from $x_1(4)$, $x_2(4)$, $x_3(4)$ to $x_1(5)$. Now go back to stage 3 and for each of the states $x_1(3)$, $x_2(3)$, $x_3(3)$ determine the optimal (minimum cost) J_0 from stage 3 to stage 5. These are shown below. (The minimum numerical values may be written at each node and encircled.)

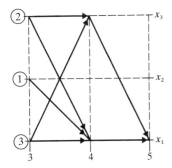

Fig. 10.6 Initial stages of calculation

Note that the paths from $x_1(3)$ form two equal optimal alternatives

$$x_1(3) \rightarrow x_3(4) \rightarrow x_1(5)$$

and

$$x_1(3) \rightarrow x_1(4) \rightarrow x_1(5)$$

each having a total cost of 3 (circled). Similarly there are two equal paths from state x_3 at stage 3. Pass back now to x_2 and by comparison of the options, bearing in mind the minimum cost paths already established from stage 3, the optimal alternatives become as Fig. 10.7, where the first column of circled figures are now the total minimum

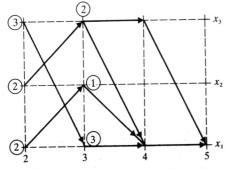

Fig. 10.7 Further stage in calculating optimal path

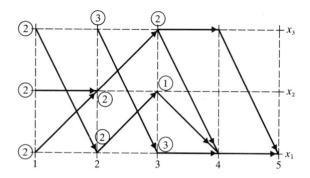

Fig. 10.8 Penultimate stage in calculating optimal path

costs from $x_1(2)$, $x_2(2)$ and $x_3(2)$ respectively to $x_1(5)$. Back to stage 1 gives Fig. 10.8. We have now arrived at the unusual, but true result on the randomly chosen individual costs, that the minimum cost paths from all three first stage states have the same cost. The overall optimum path will thus start at $x_3(0)$, go with a minimum cost of 1 to $x_1(1)$ and then follow the optimum as below (Fig. 10.9). We are still able to go through two possible alternative sequences from $x_3(3)$ to $x_1(5)$. We thus have two optimal control sequences, u_0, as follows:

Stage	$k = 0$	1	2	3	4
Alternative control sequences, $u_0(k) =$ $\begin{cases} \\ \\ \end{cases}$ $x(k+1) - x(k)$	$\begin{matrix} -2 \\ -2 \end{matrix}$	$\begin{matrix} 1 \\ 1 \end{matrix}$	$\begin{matrix} 1 \\ 1 \end{matrix}$	$\begin{matrix} -2 \\ 0 \end{matrix}$	$\begin{matrix} 0 \\ -2 \end{matrix}$

In each case the minimum total cost is 3.

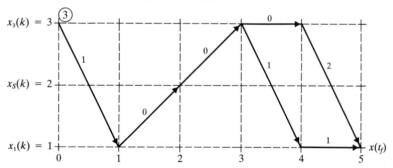

Fig. 10.9 Optimal path(s)

Pontryagin's minimum principle

A continuous and a discrete form of the Pontryagin method may be derived but we shall consider only the continuous form. Pontryagin's method provides a necessary condition for optimal control rather than a direct computation of the control itself, and its method of application is thus to some extent dependent on the type of problem posed, e.g. what are the boundary conditions. The full derivation (Pontryagin *et al.* 1962) requires extensive proofs of intermediate steps and few text books endeavour to present the details. As a result many alternative presentations arise even for graduate texts (e.g. Lapidus and Luus 1967; Koppel 1968; Takahashi, Rabins and Auslander 1970). The problem and aim in this introduction is to present the principle consistent with the level of the remainder of this book and the student's work and yet, at the same time, open the door to the more mathematical precision and notation of the methods. To endeavour to satisfy this requirement the principle is presented, rather than proved, and after the general statement, which follows the presentation of Lapidus and Luus (Lapidus and Luus 1967) more specific categories of problems are discussed. Some examples will be included.

General statement of the principle

The problem of optimization is stated in a basically different manner from that presented above, the performance index being incorporated in the state equation of the system. For the continuous-time n^{th} order dynamic system we now write

$$\dot{\mathbf{x}}(t) = \mathbf{f}(\mathbf{x}(t), \mathbf{u}(t)) \tag{10.16}$$

i.e.

$$\dot{\mathbf{x}}_i = f_i(\mathbf{x}, \mathbf{u}) \quad \text{for} \quad i = 0, 1, \ldots n$$

where the first state variable x_0 in the state vector $\mathbf{x}$ is the performance index J itself with the integral form

$$J = x_0$$
$$= \int_0^{t_f} f_0(\mathbf{x}, \mathbf{u}) \, dt, \quad x_0(0) = 0 \tag{10.17}$$

i.e. $\dot{x}_0 = f_0(\mathbf{x}, \mathbf{u})$. With the system state vector being n^{th} order the vector $\mathbf{x}$ is of order $(n+1)$. The control $\mathbf{u}$ is p-dimensional ($p \leqslant n$). The general problem is to find the optimal control $\mathbf{u}_0(t)$ which takes the system from an initial state $\mathbf{x}(0)$ to a final state $\mathbf{x}(t_f)$ at some time t_f so that J is minimized, J_0. The $\mathbf{u}$ may or may not be constrained, $\mathbf{x}(t_f)$ may be free or constrained and t_f may or may not be fixed, i.e. we have variable end conditions. Combinations of these conditions give rise to certain well-known optimization problems. Note also that because of the inclusion of the performance index in the state equation, even a scalar system will fall naturally into the vector notation used in the problem statement and so this approach is consistent with the introduction already given for the calculus of variation and dynamics programming sections where the scalar case has been established.

If the system equation contains time t explicitly then a further variable is added to the $(n+1)$ already included in equation (10.16) (i.e. $x_0, x_1, \ldots x_n$) so that

$$x_{n+1} = t$$
$$\dot{x}_{n+1} = 1, \qquad x_{n+1}(0) = 0$$

and the system equation vector $\mathbf{x}$ increases to order $(n+2)$, x_0 to x_{n+1}.

The minimum (maximum) principle provides necessary conditions for an optimal control but does not guarantee the existence of such a control. The principle is developed by considering variations in the optimal control function $\mathbf{u}_0(t)$ which may not be necessarily small but which do occur over a small interval of time.

To progress the adjoint, or costate, variables $z_i(t)$ are introduced where they are defined by

$$\dot{z}_i(t) = - \sum_{j=0}^{n} \frac{\partial f_j}{\partial x_i} z_j(t).$$

The adjoint vector $\mathbf{z}$ of the same order as $\mathbf{x}$ may then be expressed in terms of the state vector $\mathbf{x}$ and the function vector $\mathbf{f}$. If

$$\mathbf{z} = [z_0 \quad z_1 \quad \ldots \quad z_n]^T$$

and

$$\mathbf{f} = [f_0 \quad f_1 \quad \ldots \quad f_n]^T$$

then

$$\frac{\partial \mathbf{f}^T}{\partial \mathbf{x}} = \begin{bmatrix} \dfrac{\partial f_0}{\partial x_0} & \dfrac{\partial f_1}{\partial x_0} & \cdots & \dfrac{\partial f_n}{\partial x_0} \\ \cdot & \cdot & & \cdot \\ \cdot & \cdot & & \cdot \\ \cdot & \cdot & & \cdot \\ \dfrac{\partial f_0}{\partial x_n} & \dfrac{\partial f_1}{\partial x_n} & \cdots & \dfrac{\partial f_n}{\partial x_n} \end{bmatrix}$$

and

$$\frac{\partial \mathbf{f}^T}{\partial \mathbf{x}} \mathbf{z} = \begin{bmatrix} \dfrac{\partial f_0}{\partial x_0} z_0 + \dfrac{\partial f_1}{\partial x_0} z_1 + \ldots + \dfrac{\partial f_n}{\partial x_0} z_n \\ \cdot \\ \cdot \\ \cdot \\ \dfrac{\partial f_0}{\partial x_n} z_0 + \dfrac{\partial f_1}{\partial x_n} z_1 + \ldots + \dfrac{\partial f_n}{\partial x_n} z_n \end{bmatrix}$$

The ith 'row' of this vector is $\sum\limits_{j=0}^{n} \dfrac{\partial f_j}{\partial x_i} z_j$, i.e. $-z_i$ by definition so that

$$\dot{\mathbf{z}}(t) = -\frac{\partial \mathbf{f}^T}{\partial \mathbf{x}} \mathbf{z}(t). \tag{10.18}$$

For any initial conditions $\mathbf{z}(0)$ there is a unique continuous solution for the adjoint vector $\mathbf{z}$.

The state equation (10.16) and the adjoint vector $\mathbf{z}$ are combined into a scalar function using the Hamiltonian formulation (from mechanics) H. The Hamiltonian is defined by

$$H(\mathbf{x}, \mathbf{u}, \mathbf{z}) = \mathbf{z}^T \mathbf{f}$$

$$= \mathbf{f}^T \mathbf{z}$$

$$= \sum_{i=0}^{n} z_i f_i(\mathbf{x}, \mathbf{u}) \tag{10.19}$$

with the observed properties that

$$\frac{\partial H}{\partial z_i} = f_i = \dot{x}_i(t) \tag{10.20}$$

$$\frac{\partial H}{\partial x_i} = \sum_{j=0}^{n} z_j \frac{\partial f_j}{\partial x_i} \tag{10.21}$$

$$\frac{\partial H}{\partial \mathbf{z}} = \mathbf{f} = \dot{\mathbf{x}}(t) \tag{10.22}$$

$$\frac{\partial H}{\partial \mathbf{x}} = \frac{\partial \mathbf{f}^T}{\partial \mathbf{x}} \mathbf{z}. \tag{10.23}$$

From the equation introducing the adjoint variables we then obtain

$$\dot{z}_i(t) = -\frac{\partial H}{\partial x_i} \tag{10.24}$$

or

$$\dot{\mathbf{z}}(t) = -\frac{\partial H}{\partial \mathbf{x}}. \tag{10.25}$$

The vector forms, equations (10.22) and (10.25), can also be written

$$\dot{\mathbf{x}}(t) = \nabla_z H \tag{10.26}$$

$$\dot{\mathbf{z}}(t) = -\nabla_x H \tag{10.27}$$

where the notation is

$$\nabla_z H = \begin{bmatrix} \dfrac{\partial H}{\partial z_1} \\ \cdot \\ \cdot \\ \cdot \\ \dfrac{\partial H}{\partial z_n} \end{bmatrix}.$$

If $z(t)$ and $x(t)$ are fixed so that H is only a function of $u(t)$ the maximum or supremum value of H is denoted by $M(z, x)$ so that

$$M(z, x) = \sup_u H(z, x, u)$$

i.e. M is the supremum of H, for given z, x, with respect to u. A 'supremum' (infinum) value is the limiting highest (lowest) value within which a functional or variable value may fall, e.g. for

$0 < x < 1$

$\sup x = 1$, $\inf x = 0$.

The maximum (minimum) value gives the value the variable may actually reach and equal, e.g. for

$0 \leqslant x \leqslant 1$

$\max x = 1$, $\min x = 0$

and for

$0 < x \leqslant 1$

$\max x = 1$, $\inf x = 0$.

the subscript, e.g. u in $M(z, x)$ indicates the variable which is varied in order to achieve the maximization or minimization. Although the distinction is frequently glossed over and in practice the difference may not appear significant, as in the examples used here, it is a feature of the rigorous analysis.

In terms of the above the Pontryagin principle may be stated in reduced form as:

'Let $u(t)$ be an admissible control such that starting with initial conditions $x(0) = [x_0(0), x_1(0), \ldots x_n(0)]^T$ the trajectory passes through the point $x(t_f) = [x_0(t_f), x_1(t_f), \ldots x_n(t_f)]^T$ at some time t_f. If $u(t)$ is optimal, in that it minimizes $\int_0^{t_f} f_0(x, u)\, dt$ then there exists a non zero, continuous vector $z(t)$ satisfying equations (10.22) and (10.25), [(10.26) and (10.27)]. In addition

(i) for all t in the interval $[0, t_f]$, $M(z, x) = H(z, x, u)$ and
(ii) at t_f, $z_0(t_f) \leqslant 0$, $M(z(t_f), x(t_f)) = 0$.

This reduced statement of the more rigorous full Pontryagin principle enables simple cases of optimization to be considered to illustrate the methodology. For a full exposition see for example Lapidus and Luus (1967).

Note that (i) says that at all times H obtains its supremum with respect to $\mathbf{u}$. It can also be shown, (ii), that M is zero not only at t_f but at all times t in $[0, t_f]$. These, (i) and (ii), imply a constant value of M or H over the entire optimal trajectory. From this general statement of the problem and principle we may now consider more specific end conditions. These are only a limited selection from the many special or constrained cases which might be considered, e.g. constraint of some x_i only over the full trajectory, and examples will be given to illustrate the principles in use in these cases.

(a) *Minimal time problem*

The minimal time problem will be considered in more detail in a later section. However, if we wish to minimize the time t_f to reach the desired final state then put J equal to x_0 and

$$\dot{x}_0 = f_0(\mathbf{x}, \mathbf{u}) = 1$$

so that we are minimizing $\int_0^t 1 \, dt$, i.e. min t_f. the Hamiltonian may be expressed as

$$H(\mathbf{z}, \mathbf{x}, \mathbf{u}) = z_0(t) \cdot 1 + \sum_{i=1}^{n} z_i(t) f_i(\mathbf{x}, \mathbf{u})$$

$$= z_0(t) + \bar{H}(\mathbf{z}, \mathbf{x}, \mathbf{u}), \text{ say.}$$

Then

$$\dot{\mathbf{x}}(t) = \nabla_z \bar{H}$$
$$\dot{\mathbf{z}}(t) = -\nabla_x \bar{H},$$

$\mathbf{x}, \mathbf{z}$ now of order $_n$, and

$$\bar{M}(\mathbf{z}, \mathbf{x}) = \sup_{\mathbf{u}} \bar{H}(\mathbf{z}, \mathbf{x}, \mathbf{u})$$

$$= M(\mathbf{z}, \mathbf{x}) - z_0.$$

The general principle is unchanged except that

(i) for all t in the interval $[0, t_f]$, $\bar{M}(\mathbf{z}, \mathbf{x}) = \bar{H}(\mathbf{z}, \mathbf{x}, \mathbf{u})$
(ii) at time t_f, $\bar{M}(\mathbf{z}(t_f), \mathbf{x}(t_f)) \geqslant 0$.

(b) $\mathbf{x}(t_f)$ *unspecified*, t_f *fixed*

The problem now becomes a two-point boundary value problem. The system equation is still

$$\dot{\mathbf{x}}(t) = \mathbf{f}(\mathbf{x}(t), \mathbf{u}(t))$$

with given $\mathbf{x}(0)$. The performance index, a final value type, can be written as

$$J(\mathbf{x}(0), t_f) = \sum_{i=1}^{n} c_i x_i(t_f)$$

$$= \mathbf{c}^T \mathbf{x}(t_f).$$

The c_i are constants weighting the state variables of $\mathbf{x}(t_f)$. Subsequently it may be stated that if there is an optimal control $\mathbf{u}_0$ which minimizes $J(\mathbf{x}(0), t_f)$ then there exists an optimal adjoint vector satisfying the equation

$$\dot{\mathbf{z}}(t) = -\nabla_x H$$

with boundary conditions (as may be seen in the example by substituting back)

$$\mathbf{z}(t_f) = -\mathbf{c}$$

and $z_0(t)$ is such that at each instant of time along the optimal trajectory $(\mathbf{x}_0)$

$$M(\mathbf{z}, \mathbf{x}) = H(\mathbf{z}_0, \mathbf{x}_0, \mathbf{u}_0)$$

$$= \max_{\mathbf{u}} H(\mathbf{z}, \mathbf{x}, \mathbf{u})$$

and $M(\mathbf{z}, \mathbf{x})$ is a non-negative constant.

We note that there is a total of $2n$ differential equations, n in $\mathbf{x}(t)$ and n in $\mathbf{z}(t)$, with n boundary conditions at time zero in the states $\mathbf{x}(0)$ and n at time t_f in the adjoint variables, $\mathbf{z}(t_f)$. The $2n$ differential equations

$$\dot{\mathbf{x}}(t) = \nabla_z H, \quad \mathbf{x}(0) \text{ given}$$

$$\dot{\mathbf{z}}(t) = -\nabla_x H, \quad \mathbf{z}(t_f) \text{ given}$$

are known as the canonical equations.

The optimal control, i.e. that which minimizes J is also that which maximizes the Hamiltonian. (Note that for the alternative specification $\mathbf{z}(t_f) = \mathbf{c}$, then $\mathbf{u}_0$ minimizes the Hamiltonian as well as J. Both conventions may be used.)

(c) *Some $x_i(t_f)$ specified, t_f fixed.*

If the first m of the n $x_i(t)$ are specified at t_f then there are $(n+m)$ boundary conditions on $\mathbf{x}(t)$ including the n at time zero. For the $2n$ differential equations we still need $2n$ boundary conditions and the remaining $(n-m)$ are obtained from the later adjoint variables, i.e.

$$z_{m+1}(t_f) = -c_{m+1}$$
$$\cdot$$
$$\cdot$$
$$\cdot$$
$$\dot{z}_n(t_f) \quad = -c_n$$

and the theorem is otherwise the same.

It must be stressed again that all the above is of a general manner and the restrictions imposed are only examples, although important ones, of possible conditions in the optimization problem. No restriction has been placed on $\mathbf{f}(\mathbf{x}, \mathbf{u})$, e.g. as regards linearity and none has been placed on J. However, as in the other methods of optimization the problem becomes much easier to solve for linear systems and simple, linear or quadratic, or time-optimal, performance criteria. The examples covering this section will be restricted to this form of problem in order to guide the reader through the somewhat tortuous algebra.

Example Using Pontryagin's minimum principle determine the optimal control $u_0(t)$ to take the process whose state equation is

$$\dot{x} = f(x, u)$$
$$= ax + u$$

between the states $x(0)$ at time $t = 0$ and the state $x(t_f)$ at set time t_f subject to minimizing the performance index

$$J = \int_0^{t_f} (x^2 + u^2)\, dt = \int_0^{t_f} F(x, u)\, dt.$$

$x(t_f)$ is not specified.

We augment the state vector so that

$$\dot{x}_0(t) = f_0(x, u)$$
$$= F(x, u)$$
$$= x^2 + u^2$$

and hence

$$x_0 = J(x(0), t_f),$$

and

$$\dot{x}_1(t) = \dot{x}(t)$$
$$= ax + u.$$

The equations may thus be rewritten

$$\dot{x}_0 = x_1^2 + u^2$$
$$\dot{x}_1 = ax_1 + u$$

i.e.

$$\dot{\mathbf{x}} = \begin{bmatrix} \dot{x}_0 \\ \dot{x}_1 \end{bmatrix} = \mathbf{f}(\mathbf{x}, \mathbf{u})$$

$$= \begin{bmatrix} x_1^2 + u^2 \\ ax_1 + u \end{bmatrix}.$$

The initial conditions are

$$x_0(0) = 0 \qquad \text{at} \quad t = 0$$
$$x_1(0) = x(0) \quad \text{at} \quad t = 0$$

In subsection (b) above we said that in this form of problem the performance index may be written in terms of the final states $\mathbf{x}(t_f)$ i.e.

$$J(\mathbf{x}(0), \ t_f) = \sum c_i x_i(t_f)$$

$$= \mathbf{c}^T \mathbf{x}(t_f)$$
$$= c_0 x_0(t_f) + c_1 x_1(t_f).$$

Now J has also been specified as

$$J = \int_0^{t_f} (x^2 + u^2) \, dt$$

$$= \int_0^{t_f} (\dot{x}_0) \, dt.$$

As $x_0(0) = 0$

$$J = x_0(t_f).$$

Therefore by specifying $c_0 = 1$, $c_1 = 0$ the two forms of performance index coincide, minimization of $x_0(t_f)$ satisfies our initial requirement, and the alternative form of J is also justified.

Now introduce the Hamiltonian, equation (10.19).

$$H(\mathbf{x}, \mathbf{z}, \mathbf{u}) = \mathbf{z}^T \mathbf{f}$$

$$= z_0(t) f_0 + z_1(t) f_1$$
$$= z_0(t)(x_1^2 + u^2) + z_1(t)(ax_1 + u).$$

Using equation (10.25),

$$\dot{z}_i(t) = -\frac{\partial H}{\partial x_i}$$

gives

$$\dot{z}_0(t) = 0$$
$$\dot{z}_1(t) = -2x_1 z_0(t) - a z_1(t).$$

The final time values of $\mathbf{z}$ are required boundary conditions, i.e. those of $\mathbf{z}(t_f) = -\mathbf{c}$ for a linear system, so with $c_0 = 1$, $c_1 = 0$, $z_0(t_f) = -1$, $z_1(t_f) = 0$.

Because $\dot{z}_0(t) = 0$, $z_0(t)$ will have the constant value -1, i.e. as z_0 does not change with time z_0 must equal $z_0(t_f)$ at all times. H becomes

$$H = -(x_1^2 + u^2) + z_1(ax_1 + u).$$

To maximize H, $\quad \dfrac{\partial H}{\partial u} = 0,$

i.e.
$$-2u + z_1 = 0$$
$$u_0 = \frac{z_1}{2},$$

and hence the optimal trajectory is given by

$$\dot{x}_1 = ax_1 + z_1/2, \; x_1(0) = x(0)$$

But also

$$\dot{z}_1 = 2x_1 - az_1, \; z_1(t_f) = 0.$$

Solution of these two simultaneous first order differential equations with the given boundary conditions will yield $x_1(t)$, $z_1(t)$ and hence $u_0(t)$. Solution may be by first eliminating x_1. Differentiate the last equation with respect to time,

$$\ddot{z}_1 = 2\dot{x}_1 - a\dot{z}_1$$
$$= 2ax_1 + z_1 - a\dot{z}_1$$

and on substituting for x_1

$$\ddot{z}_1 = a(\dot{z}_1 + az_1) + z_1 - a\dot{z}_1$$
$$= z_1(a^2 + 1)$$

and

$$z_1 = A \exp\left(\sqrt{(1 + a^2)}t\right) + B \exp\left(-\sqrt{(1 + a^2)}t\right)$$

With

$$\dot{z}_1(0) = 2x_1(0) - az_1(0)$$

then

$$A(a + \sqrt{(1 + a^2)}) + B(a - \sqrt{(1 + a^2)}) = 2x(0)$$

and for $z_1(t_f) = 0$,

$$A \exp\left(\sqrt{(1 + a^2)}t_f\right) + B \exp\left(-\sqrt{(1 + a^2)}t_f\right) = 0$$

so that from these two equations,

$$A = \frac{2x(0) \exp(-\sqrt{(1+a^2)}t_f)}{(a+\sqrt{(1+a^2)}) \exp(-\sqrt{(1+a^2)}t_f) - (a-\sqrt{(1+a^2)}) \exp(\sqrt{(1+a^2)}t_f)}$$

$$B = \frac{2x(0) \exp(\sqrt{(1+a^2)}t_f)}{(a-\sqrt{(1+a^2)}) \exp(\sqrt{(1+a^2)}t_f) - (a+\sqrt{(1+a^2)}) \exp(-\sqrt{(1+a^2)}t_f)}.$$

Hence substitution for A and B gives the full expression for z_1. Noting that $u_0 = z_1/2$

$$u_0(t) = \frac{x(0)[\exp(-\sqrt{(1+a^2)}(t_f-t)) - \exp(\sqrt{(1+a^2)}(t_f-t))]}{(a+\sqrt{(1+a^2)}) \exp(-\sqrt{(1+a^2)}t_f) - (a-\sqrt{(1+a^2)}) \exp(\sqrt{(1+a^2)}t_f)}$$

It is interesting to compare this solution with that of the first problem. Note that in that example the end-state $x(t_f)$ was fixed, here it is free. Note also that in this case if $x(0) = 0$, as in the first example, the $u_0 = 0(t)$. This makes sense as it means no change would be created in $x(t)$ and hence $\int_0^{t_f} (x^2 + u^2) \, dt$ would be the minimum obtainable from a quadratic performance index, i.e. zero, and $x(t_f)$ would also be zero. If the actual trajectory $x(t)$ from $x(0)$ to $x(t_f)$ is required then one must use the solution for $z_1(t)$ or $u_0(t)$ above to work by substitution to $x(t)$ itself, e.g. from

$$x(t) = x_1(t)$$
$$= \tfrac{1}{2}(\dot{z}_1 + az_1)$$

and hence to $x(t_f)$.

10.4 Time-optimal control

Although falling under the general heading of optimal control and the Pontryagin maximum principle above, time-optimal control falls also into a particularly suitable category for introducing optimal control ideas. We shall thus look at it in a little more detail, but at the same time restrict consideration to a linear time-invariant system of low order. The literature of time-optimal control is extensive and there is a tendency to distinguish it from all other optimal control problems.

Purely qualitative reasoning (especially once the answer is known!) can lead to an idea of the control required to take a system from state $\mathbf{x}(0)$ to a given state $\mathbf{x}(t_f)$ in the shortest possible time. If we wish to carry out such a transition of state in a minimum time then it is logical to expect to do this by applying a maximum of control energy. In practice the control variables are limited and thus we introduce as part of our system and control definition the qualification that the control is limited, or constrained. Otherwise infinite energy, and speed, would be possible. These constraints will comprise an upper and a lower limit, the lower limit being maximum control applied in a reversed direction.

For the linear time invariant system we may write

$$\dot{\mathbf{x}}(t) = \mathbf{f}(\mathbf{x}, \mathbf{u})$$
$$= \mathbf{Ax} + \mathbf{Bu}, \ \mathbf{x}(0) \text{ given} \tag{10.28}$$

where $\mathbf{A}$ and $\mathbf{B}$ are constant matrices, $\mathbf{x}$ is of order n and $\mathbf{u}$ is constrained between the same upper and lower bounds so that

$$-\mathbf{m} \leqslant \mathbf{u} \leqslant \mathbf{m} \tag{10.29}$$

i.e. $-m_i \leqslant u_i \leqslant m_i$ for all control variables. We wish to minimize $J = \int_0^{t_f} 1 . dt$, i.e. t_f itself. We may now proceed using either the augmented state vector with $f_0(\mathbf{x}, \mathbf{u}) = 1$ or proceed using the Hamiltonian with the state vector not augmented, i.e. use the $\bar{H}$ of section 10.3(a) directly. Proceeding by the first method to illustrate the more general use of the augmented vector

$$\dot{\mathbf{x}}_a = \begin{bmatrix} \dot{x}_0 \\ \dot{\mathbf{x}} \end{bmatrix}$$
$$= \begin{bmatrix} f_0 \\ \mathbf{f} \end{bmatrix}$$
$$= \begin{bmatrix} 1 \\ \mathbf{f} \end{bmatrix}$$

and the adjoint vector $\mathbf{z}_a$ will also be of order $n+1$. The suffix a is added to distinguish between the process state equation and the augmented equation. This was not necessary for clarity before. (A common alternative notation is to augment with x_{n+1} in place of x_0.)

The Hamiltonian H is

$$H(\mathbf{z}_a, \mathbf{x}_a, \mathbf{u}) = \mathbf{z}_a^T \dot{\mathbf{x}}_a$$
$$= z_0 . 1 + \mathbf{z}^T(\mathbf{Ax} + \mathbf{Bu})$$

where $\mathbf{z}$ is the adjoint of the vector $\mathbf{x}$ and, as $\dot{\mathbf{z}}_a = -(\partial H/\partial \mathbf{x}_a)$ where $\bar{H} = \mathbf{z}^T(\mathbf{Ax} + \mathbf{Bu})$,

$$\dot{\mathbf{z}}(t) = -\frac{\partial \bar{H}}{\partial \mathbf{x}}$$

and so

$$\dot{\mathbf{z}}(t) = -\mathbf{A}^T \mathbf{z}.$$

For $\mathbf{u}$ to be time optimal $H(\mathbf{z}_a, \mathbf{x}_a, \mathbf{u})$ or $\bar{H}(\mathbf{z}, \mathbf{x}, \mathbf{u})$ must be maximized with respect to $\mathbf{u}$ for all time $0 \leqslant t \leqslant t_f$, as this is equivalent to minimizing the performance index t_f, and $\bar{H}(\mathbf{z}, \mathbf{x}, \mathbf{u})$ must be positive at time t_f.

If $\dot{\mathbf{z}}(t) = -\mathbf{A}^T \mathbf{z}$ then this may be solved to give

$$\mathbf{z}(t) = e^{-\mathbf{A}^T t} \mathbf{z}(0)$$

where $\mathbf{z}(0)$ is by definition the value of the adjoint vector at $t=0$. The Hamiltonian is then

$$
\begin{aligned}
H(\mathbf{z}_a, \mathbf{x}_a, \mathbf{u}) &= z_0 + \bar{H}(\mathbf{z}, \mathbf{x}, \mathbf{u}) \\
&= z_0 + [e^{-\mathbf{A}^T t}\mathbf{z}(0)]^T(\mathbf{Ax}+\mathbf{Bu}) \\
&= z_0 + \mathbf{z}^T(0)e^{-\mathbf{A}t}(\mathbf{Ax}+\mathbf{Bu}).
\end{aligned}
\tag{10.30}
$$

To maximize H by varying $\mathbf{u}(t)$ in equation (10.30) $\mathbf{u}(t)$ must take its maximum value if $\mathbf{z}^T(0)e^{-\mathbf{A}t}$ is positive and its minimum value if $\mathbf{z}^T(0)e^{-\mathbf{A}t}$ is negative, i.e.

$$
u_i = \begin{cases} m_i & \text{when} \quad [\mathbf{z}^T(0)e^{-\mathbf{A}t}]_i > 0 \\ -m_i & \text{when} \quad [\mathbf{z}^T(0)e^{-\mathbf{A}t}]_i < 0 \end{cases}
$$

for the optimal control vector $\mathbf{u}_0(t)$.

An example illustrating this and the subsequent generation of the 'bang-bang' controlling sequence, i.e. the switching sequence and curves will be now given.

Note that in this case the use of the Hamiltonian of either the augmented or the non-augmented system equations gives the same solution with comparable ease.

Example   The now almost classical example to illustrate time-optimal control is that of the pure inertia system without damping, i.e. the input u-output (position) x relationship is

$$M\ddot{x} = u$$

where u is an input force, (or torque), and for a mechanical system M is the mass (or moment of inertia). This may be expressed in terms of the system state variables, when, taking $M=1$ (or scaling by letting $u' = u/M$)

$$x_1 = x$$
$$\dot{x}_1 = x_2 \text{ (velocity)}$$
$$\dot{x}_2 = u \ (=\ddot{x})$$

i.e.

$$
\begin{bmatrix} \dot{x}_1 \\ \dot{x}_2 \end{bmatrix} = \begin{bmatrix} 0 & 1 \\ 0 & 0 \end{bmatrix}\begin{bmatrix} x_1 \\ x_2 \end{bmatrix} + \begin{bmatrix} 0 \\ 1 \end{bmatrix}u,
$$

or

$$\dot{\mathbf{x}} = \mathbf{Ax} + \mathbf{Bu}.$$

For some non-zero initial condition $\mathbf{x}(0)$ we wish to drive the system using the control $u(t)$ to zero position and velocity in minimum time, given that the control u is bounded at $\pm u_{\max}$ equal to $\pm$unity. If we are minimizing the time t_f then the augmenting state variable is given by $\dot{x}_0 = f_0 = 1$ and for this system we see by inspection that $\dot{x}_1 = f_1 = x_2$; $\dot{x}_2 = f_2 = u$.

We now form the adjoint variables $\mathbf{z}$ utilizing

$$\dot{z}_i(t) = -\sum_{j=0}^{n} \frac{\partial f_j}{\partial x_i} z_j(t)$$

and note that the Hamiltonian is given by

$$\begin{aligned}
H &= \mathbf{z}_a^T \mathbf{f}_a \\
&= z_0 f_0 + z_1 f_1 + z_2 f_2 \qquad &(10.19) \\
&= z_0 + \mathbf{z}^T(0) e^{-\mathbf{A}t} (\mathbf{A}\mathbf{x} + \mathbf{B}\mathbf{u}). \qquad &(10.30)
\end{aligned}$$

$\dot{x}_0$ is the cost function F and x_0 the performance index J and does not appear in f_0, f_1, f_2. Therefore $\partial f_i / \partial x_0 = 0$, $i = 0, 1, 2$, so $\dot{z}_0$ is zero and z_0 is a constant which we can put as -1 (i.e. z_0 (or c_0) is fixed as before). Since the defining relations for the adjoint variables, and the expression for the Hamiltonian, are linear in the adjoint variables we can freely select this value of z_0.

Then

$$\begin{aligned}
\dot{z}_1 &= -\frac{\partial f_1}{\partial x_1} z_1 - \frac{\partial f_2}{\partial x_1} z_2 \\
&= 0 + 0 = 0
\end{aligned}$$

and

$$\begin{aligned}
z_1 &= \text{constant, } C_1 \\
\dot{z}_2 &= -\frac{\partial f_1}{\partial x_2} \cdot z_1 - \frac{\partial f_2}{\partial x_2} \cdot z_2 \\
&= -1 \cdot z_1 \\
&= -C_1
\end{aligned}$$

and

$$z_2 = -C_1 t + C_2.$$

The Hamiltonian is then

$$\begin{aligned}
H &= -1 \cdot f_0 + C_1 f_1 + (C_2 - C_1 t) f_2 \\
&= -1 + C_1 x_2 + (C_2 - C_1 t) u.
\end{aligned}$$

(Evaluation in this way has not required evaluation of $\mathbf{z}^T(0)$ itself in equation (10.30).)

For optimal, minimum-time, control H must be maximized with respect to u. Inspection of H shows that H is maximized with respect to u if $(C_2 - C_1 t)$ is always of the same sign as u and u is at its limit of ± 1, i.e.

$$u = +1 \quad \text{if} \quad C_2 - C_1 t > 0$$
$$\quad = -1 \quad \text{if} \quad C_2 - C_1 t < 0$$

so that

$$u_0 = \text{sign}\,(C_2 - C_1 t)$$

i.e. u takes either one or the other of its extreme values. This is a 'bang-bang' control. Having determined the nature of the optimal u we may now return to the original state equation to determine the actual trajectory $\mathbf{x}(t)$ but we have to evaluate C_1, C_2 or determine the switching sequence from a consideration of the solution of $\dot{\mathbf{x}}(t)$. We see, however, that with C_1 and C_2 constants and z_2 equal to $(-C_1 t + C_2)$, z_2 can change its sign at most only once with increasing time. Hence during the optimal control sequence, $u_0 = \text{sign}\,(C_2 - C_1 t)$, the control will also only change its sign once at most. Any other control involving more switching will thus not be optimal. The switching time when u_0 changes sign will be at $t = t_s = C_2/C_1$ and from the equating of the Hamiltonian H to zero (at optimal control) we see that the constants C_1 and C_2 and hence t_s depend on the initial condition at t zero.

Solution of the state equation, with initial conditions $x_1 = x_1(0)$, $x_2 = x_2(0)$ at $t = 0$ yields

$$x_1(t) = x_1(0) + x_2(0)\,.\,t + \int_0^t (t - \tau)u(\tau)\,d\tau$$

$$x_2(t) = x_2(0) + \int_0^t u(\tau)\,d\tau.$$

Substituting the two possible values $u = +1$, $u = -1$ gives: with $u = +1$:

$$x_1(t) = x_1(0) + x_2(0)t + t^2/2$$
$$x_2(t) = x_2(0) + t$$

so that

$$x_1(t) = \frac{x_2^2(t)}{2} + \left\{ x_1(0) - \frac{x_2^2(0)}{2} \right\},$$

and with $u = -1$ similarly:

$$x_1(t) = -\frac{x_2^2(t)}{2} + \left\{ x_1(0) + \frac{x_2^2(0)}{2} \right\}.$$

These functions $x_1(x_2)$ for different $\mathbf{x}(0)$ may be plotted as parabolae on the x_1, x_2 co-ordinates (i.e.. x, $\dot{x}$ on our phase plane representation),

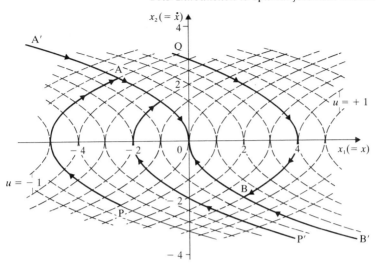

Fig. 10.10   Phase plane for time optimal control of the system $M\ddot{x} = u$

as shown in Fig. 10.10. For a given $x_1(0)$, $x_2(0)$ at P the initial state is fully defined and it is obvious that, as shown in the diagram for the chosen $x_1(0)$, $x_2(0)$ we need initial $u = +1$. To arrive at the origin with only one switching of u we must switch at A to $u = -1$. The trajectory PAO is then the minimum time trajectory from the initial state P. Starting at Q the optimal trajectory would be QBO. Obviously for the time-optimal control of this system with just one change in the sign of u we must always switch along the line A' and B' and this is the switching line of the system.

For switching at $x_2(t_s) > 0$ the equation of the switching line $A'O$ is $x_1(t) = -x_2^2/2$. For switching at $x_2(t_s) < 0$ the equation of $B'O$ is $x_1(t) = x_2^2/2$. The switching line is thus comprised of parabola sections.

10.5 Introduction to optimal feedback control

Since feedback has been a prime consideration in all previous sections of our control theory, it would seem amiss to leave optimal control without looking at the theory of optimization of feedback systems. We shall omit the discrete-time case in this introduction and look only at the application of the above methods in continuous time systems. The system will also be linear time-invariant and the cost function will be quadratic, i.e.

$$\dot{\mathbf{x}} = \mathbf{f}(\mathbf{x}, \mathbf{u})$$
$$= \mathbf{A}\mathbf{x} + \mathbf{B}\mathbf{u} \qquad (10.31)$$

and the performance index is given by the chosen form

$$J(\mathbf{x}(0), t_f) = \tfrac{1}{2} \int_0^{t_f} (\mathbf{x}^T \mathbf{Q}\mathbf{x} + \mathbf{u}^T \mathbf{R}\mathbf{u}) \, dt. \tag{10.32}$$

$\mathbf{Q}$ is symmetric positive semi-definite and $\mathbf{R}$ is symmetric positive definite so that there is always some penalty (cost) on control action. Other quadratic cost functions may be chosen leading to different forms of the solution.

Following the general procedure the variable x_0 is defined so that

$$\dot{x}_0 = f_0(\mathbf{x}, \mathbf{u})$$
$$= \tfrac{1}{2}(\mathbf{x}^T \mathbf{Q}\mathbf{x} + \mathbf{u}^T \mathbf{R}\mathbf{u})$$

and hence the performance criterion may be expressed by

$$J(\mathbf{x}(0), t_f) = \int_0^{t_f} f_0(\mathbf{x}, \mathbf{u}) \, dt$$
$$= x_0(t_f), \quad x_0(0) = 0.$$

From equation (10.19) the Hamiltonian is

$$H = \mathbf{f}_a^T(\mathbf{x}, \mathbf{u})\mathbf{z}_a$$

where $\mathbf{f}_a$ and $\mathbf{z}_a$ are the augmented state vector and adjoint vector, so that

$$H = f_0 z_0 + \mathbf{z}^T \mathbf{f}(\mathbf{x}, \mathbf{u})$$
$$= \tfrac{1}{2}z_0(\mathbf{x}^T \mathbf{Q}\mathbf{x} + \mathbf{u}^T \mathbf{R}\mathbf{u}) + \mathbf{z}^T(\mathbf{A}\mathbf{x} + \mathbf{B}\mathbf{u}). \tag{10.33}$$

The initial conditions specified are $x_0(0) = 0$ and $\mathbf{x}(0)$. The adjoint boundary conditions are not specified at time zero but above, section 9.3, it was seen that if the performance index J can be expressed in terms of the augmented vectors, as

$$J(\mathbf{x}(0), t_f) = \mathbf{c}^T \mathbf{x}(t_f)$$

then

$$\mathbf{z}_a(t_f) = -\mathbf{c}.$$

From our relationship between J and $x_0(t_f)$ we see that c_0 is unity so that in this case

$$z_0(t_f) = -1$$

and hence

$$H = -\tfrac{1}{2}(\mathbf{x}^T \mathbf{Q}\mathbf{x} + \mathbf{u}^T \mathbf{R}\mathbf{u}) + \mathbf{z}^T(\mathbf{A}\mathbf{x} + \mathbf{B}\mathbf{u})$$

The optimizing $\mathbf{u}$, $\mathbf{u}_0$, is that which maximizes the Hamiltonian. (Note that letting $\mathbf{z}_a(t_f) = \mathbf{c}$ leads to minimizing the Hamiltonian, see above.)

Differentiation of H with respect to $\mathbf{u}$ yields for a maximum in H

$$\frac{\partial H}{\partial \mathbf{u}} = -\mathbf{R}\mathbf{u} + \mathbf{B}^T\mathbf{z}$$

$$= \mathbf{0}$$

i.e.

$$\mathbf{u}_0 = \mathbf{R}^{-1}\mathbf{B}^T\mathbf{z}$$

and

$$\frac{\partial^2 H}{\partial \mathbf{u}^2} = -\mathbf{R}$$

so that H is a maximum and its second derivative is negative, $\mathbf{R}$ being defined as positive definite. (It is assumed also that $\mathbf{R}$ is non-singular i.e. its inverse exists.) [Note again that it is the form of definition of the adjoint vector $\mathbf{z}_a(t_f)$ which is important. However, consistent working with either $\mathbf{z}_a(t_f) = -\mathbf{c}$ or $\mathbf{z}_a(t_f) = \mathbf{c}$ will lead to the same conclusion.]

If the control vector $\mathbf{u}$ is formed by state feedback, i.e.

$$\mathbf{u}_0(t) = -\mathbf{K}(t)\mathbf{x}(t)$$

and if the adjoint $\mathbf{z}$ and state vector $\mathbf{x}$ are related by

$$\mathbf{z}(t) = -\mathbf{M}(t)\mathbf{x}(t) \tag{10.34}$$

where $\mathbf{M}(t)$ is a symmetric positive definite, but as yet unknown, matrix (Riccati transformation) then,

$$\mathbf{u}_0 = \mathbf{R}^{-1}\mathbf{B}^T\mathbf{z}$$

$$= -\mathbf{R}^{-1}\mathbf{B}^T\mathbf{M}(t)\mathbf{x}(t)$$

and

$$\mathbf{K}(t) = \mathbf{R}^{-1}\mathbf{B}^T\mathbf{M}(t). \tag{10.35}$$

$\mathbf{K}(t)$ is the feedback control law that we are seeking in order to make the control $\mathbf{u}$ the optimal, $\mathbf{u}_0$. Hence we need to determine this $\mathbf{K}(t)$ and we do this by first finding $\mathbf{M}(t)$ and then using equation (10.35) to find $\mathbf{K}(t)$ itself. To do this we return to the canonical equations,

$$\dot{\mathbf{z}}(t) = -\frac{\partial H}{\partial \mathbf{x}}$$

$$= -(-\mathbf{Q}\mathbf{x} + \mathbf{A}^T\mathbf{z}).$$

Also since

$$\dot{\mathbf{x}}(t) = \frac{\partial H}{\partial \mathbf{z}}$$

$$= \mathbf{A}\mathbf{x} + \mathbf{B}\mathbf{u}$$

so

$$\dot{x}(t) = Ax(t) - BR^{-1}B^T M(t)x(t). \tag{10.36}$$

Differentiating equation (10.34) and using equation (10.36),

$$\dot{z}(t) = -\dot{M}(t)x(t) - M(t)\dot{x}(t)$$
$$= -\dot{M}(t)x(t) - M(t)[A - BR^{-1}B^T M(t)]x(t).$$

Equating this to the expression for $\dot{z}(t)$ and using equation (10.34) also yields

$$\dot{M}(t) + M(t)A - M(t)BR^{-1}B^T M(t) + Q + A^T M(t) = 0. \tag{10.37}$$

Solution of equation (10.37) for $M(t)$ and substitution into (10.35) gives the feedback law $K(t)$. Equation (10.37) is a matrix Riccati equation. [If $z_a(t_f) = c$ and $z(t) = M(t)x(t)$ had been used then $u_0 = -R^{-1}B^T z$, $K(t) = R^{-1}B^T M(t)$ and the matrix Riccati equation is still

$$\dot{M}(t) + M(t)A - M(t)BR^{-1}B^T M(t) + Q + A^T M(t) = 0.$$

Hence $K(t)$ becomes the same and u_0 is the same also.] The boundary conditions for $M(t)$ required in the solution of the Riccati equation are formed at t_f from the relationships between $z(t_f)$ and J, (much simplified if, as here, $t_f \to \infty$). Solution normally requires numerical techniques (e.g. Kalman 1963, Kalman and Englar 1963).

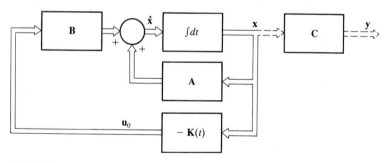

Fig. 10.11 Basic optimal feedback system

The multivariable optimal feedback system may be represented by Fig. 10.11, corresponding to the conventions above. If $t_f \to \infty$, the optimal gain control law becomes constant, $\dot{M}(t)$ is zero, $M(t)$ becomes a constant matrix and the Riccati equation is reduced to

$$M(t)A - M(t)BR^{-1}B^T M(t) + Q + A^T M(t) = 0. \tag{10.38}$$

Example For the pure inertia system problem above determine the optimal linear state-feedback controller for the given quadratic performance index below.

The state equation is

$$\begin{bmatrix} \dot{x}_1 \\ \dot{x}_2 \end{bmatrix} = \begin{bmatrix} 0 & 1 \\ 0 & 0 \end{bmatrix}\begin{bmatrix} x_1 \\ x_2 \end{bmatrix} + \begin{bmatrix} 0 \\ 1 \end{bmatrix} u$$

and we wish to determine the gain matrix $\mathbf{K}(t)$ for the feedback controller

$$\mathbf{u}_0(t) = -\mathbf{K}(t)\mathbf{x}(t).$$

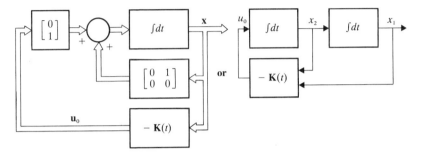

Fig. 10.12 Optimal state feedback controller for undamped second order system

This system is shown by the block diagrams, Fig. 10.12. The performance index which we require to be minimized is now

$$J = \int_0^\infty \tfrac{1}{2}(\mathbf{x}^T\mathbf{Q}\mathbf{x} + \mathbf{u}^T\mathbf{R}\mathbf{u})\,dt$$

$$= \int_0^\infty F(\mathbf{x}, \mathbf{u})\,dt.$$

The control vector $\mathbf{u}$ is

$$\mathbf{u}_0(t) = -\mathbf{K}(t)\mathbf{x}(t)$$

where

$$\mathbf{K}(t) = \mathbf{R}^{-1}\mathbf{B}^T\mathbf{M}(t) \tag{10.35}$$

and $\mathbf{M}(t)$ is the solution of Riccati equation (10.38),

$$\mathbf{M}(t)\mathbf{A} - \mathbf{M}(t)\mathbf{B}\mathbf{R}^{-1}\mathbf{B}^T\mathbf{M}(t) + \mathbf{Q} + \mathbf{A}^T\mathbf{M}(t) = \mathbf{0}.$$

Let us assume that we are only interested in the output $x_1(=x)$ and the cost function is

$$F = \tfrac{1}{2}(x^2 + \sigma u^2)$$

then

$$\mathbf{Q} = \begin{bmatrix} 1 & 0 \\ 0 & 0 \end{bmatrix}; \ \mathbf{R} = [\sigma]$$

$$\mathbf{A} = \begin{bmatrix} 0 & 1 \\ 0 & 0 \end{bmatrix}; \ \mathbf{B} = \begin{bmatrix} 0 \\ 1 \end{bmatrix}.$$

Substitution in equation (10.38) gives for the 2×2 $\mathbf{M}(t)$ matrix

$$\mathbf{M} \begin{bmatrix} 0 & 1 \\ 0 & 0 \end{bmatrix} - \mathbf{M} \begin{bmatrix} 0 \\ 1 \end{bmatrix} \frac{1}{\sigma} [0 \ \ 1] \mathbf{M} + \begin{bmatrix} 1 & 0 \\ 0 & 0 \end{bmatrix} + \begin{bmatrix} 0 & 0 \\ 1 & 0 \end{bmatrix} \mathbf{M} = \mathbf{0}.$$

Writing $\mathbf{M} = \begin{bmatrix} m_{11} & m_{12} \\ m_{12} & m_{22} \end{bmatrix}$, we have

$$\begin{bmatrix} 0 & m_{11} \\ 0 & m_{12} \end{bmatrix} - \frac{1}{\sigma} \begin{bmatrix} m_{12}^2 & m_{12}m_{22} \\ m_{12}m_{22} & m_{22}^2 \end{bmatrix} + \begin{bmatrix} 1 & 0 \\ 0 & 0 \end{bmatrix} + \begin{bmatrix} 0 & 0 \\ m_{11} & m_{12} \end{bmatrix} = \begin{bmatrix} 0 & 0 \\ 0 & 0 \end{bmatrix}$$

and hence

$$m_{11} - m_{12}m_{22}/\sigma = 0$$
$$- m_{12}^2/\sigma + 1 = 0$$
$$m_{12} - m_{22}^2/\sigma + m_{12} = 0.$$

For **M** to be positive definite

$$m_{12} = \sqrt{\sigma} = \sigma^{0.5}$$
$$m_{22} = \sqrt{\sigma} . \sqrt{2} m_{12} = \sqrt{2} . \sigma^{0.75}$$
$$m_{11} = \sqrt{2} . \sigma^{0.25}$$

i.e.

$$\mathbf{M}(t) = \begin{bmatrix} \sqrt{2} . \sigma^{0.25} & \sigma^{0.5} \\ \sigma^{0.5} & \sqrt{2} . \sigma^{0.75} \end{bmatrix}$$

and

$$\mathbf{K}(t) = \mathbf{R}^{-1} \mathbf{B}^T \mathbf{M}(t)$$

$$= \sigma^{-1} [0 \ \ 1] \begin{bmatrix} \sqrt{2} . \sigma^{0.25} & \sigma^{0.25} \\ \sigma^{0.5} & \sqrt{2} . \sigma^{0.75} \end{bmatrix}$$

$$= \frac{1}{\sigma} [\sigma^{0.5} \ \ \sqrt{2} . \sigma^{0.75}]$$

Then

$$u_0 = -\mathbf{K}(t)\mathbf{x}$$

$$= -\frac{1}{\sigma} [\sigma^{0.5} \ \ \sqrt{2} . \sigma^{0.75}] \begin{bmatrix} x_1 \\ x_2 \end{bmatrix}$$

$$= -\sigma^{-0.5} x_1 - \sqrt{2} . \sigma^{-0.25} x_2.$$

Note that the effect of a higher penalty on the control, i.e. higher σ becomes very obvious here, the optimal controller gains being inversely proportional to positive powers of the control cost.

10.6 Summary

It has been the intention of this chapter to introduce various aspects of optimal control, which are illustrated by example, so that the person new to the concepts may both appreciate some of the methods available for the study of the optimal control of dynamic systems and also be more able to utilize if the need arises fuller texts on the subject. To keep the overall presentation short some of the content has been presented in a condensed form and it is inevitable that some questions remain unanswered and not all variations, e.g. in convention, can be fully covered.

In order that a control may be classed as optimal it must reduce some performance index J to a minimum (or maximum). This performance index is itself a function, e.g. an integral, of a cost function F which is a function both of the controlled state $\mathbf{x}$ and of the magnitudes of the control vector $\mathbf{u}$. Thus both process deviation and extreme control actions are penalized. Methods of use in the establishing of optimal controls include the calculus of variations, dynamic programming and Pontryagin's minimum (maximum) principle. The control problems themselves fall into various classifications including those with a fixed final time, fixed or partially fixed final states, those with constrained or unconstrained states and controls and those where it is desired to produce a final end state in a minimum time, time-optimal control. Optimal control, like other controls, may be open or closed loop in nature and optimal feedback control based on state variable feedback is a further category of optimal control.

Within optimal control methods the theory tends to become more complex than that which we have experienced so far in this text and perhaps even more important is the difficulty of actually applying the control methods. This is because as the order of the state vector becomes even of low to medium order the computational requirements become very great and may render the theoretical calculations alone even unpracticable. Actual implementation of control to the standard set by the optimal control theory may prove to be not feasible in practice and sub-optimal controllers, which may however be compared with the computed optimal, will then be the more beneficial practical alternative. Some of the potential complexity may be observed from problems which appear trivial but involve a fair degree of manipulation. Similarly it can be seen that some applications are simpler than might appear at first sight.

Chapter 11
Introduction to adaptive and stochastic systems

11.1 Introduction

So far all consideration has been given to systems where equations may be established on a knowledge of the plant and whose dynamics and impulse characteristics are known. Such systems may be referred to as deterministic and in contrast we have stochastic systems where the inputs and possibly system parameters may have components which vary randomly with time. This randomness may not be complete but may be of a nature so that the basic signal is said to be contaminated with an unwanted noise signal. These systems, one aspect of which we shall look at briefly here, variables which are described by their statistical properties, e.g. their expected mean value. Some mention of such systems is desirable but it is important to realize that only the briefest coverage is given here to illustrate the nature of the problem and one particular aspect, the Kalman filter, of its solution.

Another development from the purely deterministic aspects of control systems is the adaptive controller which assesses plant and environment variations and then changes its control algorithm correspondingly. Such systems usually require considerable on-line computational capabilities and may include various aspects of the stochastic systems work. Once again, only the nature of the system will be illustrated here.

Both of these extensions in control are allied closely with the concepts of optimal control of Chapter 10.

11.2 Adaptive control

As indicated above the adaptive control system changes according to changes in the environment and plant, i.e. it adapts itself so as to maintain satisfactory control, which is usually judged by some performance index. To be of use the controller must adapt as rapidly as possible in comparison with the parameter changes of the plant or environment so that it is always at, or nearly at, its most suitable performance. Various degrees of adaption may fall within the rather ill-defined area of adaptive control, from feedback controllers supplemented by feedforward elements to full optimization procedures. Some degree of adaptive capacity is sought when it is known that plant or environment will be subject to considerable variations outside of the usual controllable inputs which have so far been discussed. Some definitions of adaptive systems specifically include the requirement that the system is 'self-organizing' (Mishkin *et al.* 1961) and this rules out some of the direct model reference systems. Other sources would include the latter within the general adaptive area of control. In the model reference control system the control signal is generated according to the difference between the output of a model and the output of the real plant (Fig. 11.1).

As the adaptive controller requires a performance index it is important that the chosen cost function should reflect the effect of system parameters which are likely to vary, so that the adaptive nature can be realized to the full. The controller may also require secondary performance criteria to ensure that the dynamic, transient, response of

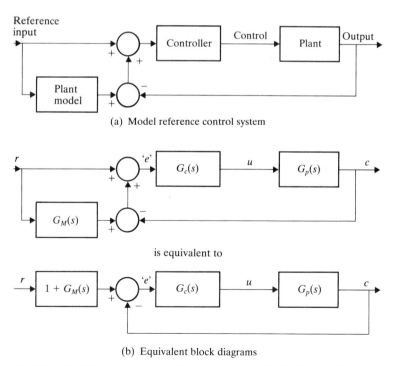

(a) Model reference control system

is equivalent to

(b) Equivalent block diagrams

Fig. 11.1 Model reference system (a) and equivalent block diagrams (b)

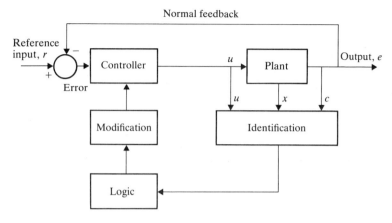

Fig. 11.2   An adaptive control scheme (Elgerd 1967)

the system remains good in the face of maintaining the main performance index as conditions undergo large variations. Given a suitable performance criteria the adaptive control system has three major functions to perform. These are the identification of the plant dynamics, the decision in selecting the required control, and the subsequent modification of the controller according to the decision that is taken. These three functions are carried out continuously or at selected time internals to allow for unpredictable external influences and unpredictable or unexpected (and hence not possible for consideration within a model reference scheme) changes in plant parameters. The adaptive control scheme, shown in Fig. 11.2, is a feedback system and hence is subject to the considerations of stability of other feedback systems. The 'normal feedback' shown may be incorporated within the adaptative feedback path but the action of the adaptive loop is to modify the controller in its action on the error signal.

Consider now the three functions of the adaptive control system:

Identification

The identification relies on continuous or frequent plant measurements, i.e. at intervals much less than the 'period' of likely variations and at least of a similar order to the appropriate plant time constants. Simultaneously with speed of measurement there must be adequate accuracy of measurement and some prior knowledge of the plant is usually essential. The plant inputs may themselves serve as test signals for identification or additional sinusoidal or stochastic signals input and cross-correlation techniques used to identify the system, i.e. establish the input–output response characteristic of the plant, which may be time varying. The principal task of the identification function is to determine the performance index, compare it with that desired, and decide if the difference calls for a full evaluation of the plant parameters and adjustment of the controller. The use of cross-correlation techniques and stochastic signals enables low magnitude signals to be used which do not themselves adversely affect the plant output. Details of such correlation techniques may be found in the texts (Mishkin *et al.* 1961; Yore and Takahashi 1967).

Decision

Based on the identification of the plant and evaluation of the performance index and its comparison with the required, optimal index, the decision as to the form the modification to the controller should take must be decided. A threshhold deviation limit is set after which some change is required which may be a simple logical change of controller parameters or even a change in controller strategy in acute cases. The decision-making and modification-calculating processes must be rapid and are generally faster than the identification stage once the alternative control patterns have been decided upon.

Modification

The final function is the implementation, or actuation, resulting from the decision function stage. This may be effected by changing the controller parameters, and hence plant input, or by digital calculation of an optimal control signal based on the plant identification and performance index. The latter places a higher requirement on on-line computing effort and components.

It can be seen that adaptive control systems rely both on statistical methods of analysis and on optimizing procedures, and even on the 'learning' aspects of control where the adaptive step forced by some deviation is retained for use in a similar recurring circumstance. It is also a comparatively expensive control technique relying on significant knowledge and computation and its use is only justified where high performance 'plant' of various forms are subject to large plant and environmental variations.

11.3 Stochastic control systems

In this section a brief look is taken at linear continuous control systems subject to random disturbances and operating to a quadratic cost function. A full treatment, or even a detailed introduction, takes more space than it is wished to take up in this general introductory text. Only those features of immediate interest to the specific problem will be mentioned and for a general study of statistical theory and stochastic systems, one is referred to other texts (e.g. Meditch 1969; Lee 1960; Papoulis 1965; Åström 1970).

Some definitions of probability functions

Let us start by defining certain terms and functions which are basic to the study of stochastic systems. The characteristic of these systems which is of importance is that certain variables contain randomly fluctuating components and precise values cannot therefore be given to the variables at any instant of time. However, we can express a probability that the variable will take on certain values and we can estimate what the value of these and other dependent variables will be. From these estimates control signals may be generated. The problem is thus one of estimation followed by control.

A random variable X is the outcome of an 'experiment' in a sample space which is made up of all possible outcomes. That is, we know that X may have discrete values or may take values which vary continuously over the sample space, e.g. any number from 1 to 10, or any point in a fixed length. X may be a scalar or we may have a vector $\mathbf{X}$, a vector valued random variable.

The probability distribution function $F(x)$ is the scalar-valued function that specifies the probability that the random variable X takes on a value less than or equal to a specific value x,

$$F(x) = P(X \leqslant x).$$

For a vector-valued random variable

$$F(\mathbf{x}) = P(\mathbf{X} \leqslant \mathbf{x})$$
$$= P(X_1 \leqslant x_1, X_2 \leqslant x_2, \ldots X_n \leqslant x_n)$$

where P means 'the probability that'.

The probability density function of X, $f(x)$ is defined by

$$f(x) = \frac{dF(x)}{dx}$$

or if $\mathbf{X}$ is vector valued

$$f(\mathbf{x}) = \frac{\partial F^n(\mathbf{x})}{\partial x_1 \, \partial x_2 \ldots \partial x_n}.$$

The probability that X lies between two values is given by

$$P(x_1 < X \leqslant x_2) = \int_{x_1}^{x_2} f(x) \, dx$$
$$= F(x_2) - F(x_1)$$

and over the full range of x, i.e. if x_1 and x_2 are the limits of x, $P(x_1 < X \leqslant x_2) = 1$. For the vector valued random variable

$$P(\mathbf{x}_1 < \mathbf{X} \leqslant \mathbf{x}_2) = \int_{x_{1,1}}^{x_{1,2}} \int_{x_{2,1}}^{x_{2,2}} \ldots \int_{x_{n,1}}^{x_{n,2}} f(\mathbf{x}) \, dx_1 \, dx_2 \ldots dx_n$$

and

$$F(\mathbf{x}) = P(-\infty < \mathbf{X} \leqslant \mathbf{x}_2)$$
$$= P(\mathbf{X} \leqslant \mathbf{x}_2).$$

The expectation, expected value, mean or first moment of a function, g, of the vector-valued random variable $\mathbf{X}$ is defined as

$$E[g(\mathbf{X})] = \int_{-\infty}^{\infty} \cdots \int_{-\infty}^{\infty} g(\mathbf{x})f(\mathbf{x}) \, dx_1 \ldots dx_n$$

and if g is a function of two vector-valued random variables $g(\mathbf{X}, \mathbf{Y})$

$$E[g(\mathbf{X}, \mathbf{Y})] = \int_{-\infty}^{\infty} \cdots \int_{-\infty}^{\infty} g(\mathbf{x}, \mathbf{y})f(\mathbf{x}, \mathbf{y}) \, dx_1 \ldots dx_n \, dy_1 \ldots dy_m.$$

Thus the expected value or mean of the scalar variable X, i.e. when $g(X) = x$ is,

$$E[X] = \int_{-\infty}^{\infty} xf(x) \, dx$$
$$= \bar{x}.$$

The covariance matrix of $\mathbf{X}$ is

$$E[(\mathbf{X} - \bar{\mathbf{x}})(\mathbf{X} - \bar{\mathbf{x}})^T] = \int_{-\infty}^{\infty} \cdots \int_{-\infty}^{\infty} (\mathbf{x} - \bar{\mathbf{x}})(\mathbf{x} - \bar{\mathbf{x}})^T f(\mathbf{x}) \, dx_1 \ldots dx_n.$$

It is also known as the central second moment and for a scalar X we have

$$E[(X = \bar{x})^2] = \int_{-\infty}^{\infty} (x - \bar{x})^2 f(x) \, dx$$
$$= \sigma^2$$

and this is the variance of X. This is a measure of the scatter of X, the greater the scatter the greater is σ^2. The symbol σ stands for the standard deviation.

Correlation: Two variables are said to be uncorrelated if

$$E[X, Y] = E[X] \cdot E[Y]$$

and correlated otherwise. Correlation is a measure of the association between two random variables. X and Y may be different variables from different sample spaces or they may be the same variable but separated by a period of time, e.g. $X = X(t)$, $Y = X(t + \tau)$. If a single variable is present the process of correlation is auto-correlation, otherwise with two variables it is cross-correlation. If $\mathbf{X}$ is a vector-valued random variable the covariance matrix is diagonal if its elements

$x_1, \ldots x_n$ are uncorrelated. If $\mathbf{X}$, $\mathbf{Y}$ are uncorrelated random vectors then

$$E[(\mathbf{X}-\bar{\mathbf{x}})(\mathbf{Y}-\bar{\mathbf{y}})^T] = 0.$$

The above definitions are based on the possible values of a random variable at any instant of time and time is not explicitly used in the definitions. That is, the definitions are based on the assumption that the variable may take different values, i.e. there exists an ensemble of possible values or functions. Let us now consider a random variable which varies with time, so that the functions above, instead of being formed over the ensemble set of functions is formed over a period of time. This may be illustrated by Fig. 11.3. The ensemble mean at time t is calculated from the value of each value of the separated functions at time t. The time-mean is obtained from the average value of one

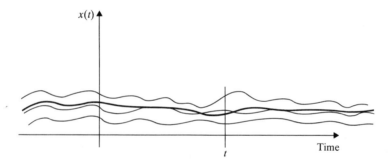

Fig. 11.3 Ensemble average and time average

function only, e.g. the heavier line over a selected period of time. If we can equate the time average mean value with the ensemble average mean the process is ergodic, and for this to be so the time averages must remain independent of time, i.e. the probability distribution function F is independent of time. The variable is said to be stationary or it is said that we have a stationary stochastic process when the probability laws governing the process remain unchanged with varying time. If these assumptions are sufficiently closely held to in practice then the study of processes involving random disturbances is greatly facilitated and we have, for example,

$$E[g(\mathbf{X})] = \int_{-\infty}^{\infty} \ldots \int_{-\infty}^{\infty} g(\mathbf{x})f(\mathbf{x}) \, dx_1 \ldots dx_n$$

$$= \lim_{T \to \infty} \frac{1}{2T} \int_{-T}^{T} g(\mathbf{x}) \, dt$$

and in particular the mean value

$$E[X] = \int_{-\infty}^{\infty} xf(x)\, dx$$

$$= \lim_{T \to \infty} \frac{1}{2T} \int_{-T}^{T} x(t)\, dt.$$

Returning to the concept of correlation it is seen that the mean value of the product $X(t)X(t+\tau)$ is

$$E[X(t)X(t+\tau)] = \lim_{T \to \infty} \frac{1}{2T} \int_{-T}^{T} x(t)x(t+\tau)\, dt.$$

This is the auto-correlation function $\phi_{xx}(\tau)$.

For two variables $X(t)$, $Y(t)$ the cross-correlation function $\phi_{xy}(\tau)$ is

$$E[X(t)Y(t+\tau)] = \lim_{T \to \infty} \frac{1}{2T} \int_{-T}^{T} x(t)y(t+\tau)\, dt.$$

The time τ is the interval in time between the two samples of the random variable(s). For stationary processes $E[X(t)Y(t+\tau)]$ and similar functions are functions only of the separation τ and may be written $E[X, Y(\tau)]$.

From these definitions consider one form of random variable and process in particular. By definition a stochastic process is said to be Gaussian or normal if, for the scalar case, the probability density function is given by

$$f(x) = \frac{1}{\sigma\sqrt{2\pi}} \exp\left(\frac{-(x-\bar{x})^2}{2\sigma^2}\right),$$

where σ^2 is the variance and $\bar{x}$ the mean value of X (Fig. 11.4).

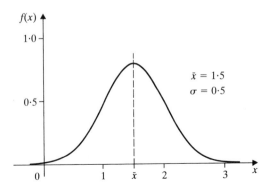

Fig. 11.4 Probability distribution function $f(x)$ for Gaussian random variable

For the set of m time points $t_1, \ldots t_m$ the set of random n^{th} order vector valued random variables is jointly Gaussian if the joint probability density function is

$$f(\mathbf{x}^*) = \frac{1}{\sqrt{((2\pi)^{nm}|\mathbf{P}^*|)}} \exp\left(\frac{-(\mathbf{x}^* - \bar{\mathbf{x}}^*)\mathbf{P}^{*-1}(\mathbf{x}^* - \bar{\mathbf{x}}^*)}{2}\right)$$

where $\mathbf{x}^*$ is the nm vector $[x_1 \ldots x_{nm}]^T$ and $\mathbf{P}^*$ is the $nm \times nm$ matrix whose ij element is

$$P_{ij} = E[(\mathbf{x}(t_i) - \bar{\mathbf{x}}(i))(\mathbf{x}(t_j) - \bar{\mathbf{x}}(j))^T].$$

For a further exposition on such functions other texts should be consulted (Meditch 1967; Lee 1960; Laning and Battin 1956).

The Gaussian distribution is of special importance in that it is analytically convenient to manage, may be approximated to in practice, and a Gaussian process which undergoes any linear transition, e.g. between a linear system input to its output, gives rise to a further Gaussian process. If in addition the mean is zero, which can usually be achieved by choice of variable, then further simplification occurs. A Gaussian white process has the additional properties that for a vector-valued variable it has the auto-correlation function

$$\phi_{xx}(\tau) = E[\mathbf{X}(t)\mathbf{X}^T(t + \tau)]$$
$$= \mathbf{P}\,\delta(\tau)$$

$\delta(\tau)$ is the Dirac delta function. The elements of $\mathbf{P}$ are given by $P_{ij} = E[x_i(t)x_j(t + \tau)]$. For the scalar case

$$\phi_{xx}(\tau) = E[x(t)x(t + \tau)]$$
$$= \sigma^2\delta(\tau)$$

where σ^2 is the variance again. Because of the delta function in the white noise auto-correlation function such a signal cannot be in practice produced but must be approximated to.

If the individual elements X_i of $\mathbf{X}$ are uncorrelated then $\mathbf{P}$ becomes diagonal.

Two random vectors which are independent, i.e. $E[\mathbf{XY}] = E[\mathbf{X}]E[\mathbf{Y}]$, are also uncorrelated. The reverse is not necessarily true unless the random vectors are Gaussian in which case 'independence' or 'uncorrelated' always infers the other property also.

We are now in a position, possibly with further support, to return to our original problem.

The Kalman filter in optimal feedback control

The original work of Kalman (Kalman 1960; see also Meditch 1969) has received much attention, extension, and has become the focus of much of the work in the field of stochastic systems. The treatment of it in isolation presents some problems. With the background of optimal systems of Chapter 10, this section will follow closely the presentation of Schultz and Melsa (1969). It will be seen that the definitions above play an important role in defining the characteristics of the system input but otherwise little direct further knowledge of statistics and stochastic process is required. Although the presentation here is for continuous systems the initial derivation was in terms of discrete systems, the continuous system being treated as a limiting case. This continuous case ties in, however, with the greater emphasis given in this text to the study of continuous systems, and is probably thus more suitable as an introduction.

The previously deterministic inputs to our system are now replaced or added to by inputs contaminated with noise, i.e. a superimposed random variation. As well as occurring in the system plant inputs it also is considered as contaminating the plant measurements. The purpose of the Kalman filter is to estimate, on the basis of the noisy measurements, the state variable values of the plant which is subject to stochastic inputs. Using the estimated states an optimal feedback control system is sought.

For a linear time-invariant system the following change in system equations are made.

$$\dot{\mathbf{x}} = \mathbf{Ax} + \mathbf{Bu} \rightarrow \dot{\mathbf{x}} = \mathbf{Ax} + \mathbf{B}(\mathbf{u} + \mathbf{v}) \qquad (11.1)$$

$$\mathbf{y} = \mathbf{Cx} \qquad \rightarrow \tilde{\mathbf{y}} = \tilde{\mathbf{C}}\mathbf{x} + \mathbf{w}. \qquad (11.2)$$

The noise vectors are $\mathbf{v}(t)$ and $\mathbf{w}(t)$. It is assumed that they are both Gaussian stationary processes with zero mean and $\mathbf{v}(t)$ and $\mathbf{w}(t)$ are independent. Then the cross-correlation function

$$\phi_{\mathbf{vw}}(\tau) = E[\mathbf{v}(t)\mathbf{w}^T(t+\tau)]$$
$$= 0.$$

If the disturbances are also white noise the auto-correlation functions are

$$\phi_{\mathbf{vv}}(\tau) = E[\mathbf{v}(t)\mathbf{v}^T(t+\tau)]$$
$$= \tilde{\mathbf{Q}}\,\delta(\tau)$$
$$\phi_{\mathbf{ww}}(\tau) = E[\mathbf{w}(t)\mathbf{w}^T(t+\tau)]$$
$$= \tilde{\mathbf{R}}\delta(\tau).$$

The white noise assumption is not a practical reality but is suitable for low frequency control system analysis of practical systems.

The output vector $\tilde{\mathbf{y}}$ ($\mathbf{y}$ tilde) contains the measurement outputs, i.e. those which it is possible to measure directly in the system, but these are not necessarily those which are being controlled, e.g. measurement of a secondary shaft position may be made but we are interested in controlling output shaft position. The actual controlled

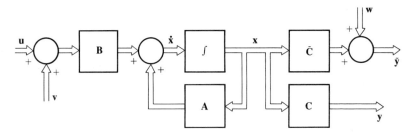

Fig. 11.5 General state space representation with input and measurement noise

outputs, e.g. the output shaft position, are retained by our original notation

$$\mathbf{y} = \mathbf{C}\mathbf{x}. \tag{11.3}$$

Because of our specification of a linear system and Gaussian noise all noise can be shown, without loss of generality, as entering the system with the input **u**. The general system representation is then as Fig. 11.5. (Compare with Fig. 7.7 without feedback.) It becomes clearer from this diagram that given the noisy measurements $\tilde{\mathbf{y}}$ we need to find **x**, the state, and hence the sought output **y**. To do this we need feedback from the noisy measurements to determine the control **u**. This **u** is obtained by estimating the state **x** from $\tilde{\mathbf{y}}$ and then using the optimal control law

$$\mathbf{u}_0 = -\mathbf{K}\hat{\mathbf{x}}(t).$$

The estimate of the state determined from the noisy measurements is denoted by $\hat{\mathbf{x}}$. The Kalman filter, or estimator, is used to provide these estimates of the state variables. This process of estimation is also known as state, or state variable, reconstruction (Fig. 11.6). Because of the linear system and the stipulation of Gaussian noise any random signal in the system will be Gaussian also and completely specified by its mean and its variance. The Kalman filter provides an unbiased estimate with minimum variance and gives the best estimate of the state vector. Combination of this best estimate with optimal control gives an overall optimal system, but the estimation and control may be worked upon separately thus preventing the overall combined problem from becoming of greatly increased complexity.

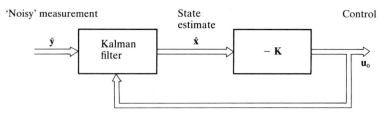

Fig. 11.6 Use of the Kalman filter to construct state estimate and control input

The notation used in the following is chosen to make this section's notation consistent with that of section 10.5 on optimal feedback control. Unfortunately there appears to be no generally accepted standard notation. With this in mind a solution to the problem of estimating $\mathbf{x}$ from $\bar{\mathbf{y}}$, the Kalman filter, may be written in the form of the dynamics of a linear system:

$$\dot{\hat{\mathbf{x}}} = \mathbf{A}\hat{\mathbf{x}} + \bar{\mathbf{K}}(\bar{\mathbf{y}} - \tilde{\mathbf{C}}\hat{\mathbf{x}}) + \mathbf{B}\mathbf{u} \qquad (11.4)$$

where

$$\hat{\mathbf{x}} = \hat{\mathbf{x}}(t), \ \bar{\mathbf{y}} = \hat{\mathbf{y}}(t), \ \mathbf{u} = \mathbf{u}(t)$$

and

$$\bar{\mathbf{K}} = \bar{\mathbf{R}}^{-1}\tilde{\mathbf{C}}\bar{\mathbf{M}}. \qquad (11.5)$$

$\bar{\mathbf{M}}(t)$ is the steady state solution of a matrix Riccati equation,

$$-\dot{\bar{\mathbf{M}}} + \bar{\mathbf{M}}\mathbf{A}^T - \bar{\mathbf{M}}\tilde{\mathbf{C}}^T\bar{\mathbf{R}}^{-1}\tilde{\mathbf{C}}\bar{\mathbf{M}} + \mathbf{B}\bar{\mathbf{Q}}\mathbf{B}^T + \mathbf{A}\bar{\mathbf{M}} = 0 \qquad (11.6)$$

with initial condition $\bar{\mathbf{M}}(0) = E[\mathbf{x}(0)\mathbf{x}^T(0)]$. The initial state $\mathbf{x}(0)$ is taken as a Gaussian random variable with zero mean, $E[\mathbf{x}(0)] = 0$. If only the steady state value of $\bar{\mathbf{M}}(t)$ is required (i.e. as $t_f \to \infty$), this value is independent of the initial condition $\bar{\mathbf{M}}(0)$. The block diagram of this 'system' is shown in Fig. 11.7. Noting that the operation $\tilde{\mathbf{C}}\hat{\mathbf{x}}$ gives an estimated $\hat{\mathbf{y}}$, an 'error' is generated $\bar{\mathbf{y}} - \hat{\mathbf{y}}$ and with $\bar{\mathbf{K}}$ as the input matrix

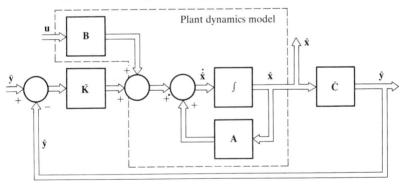

Fig. 11.7 Diagram of Kalman filter showing contained system model. (The two central summations are shown separately to illustrate equation (11.4) more clearly)

this error signal is used as an input to the original system with the same matrix $\mathbf{A}$. There is thus considerable similarity of the proposed filter to the original system and the problems of solution of the system equations are similar to those of the optimal control problem. Note also the similarity to our earlier treatment of observers. This similarity may be high-lighted in the following way;

For the system

$$\dot{\mathbf{x}} = \mathbf{A}\mathbf{x} + \mathbf{B}\mathbf{u} \qquad (10.31)$$

with the performance index

$$J = \tfrac{1}{2} \int_0^{t_f} (\mathbf{x}^T \mathbf{Q} \mathbf{x} + \mathbf{u}^T \mathbf{R} \mathbf{u}) \, dt \tag{10.32}$$

the optimal control vector was

$$\mathbf{u}_0 = -\mathbf{K}\mathbf{x}$$

with

$$\mathbf{K} = \mathbf{R}^{-1} \mathbf{B}^T \mathbf{M} \tag{10.35}$$

and $\mathbf{M}$ the solution of

$$\dot{\mathbf{M}} + \mathbf{M}\mathbf{A} - \mathbf{M}\mathbf{B}\mathbf{R}^{-1}\mathbf{B}^T\mathbf{M} + \mathbf{Q} + \mathbf{A}^T\mathbf{M} = 0. \tag{10.37}$$

By direct substitution if we define a system by the equation

$$\dot{\mathbf{x}} = \mathbf{A}^T \mathbf{x} + \tilde{\mathbf{C}}^T \mathbf{u} \tag{11.7}$$

with the performance index

$$J = \tfrac{1}{2} \int_0^{t_f} (\mathbf{x}^T (\mathbf{B}\bar{\mathbf{Q}}\mathbf{B}^T)\mathbf{x} + \mathbf{u}^T \bar{\mathbf{R}} \mathbf{u}) \, dt \tag{11.8}$$

then the optimal feedback control

$$\mathbf{K} = \bar{\mathbf{R}}^{-1} \tilde{\mathbf{C}} \mathbf{M} \tag{11.9}$$

amd $\mathbf{M}$ is the solution of

$$\dot{\mathbf{M}} + \mathbf{M}\mathbf{A}^T - \mathbf{M}\tilde{\mathbf{C}}^T\bar{\mathbf{R}}^{-1}\tilde{\mathbf{C}}\mathbf{M} + \mathbf{B}\bar{\mathbf{Q}}\mathbf{B}^T + \mathbf{A}\mathbf{M} = 0. \tag{11.10}$$

Comparison of equations (11.9) and (11.10) with equations (11.5) and (11.6) for the Kalman filter show that the only difference is in the sign of $\dot{\mathbf{M}}$ in equations (11.6) and (11.10). That is, the filter and optimal control problem are identical if we consider time as running backwards in the filter, thus changing the sign of the differential term. As stated in section 10.5 the boundary condition for equation (10.37) and hence (11.10) is $\mathbf{M}(t_f)$ and with $t_f \to \infty$ the condition $\mathbf{M}(\infty) = 0$ is used. Thus for equivalence in solving equation (11.6) forward integration with $\bar{\mathbf{M}}(0) = 0$ is used and the steady state solution of equations (10.37) and (11.6) are then identical.

The determination of $\bar{\mathbf{K}}$ may thus be made exactly equivalent to the determination of the optimal controller $\mathbf{K}$ for the system of equations (11.7) and (11.8), with time running backwards. This system is the dual of the original system and the problem of determining its controller is Kalman's dual control problem. Thus solution of the optimal linear feedback control problem now infers solution of the optimal filter problem. For the steady state solution of $\mathbf{M}$ or $\dot{\mathbf{M}}$, the time derivative is zero and both forms of the Riccati equation are the same and purely algebraic.

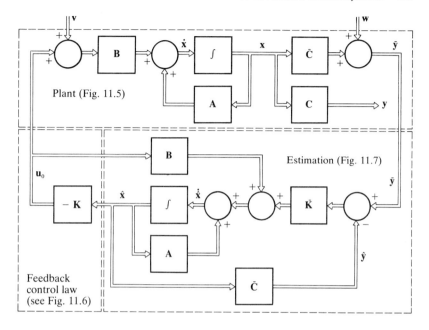

Fig. 11.8 The plant-estimation-control relationship

The complete estimation and control problem and its solution may be represented by Fig. 11.8, the defining equations being the equations (11.1), (11.2), (11.3) for the system itself and (10.32) for the performance index $(t_f \to \infty)$ the solution being given by (10.35) and (10.37) for optimal control and (11.5), (11.6) for the Kalman filtering. (Equations (10.37) and (11.6) become the same for infinite time, steady state, solution.) Notice again the composite similarity to the combined deterministic state feedback controller (section 7.7).

Example The very nature of the problem and the form of the solution make meaningful hand calculations examples difficult to obtain. On the other hand it may be felt that for these very reasons a worked example is a necessity. We shall base our example on the pure inertial system, $\ddot{x} = u$, used in an example in section 10.5, but where there is a random disturbance input v and disturbance w_1, w_2 on the measurements $y_1(=x_1)$ and $y_2(=x_2)$. Then

$$\dot{x}_1 = x_2$$
$$\dot{x}_2 = u + v$$
$$\begin{bmatrix} \dot{x}_1 \\ \dot{x}_2 \end{bmatrix} = \begin{bmatrix} 0 & 1 \\ 0 & 0 \end{bmatrix} \begin{bmatrix} x_1 \\ \dot{x}_2 \end{bmatrix} + \begin{bmatrix} 0 \\ 1 \end{bmatrix} [u+v]$$

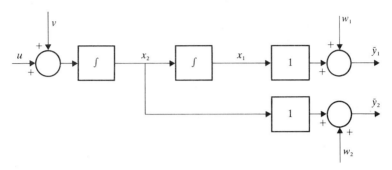

Fig. 11.9 Second order system with plant and measurement noise

i.e.

$$\dot{\mathbf{x}} = \mathbf{Ax} + \mathbf{B}(\mathbf{u} + \mathbf{v})$$

and

$$\tilde{y}_1 = x_1 + w_1$$
$$\tilde{y}_2 = x_2 + w_2 \;\}$$

$$\begin{bmatrix} \tilde{y}_1 \\ \tilde{y}_2 \end{bmatrix} = \begin{bmatrix} 1 & 0 \\ 0 & 1 \end{bmatrix} \begin{bmatrix} x_1 \\ x_2 \end{bmatrix} + \begin{bmatrix} w_1 \\ w_2 \end{bmatrix}$$

i.e.

$$\tilde{\mathbf{y}} = \tilde{\mathbf{C}}\mathbf{x} + \mathbf{w}, \quad \tilde{\mathbf{C}} = \mathbf{I}.$$

If the disturbances are stationary Gaussian processes so that

$$E[vv(\tau)] = \bar{\mathbf{Q}}\,\delta(\tau)$$
$$= [1]\,\delta(\tau)$$
$$= \delta(\tau)$$

and

$$E[\mathbf{ww}^T(\tau)] = \bar{\mathbf{R}}\,\delta(\tau)$$
$$= \begin{bmatrix} 2 & 0 \\ 0 & 1 \end{bmatrix}\delta(\tau),$$

so

$$\bar{\mathbf{R}}^{-1} = \begin{bmatrix} 0{\cdot}5 & 0 \\ 0 & 1 \end{bmatrix}$$

i.e. so that w_1, w_2 are uncorrelated, determine the matrix $\bar{\mathbf{K}}$ of the Kalman filter.

  (i) Initially let us use equations (11.5) and (11.6) directly to obtain $\bar{\mathbf{K}}$. Substitution of $\bar{\mathbf{K}}$ in equation (11.4) would then enable us to obtain the solution of this equation to give the best estimate $\hat{x}$ of the state for given measurements $\tilde{y}$ and input u. For the steady state filter we solve the reduced Riccati equation ($\dot{\bar{M}} = 0$),

$$\bar{M}A^T - \bar{M}\tilde{C}^T\bar{R}^{-1}\tilde{C}\bar{M} + B\bar{Q}B^T + A\bar{M} = 0. \tag{11.6a}$$

To avoid barring all elements of $\bar{M}$ replace the bar by † so that

$$\begin{bmatrix} m_{11} & m_{12} \\ m_{12} & m_{22} \end{bmatrix}^\dagger \begin{bmatrix} 0 & 0 \\ 1 & 0 \end{bmatrix} - \begin{bmatrix} m_{11} & m_{12} \\ m_{12} & m_{22} \end{bmatrix}^\dagger \begin{bmatrix} 1 & 0 \\ 0 & 1 \end{bmatrix} \begin{bmatrix} 0\cdot5 & 0 \\ 0 & 1 \end{bmatrix} \begin{bmatrix} 1 & 0 \\ 0 & 1 \end{bmatrix} \begin{bmatrix} m_{11} & m_{12} \\ m_{12} & m_{22} \end{bmatrix}^\dagger$$

$$+ \begin{bmatrix} 0 \\ 1 \end{bmatrix} [1] \begin{bmatrix} 0 & 1 \end{bmatrix} + \begin{bmatrix} 0 & 1 \\ 0 & 0 \end{bmatrix} \begin{bmatrix} m_{11} & m_{12} \\ m_{12} & m_{22} \end{bmatrix}^\dagger = 0$$

i.e.

$$\begin{bmatrix} m_{12} & 0 \\ m_{22} & 0 \end{bmatrix}^\dagger - \begin{bmatrix} 0\cdot5m_{11}^2 + m_{12}^2 & \vdots & 0\cdot5m_{11}m_{12} + m_{12}m_{22} \\ \hdashline 0\cdot5m_{11}m_{12} + m_{11}m_{12} & \vdots & 0\cdot5m_{12}^2 + m_{22}^2 \end{bmatrix}^\dagger$$

$$+ \begin{bmatrix} 0 & 0 \\ 0 & 1 \end{bmatrix} + \begin{bmatrix} m_{12} & m_{22} \\ 0 & 0 \end{bmatrix}^\dagger = 0$$

and hence (with $\bar{m}_{ij}$ the elements of $\bar{M}$ (i.e. $M^\dagger$))

$$\bar{m}_{12} - 0\cdot5\bar{m}_{11}^2 - \bar{m}_{12}^2 + \bar{m}_{12} = 0$$
$$-0\cdot5\bar{m}_{11}\bar{m}_{12} - \bar{m}_{12}\bar{m}_{22} + \bar{m}_{22} = 0$$
$$-0\cdot5\bar{m}_{12}^2 - \bar{m}_{22}^2 + 1 = 0$$

giving three equations and three unknowns. At once the problems of solution become apparent if only hand calculations are to be used. Elimination procedures, bearing in mind that $\bar{M}$ is to be positive definite, i.e.

$$|m_{11}|^\dagger \geqslant 0, \quad \begin{vmatrix} m_{11} & m_{12} \\ m_{12} & m_{22} \end{vmatrix}^\dagger > 0$$

gives

$$\bar{m}_{11} = 1\cdot288$$
$$\bar{m}_{12} = 0\cdot586$$
$$\bar{m}_{22} = 0\cdot915.$$

Using equation (11.5) the Kalman filter matrix $\bar{K}$ is

$$\bar{K} = \bar{R}^{-1}\tilde{C}\bar{M}$$

$$= \begin{bmatrix} 0\cdot5 & 0 \\ 0 & 1 \end{bmatrix} \begin{bmatrix} 1 & 0 \\ 0 & 1 \end{bmatrix} \begin{bmatrix} 1\cdot288 & 0\cdot586 \\ 0\cdot586 & 0\cdot915 \end{bmatrix}$$

$$= \begin{bmatrix} 0\cdot644 & 0\cdot293 \\ 0\cdot586 & 0\cdot915 \end{bmatrix}.$$

(ii) To check this result let us proceed via the dual control problem. The dual system is given by

$$\dot{\mathbf{x}} = \mathbf{A}^T \mathbf{x} + \tilde{\mathbf{C}}^T \mathbf{u}$$
$$= \mathbf{A}^* \mathbf{x} + \mathbf{B}^* \mathbf{u}$$

and this is to be optimized with respect to the quadratic performance index

$$J = \tfrac{1}{2} \int_0^\infty (\mathbf{x}^T (\mathbf{B}\bar{\mathbf{Q}}\mathbf{B}^T)\mathbf{x} + \mathbf{u}^T \bar{\mathbf{R}}\mathbf{u})\, dt$$
$$= \tfrac{1}{2} \int_0^\infty (\mathbf{x}^T \mathbf{Q}^* \mathbf{x} + \mathbf{u}^T \mathbf{R}^* \mathbf{u})\, dt.$$

The Riccati equation to be solved is now equation (10.38)

$$\mathbf{M}\mathbf{A}^* - \mathbf{M}\mathbf{B}^* \mathbf{R}^{*-1} \mathbf{B}^{*T} \mathbf{M} + \mathbf{Q}^* + \mathbf{A}^{*T}\mathbf{M} = 0$$

and

$$\mathbf{K} = \mathbf{R}^{*-1} \mathbf{B}^{*T} \mathbf{M}. \tag{10.35}$$

The $*$ is introduced to distinguish the $\mathbf{A}$, $\mathbf{B}$ etc. from those of the original system.

Now

$$\mathbf{A}^* = \begin{bmatrix} 0 & 0 \\ 1 & 0 \end{bmatrix}, \qquad \mathbf{B}^* = \begin{bmatrix} 1 & 0 \\ 0 & 1 \end{bmatrix}$$

$$\mathbf{R}^* = \begin{bmatrix} 2 & 0 \\ 0 & 1 \end{bmatrix}; \qquad \mathbf{R}^{*-1} = \begin{bmatrix} 0\cdot5 & 0 \\ 0 & 1 \end{bmatrix}$$

$$\mathbf{Q}^* = \mathbf{B}\bar{\mathbf{Q}}\mathbf{B}^T = \begin{bmatrix} 0 \\ 1 \end{bmatrix}[1][0 \quad 1]$$
$$= \begin{bmatrix} 0 & 0 \\ 0 & 1 \end{bmatrix}$$

so that the reduced Ricatti equation is

$$\begin{bmatrix} m_{11} & m_{12} \\ m_{12} & m_{22} \end{bmatrix}\begin{bmatrix} 0 & 0 \\ 1 & 0 \end{bmatrix} - \begin{bmatrix} m_{11} & m_{12} \\ m_{12} & m_{22} \end{bmatrix}\begin{bmatrix} 1 & 0 \\ 0 & 1 \end{bmatrix}\begin{bmatrix} 0\cdot5 & 0 \\ 0 & 1 \end{bmatrix}\begin{bmatrix} 1 & 0 \\ 0 & 1 \end{bmatrix}\begin{bmatrix} m_{11} & m_{12} \\ m_{12} m_{22} \end{bmatrix}$$
$$+ \begin{bmatrix} 0 & 0 \\ 0 & 1 \end{bmatrix} + \begin{bmatrix} 0 & 1 \\ 0 & 0 \end{bmatrix}\begin{bmatrix} m_{11} & m_{12} \\ m_{12} & m_{22} \end{bmatrix} = 0.$$

Immediately we see the equivalence between this equation and the one in $\bar{\mathbf{M}}$ and

$$m_{11} = 1\cdot288$$
$$m_{12} = 0\cdot586$$
$$m_{22} = 0\cdot915$$

and hence

$$\mathbf{K} = \mathbf{R}^{*-1}\mathbf{B}^{*T}\mathbf{M}$$
$$= \begin{bmatrix} 0{\cdot}644 & 0{\cdot}293 \\ 0{\cdot}586 & 0{\cdot}915 \end{bmatrix}$$
$$= \bar{\mathbf{K}}.$$

Although the cross check produced by using the dual control system appears somewhat trivial here, it does underline the relationship between the optimal control and estimation. With higher order systems and less simple matrices the numerical similarity is not quite so easy to detect and the computation required increases greatly. The determination of $\mathbf{K}(t)$ at its steady state value $\mathbf{K}(\infty)$ is also a significant simplification.

11.4 Summary

Within this chapter a brief introduction has been given to adaptive control and to one aspect of stochastic systems. This has required the definition of certain statistical terms. The adaptive control should be capable of self-organizing in the sense that for significant changes in the system parameters or environment the controller algorithm may be changed, including a possible restructuring of the whole control policy. The identification, decision and modification functions of adaptive control require considerable on-line computation.

When random inputs affect the plant and measurement a noise problem is presented and it is necessary to estimate the system state from measurements which contain contaminating noise. The Kalman filter gives an unbiased minimum variance estimate when the noise is taken to be white. The optimal estimation and control problem may be solved using the concept of the dual control problem. This utilizes the optimal feedback control laws in the estimation part of the problem. The solution is effectively two part, that of optimal estimation followed by optimal control based on the estimated state variables.

Problems

The following short selection of problems is intended to test the student's basic understanding of the text material.

1. Key variables in a system are related by the equations

$$\dot{x}_1 = -3x_1^2 + 0\cdot 6x_2$$
$$\dot{x}_2 = -x_1^2 + 0\cdot 5\sqrt{x_2}.$$

Determine the values of x_1 and x_2 at which steady states are possible, i.e. at which both $\dot{x}_1$ and $\dot{x}_2$ are zero, and hence derive the linearized relationship for deviations in $\dot{x}_1$ and $\dot{x}_2$ following a small disturbance from each of the possible steady states. Use the Laplace transformation to evaluate $x_1(t)$, $x_2(t)$ from an initial state $\mathbf{x}(0)$ when the system is subject to a step input of magnitude $0\cdot 1$ so that the second state equation becomes

$$\dot{x}_2 = -x_1^2 + 0\cdot 5\sqrt{x_2} + 0\cdot 1.$$

2. Using the Laplace transform methods solve the differential equations

$$\ddot{x} + 6\dot{x} + 0\cdot 2x = 1, \qquad \dot{x}(0) = 0\cdot 3, \ x(0) = 3\cdot 6$$
$$\ddot{x} - 3\dot{x} + 3x = -2\cdot 4, \qquad \dot{x}(0) = x(0) = 0$$
$$\dddot{x} + 4\ddot{x} + \dot{x} - 3x = 0, \qquad \ddot{x}(0) = -0\cdot 2, \ \dot{x}(0) = 0\cdot 1, \ x(0) = 3$$
$$\dddot{x} + 3\ddot{x} - \dot{x} + 2x = 3, \qquad \ddot{x}(0) = \dot{x}(0) = x(0) = 0.$$

3. Given that

$$\mathbf{A} = \begin{bmatrix} 3 & 1 & 0\cdot 5 \\ 2 & -1 & -3 \\ -1 & 0\cdot 1 & 2 \end{bmatrix}, \qquad \mathbf{B} = \begin{bmatrix} 1 & 2\cdot 5 & 0\cdot 1 \\ 3 & 2 & -1 \\ -2 & -0\cdot 5 & 1 \end{bmatrix}$$

evaluate

$$\mathbf{A}^{-1}, \ \mathbf{B}^{-1}, \ \mathbf{A}^T, \ \mathbf{B}^T, \ \mathbf{AB}, \ \mathbf{A}^{-1}\mathbf{B}, \ (\mathbf{AB})^{-1}, \ (\mathbf{AB})^T.$$

If

$$\mathbf{C} = \begin{bmatrix} t & 0{\cdot}5t^2 \\ 3t & -0{\cdot}5t \end{bmatrix}, \quad \mathbf{D} = \begin{bmatrix} t & -t^{-1} \\ -0{\cdot}5t^{-1} & t^2 \end{bmatrix}$$

evaluate

$$\frac{d(\mathbf{C}^{-1})}{dt}, \quad \frac{d(\mathbf{CD})}{dt}.$$

4. An electromechanical indicating device is represented by the equation

$$0{\cdot}01\ddot{x} + 0{\cdot}12\dot{x} + x = 10u,$$

x being indicating pointer position and u an input signal. If the initial input and deflection are both zero, plot the response of the instrument to a ramp input $u = 3t$ from time t equals zero to time t equal to $0{\cdot}4$ seconds followed by a steady value of $1{\cdot}2$ units. Comment on the suitability of the instrument for measuring changes of this and both higher and lower rates of change of signal.

5. A floating jetty is hinged at mean water level at the quay side. If the jetty length is short compared with the wavelength of the sea waves, which may be approximated as a simple sine function, derive an expression for the deflection of the jetty caused by a low amplitude swell. Making your own assumptions as to relative magnitudes, damping etc., estimate the wave frequency which would cause maximum jetty deflection at constant wave amplitude and say whether this is likely to occur.

6. What is the transition matrix $e^{\mathbf{A}t}$ for the system equation

$$\begin{bmatrix} \dot{x}_1 \\ \dot{x}_2 \\ \dot{x}_3 \end{bmatrix} = \begin{bmatrix} -2 & 0 & 0 \\ 1 & 1 & 0 \\ -1 & 1 & -1 \end{bmatrix} \begin{bmatrix} x_1 \\ x_2 \\ x_3 \end{bmatrix}?$$

Note the similarity in form between the **A** matrix and $e^{\mathbf{A}t}$ and explain why although the eigenvalues suggest an unstable system not all the states exhibit unstable behaviour.

7. A dynamic system is described by the equations

$$\begin{bmatrix} \dot{x}_1 \\ \dot{x}_2 \end{bmatrix} = \begin{bmatrix} -3 & 1 \\ 1 & -3 \end{bmatrix} \begin{bmatrix} x_1 \\ x_2 \end{bmatrix} + \begin{bmatrix} 1 \\ 2 \end{bmatrix} u$$

If the initial state at time zero is $x_1(0) = 1$, $x_2(0) = 2$ determine $\mathbf{x}(t)$ as functions of time if the system is released from this initial state simultaneously with the imposition of a forcing step input u of unit magnitude. Use both the inverse Laplace transform method and the modal eigenvector methods.

8. For the system illustrated in Fig. 3.5 reduce the block diagram to its simplest form to determine the overall transfer function between the input $Z_2(s)$ and the output $X(s)$. Take both $R(s)$ and $Z_1(s)$ as zero.

Take

$$G_1(s) = \frac{1}{s(s+1)}, \qquad G_2(s) = \frac{s+2}{(s+3)(s+0 \cdot 5)}, \qquad G_3(s) = \frac{1}{s},$$

$$H_1(s) = \frac{s}{(s+5)}, \qquad H_2(s) = 3.$$

Repeat using the signal flow graph equivalent and Mason's rule to confirm the result. Obtain the inverse of the overall expression for $x(t)$ if z_2 is a unit impulse.

9. Two masses are connected by means of a viscous damper and each is separately restrained by springs to move in the same straight line. Using the separate positions and velocities of the masses as state variables show that their motion may be represented by a fourth order state equation.

10. Two unreactive liquids each of unit specific gravity and unit specific heat flow continuously into an open mixing vessel which is cooled by water flowing through an external jacket. It may be assumed that there is perfect mixing both within the vessel and of the water within the jacket. From the base of the mixing vessel the liquid flows through a short pipe and then upwards in essentially plug flow through a second cylindrical vessel and discharges, at a height lower than the mixing vessel normal liquid level, to further processing.

The two inlet flows have temperatures and flowrates of T_1, Q_1, and T_2, Q_2 respectively. The mixing vessel is of volume V_m, temperature T_m and the second vessel is of volume V_s. The water jacket has volume V_j.

The process is subject to disturbances in only the temperature T_1, and in an attempt to keep the temperature T_p of the liquid leaving the second vessel constant the cooling water flow, of temperature T_c, is controlled by means of a controller-valve combination with a combined transfer function $G_c(s)$ so that

$$F_w(s) = G_c(s) T_p(s)$$

where $F_w(s)$ and $T_p(s)$ are the Laplace transforms of the deviations in water flowrate and final temperature T_p respectively.

Draw a diagram of the system. If lower-case letters refer to deviations from a nominal operating point establish the set of equations for the closed loop system relating t_p to t_1, t_j, t_m, f_w following an initial deviation t_1. Without solving these equations individually draw the transfer function block diagram relating $T_p(s)$ to $T_1(s)$, stating any assumptions you make. Assume a constant heat transfer coefficient between process fluid and coolant and a constant effective heat transfer area. Suggest improvements to the given control scheme.

11. Draw the Bode and Nyquist plots for the transfer function

$$G(s) = \frac{(s+0 \cdot 2)}{s(s^2+s+0 \cdot 5)(s+2)},$$

using the corner frequencies and asymptotes in the construction of the Bode plot. Hence sketch the magnitude-phase shift plot for the same open loop system.

12. A plant has a forward transfer function of $\dfrac{10}{s^2+3s+12}$. Compare the open loop transient response, to a unit step input, with that obtained using a proportional feedback controller of gain $k_p = 9$, the addition of integral action with $k_p = 9$, $T_i = 2$, and with derivative action $k_p = 9$, $T_d = 0.5$. Comment on the changes in response and select the 'most suitable' controller.

13. Investigate with the Routh–Hurwitz criteria, the stability of the systems with the closed loop characteristic equations

$$4s^3 + 4s^2 + s + 2 = 0$$
$$s^4 + (3+K)s^3 + s^2 + Ks + 1 = 0$$
$$s^3 + s^2 + Ks + K = 0$$
$$s^5 + s^4 + 3s^3 + s^2 + 6s + K = 0.$$

14. A proportioned plus integral action controller is in series with a plant which has unity gain and time constants of 0.25 and 1 minute. If k_p for the controller is also unity determine the range of integral action time T_i which will maintain stability. If an additional gain K is introduced in the plant what is the relationship between T_i and K for stability to be maintained. For $T_i = 8$ minutes sketch the roots loci as K varies.

15. A plant with the open loop transfer function $(s+4)/s(s+1)$ is controlled by a three mode, P.I.D. controller with the transfer function $k_p\left(1 + \dfrac{1}{T_i s} + T_d s\right)$. For the following controller settings draw, and interpret, the roots loci as the controller gain is varied;

k_p	variable, $T_i = \infty$,	$T_d = 0$
k_p	variable, $T_i = 0.5$,	$T_d = 0$
k_p	variable, $T_i = \infty$,	$T_d = 0.5$
k_p	variable, $T_i = 0.5$,	$T_d = 0.5$.

16. Draw the Bode plot for the open loop transfer function

$$G(s) = \frac{0.2}{s(s+0.5)(s+1.5)}.$$

Determine the closed loop gain and phase margins and hence determine the maximum additional gain factor which could be introduced to the system for stability to be maintained. Show this new $G(s)$ on the Bode plot.

17. Repeat the above question using the Nyquist in place of the Bode plots.

18. The output from a second order plant, $G(s) = \dfrac{2}{(1+0\cdot5s)(1+2s)}$ is delayed by a distance velocity lag of duration 1 second before measurement and being used with a proportional feedback controller of gain k_p. Compare the limiting values of k_p which are predicted if stability is to be maintained if (a) the delay is given its true transfer function e^{-s}, (b) it is approximated by the simple lag $\dfrac{1}{1+s}$ and (c) if the approximation to a double pole $\dfrac{1}{(1+0\cdot5s)^2}$ is used. Use the Bode and/or Nyquist plots in your determination.

19. A process plant has two reference inputs and two measured outputs. It is desired to design a non-interacting control system so that each of the outputs may be changed by using just one of the reference inputs. If the plant transfer function matrix (see Fig. 6.18) $G_p(s)$ is

$$\begin{bmatrix} \dfrac{1}{1+0\cdot5s} & \dfrac{1}{1+s} \\[2mm] \dfrac{1}{1+0\cdot6s} & \dfrac{1}{1+1\cdot5s} \end{bmatrix}$$

the measurement and actuator transfer functions are all unity, and the two major controllers $G_{c11}(s)$ and $G_{c22}(s)$ have transfer functions of unity and $(1+0\cdot4s)$ respectively complete the design for the non-interacting control system, commenting on any practical difficulties.

20. Check on the controllability and observability of the system $\dot{x} = Ax + Bu$; $y = Cx$ where A, B and C have the following forms

(a) $A = \begin{bmatrix} 2 & 1 \\ -1 & 2 \end{bmatrix}$, $B = \begin{bmatrix} 3 \\ 2 \end{bmatrix}$, $C = [2 \quad 3]$

(b) $A = \begin{bmatrix} 3 & 1 \\ -1 & -2 \end{bmatrix}$, $B = \begin{bmatrix} 2 \\ 1 \end{bmatrix}$, $C = [1 \quad 2]$

(c) $A = \begin{bmatrix} 2 & 8 \\ 2 & 2 \end{bmatrix}$, $B = \begin{bmatrix} 2 \\ 1 \end{bmatrix}$, $C = [1 \quad 2]$.

21. A system is represented by the state equations

$$\begin{bmatrix} \dot{x}_1 \\ \dot{x}_2 \end{bmatrix} = \begin{bmatrix} 0 & 2 \\ -1 & -1 \end{bmatrix} \begin{bmatrix} x_1 \\ x_2 \end{bmatrix} + \begin{bmatrix} 2 & 0 \\ 1 & 1 \end{bmatrix} \begin{bmatrix} u_1 \\ u_2 \end{bmatrix}$$

$$y = Ix.$$

Determine the feedback control law so that the closed loop A_c matrix is

$$\begin{bmatrix} -1 & 0\cdot5 \\ -1 & -2 \end{bmatrix}.$$

22. A third order system may be represented by the system equations

$$\dot{x} = Ax + Bu$$
$$y = Cx$$

with

$$A = \begin{bmatrix} -2 & 3 & -1 \\ -1 & 0 & -1 \\ -2.25 & 0 & -3 \end{bmatrix}, \qquad C = [1 \quad 0.5 \quad 1].$$

Establish an observer vector **T** to place the error eigenvectors at -20, -20, -20. (System eigenvectors are at -3, -1.5, -0.5.)

23. Show by use of a quadratic Liapunov function that the following nonlinear equations represent a stable in the large system

$$\dot{x}_1 = -3x_1 + x_2^2 + x_3$$
$$\dot{x}_2 = -x_1 x_2 - x_2 x_3$$
$$\dot{x}_3 = -x_1 + x_2^2 - x_3^3.$$

24. The dynamics of a plant are given by the differential equation

$$\ddot{x} + 4\dot{x} + 3x = u$$

where u is the forcing input and x an output position. By introducing state feedback it is desired to speed up the system dynamics and remove offset in the steady state following a change in u. Choose state variables for such a system and design a controller such that the closed loop system eigenvalues will be at -6, $-6 \pm j5$.

25. A combination of nonlinear elements may be represented by the following input–output characteristic:

$$
\begin{aligned}
\text{input } x &< -2, & y &= -1 \\
-2 \leqslant x &< -0.5, & y &= \tfrac{2}{3}x + \tfrac{1}{3} \\
-0.5 \leqslant x &< 0.5, & y &= 0 \\
0.5 \leqslant x &< 2, & y &= \tfrac{2}{3}x - \tfrac{1}{3} \\
2 \leqslant x, & & y &= 1.
\end{aligned}
$$

Derive an expression(s) for the describing function for this nonlinearity and plot it against X, the amplitude of a sinusoidal forcing function.

26. The above nonlinearity is put in series with the linear plant having the transfer function $\dfrac{1}{s(s+1)}$ and there is unity negative feedback about the combined elements. Draw a phase plane portrait for the unforced system using the method of isoclines and show a number of trajectories starting from various regions of the phase plane.

27. Establish the z-transform for the following functions of time,

$$u(t) = 1$$
$$u(t) = t^2$$
$$u(t) = te^{-at}$$
$$u(t) = t^2 e^{-at}.$$

28. The input to a simple plant with transfer function $\dfrac{1}{s(s+1)}$ comes from a sampler and zero order hold, the sampling interval being T. If the output is sampled at the same time as the input, plot the sampled output after the plant is subject, prior to the sampler, to a ramp of slope 2.

29. A lightly damped ($\zeta = 0.4$) inertial system has an input u and a natural frequency of 1 rad sec^{-1}. Taking output position and velocity as state variables and a sampling interval of 1 second form the discrete state equations.

A digital state feedback controller, with the same sampling time, is installed so that a closed loop system is formed with dynamics equivalent to those of a continuous system with a damping factor ζ of unity and an undamped natural frequency of 2 rad sec^{-1}. What are the required controller feedback gains? (Proceed by way of the closed loop characteristic polynomial for the discrete system.)

30. Using the calculus of variations determine the optimal control $u_0(t)$ to take a process variable x between two known values $x(0) = 0$ and $x(t_f)$ at a fixed time t_f. The process state equation is

$$\dot{x} = 0.5x + u.$$

Take the cost function as $F(x, u) = x^2 + 0.5u^2$, i.e. we wish to minimize $\int_0^{t_f} (x^2 + 0.5u^2)\, dt$. If $x(t_f)$ is unity and t_f is ten seconds plot $u_0(t)$ and $x_0(t)$.

31. Repeat the problem of dynamic programming illustrated in Fig. 10.5 but using the following combinations of initial and final states

(a) $x(0) = x_2$ $x(5) = x_2$
(b) $x(0) = x_1$ $x(5) = x_3$

with the additional cost information

$$x_2(0) \to x_1(1) = 2 \text{ units}, \qquad x_1(4) \to x_2(5) = 1 \text{ unit}$$
$$x_2(0) \to x_2(1) = 1, \qquad x_2(4) \to x_2(5) = 3$$
$$x_2(0) \to x_3(1) = 1, \qquad x_3(4) \to x_2(5) = 2$$

and

$$x_1(0) \to x_1(1) = 2, \qquad x_1(4) \to x_3(5) = 2$$
$$x_1(0) \to x_2(1) = 0, \qquad x_2(4) \to x_3(5) = 1$$
$$x_1(0) \to x_3(1) = 3, \qquad x_3(4) \to x_3(5) = 0.$$

32. Using Pontryagin's maximum principle determine the optimal control $u_0(t)$ to take the linear process described by the equation

$$\dot{x} = 0.1x + 0.5u$$

from the state $x(0) = 2$ at $t = 0$ to the final state $x(t_f) = 1$ at $t_f = 4$ seconds so that the performance index

$$J = \int_0^{t_f} (x^2 + 0.1u^2)\, dt$$

is minimized.

33. For the pure inertia problem (Fig. 10.12) use the Riccati equation to determine the optimal steady state $(t_f \to \infty)$ state feedback controller when

$$J = \int_0^\infty \tfrac{1}{2}(\mathbf{x}^T \mathbf{Q} \mathbf{x} + \mathbf{u}^T \mathbf{R} \mathbf{u})\, dt$$

and

$$\mathbf{Q} = \begin{bmatrix} 1 & 0 \\ 0 & 1 \end{bmatrix}, \qquad \mathbf{R} = [\sigma].$$

The state equation is

$$\begin{bmatrix} \dot{x}_1 \\ \dot{x}_2 \end{bmatrix} = \begin{bmatrix} 0 & 1 \\ 0 & 0 \end{bmatrix} \begin{bmatrix} x_1 \\ x_2 \end{bmatrix} + \begin{bmatrix} 0 \\ 1 \end{bmatrix} u.$$

References

Adler, R. J. and Hovorka, R. B. (1961) 'A finite-stage model for highly asymmetric residence-time distributions', *Second Joint Automatic Control Conf.*, Denver, Colorado.

Aris, R., Nemhauser, G. L., Wilde, D. J. (1964) 'Optimization of multistage cyclic and branching systems by serial procedures', *A.I.Ch.E.J.* **10**, 913.

Åstrom, K. J. (1970) *Introduction to Stochastic Control Theory*, Academic Press, N.Y.

Åstrom, K. J. and Eykhoff, P. (1971) 'System identification – a survey', *Automatica* **7**, 123.

Athans, M. and Falb P. L. (1966) *Optimal Control – an Introduction to the Theory and its Applications*, McGraw Hill, N.Y.

Bateman, H. (1954) *Tables of Integral Transforms*, Vol. 1, McGraw-Hill, N.Y.

Bellman, R. (1961) *Adaptive Control Processes*, Princeton U.P.

Bellman, R. (1962) *Applied Dynamic Programming*, Princeton U.P.

Bhavani, K. H. and Chen, K. (1966) 'Optimization of time-dependent systems by dynamic programming', *J.A.C.C. Proc.*

Birkhoff, G. and MacLane, S. (1941) *Survey of Modern Algebra*, Macmillan, N.Y.

Cholette, A. and Cloutier, L. (1959) 'Mixing efficiency determination for continuous flow systems', *Can. J. Chem. Eng.* **37**, 105.

Cohen, G. H. and Coon, G. A. (1953) 'Theoretical consideration of retarded control', *Trans ASME* **75**, 827.

Coon, G. A. (1956a) 'How to find controller settings from process characteristics', *Control Eng.* **3** (5), 66.

Coon, G. A. (1956b) 'How to set three term controllers', *Control Eng.* **3** (6), 71.

Danckwerts, P. V. (1953) 'Continuous flow systems. Distribution of residence times', *Chem. Eng. Sci.* **2**, 1.

D'Azzo, J. J. and Houpis, C. H. (1960) *Control System Analysis and Synthesis*, McGraw-Hill, N.Y.

Douglas, J. M. (1972) *Process Dynamics and Control*, Vols. 1 and 2, Prentice-Hall, N.J.

Dransfield, P. and Haber, D. F. (1973) *Introducing Root Locus*, Cambridge U.P.

Dreyfus, S. E. (1965) *Dynamic Programming and the Calculus of Variations*, Academic Press, N.Y.

Elgerd, O. I. (1967) *Control Systems Theory*, McGraw-Hill, N.Y.

Evans, W. R. (1948) 'Graphical analysis of control systems', *AIEE Trans.* Pt. II, 547.

Evans, W. R. (1954) *Control System Dynamics*, McGraw-Hill, N.Y.

Ellis, J. K. and White, G. W. T. (1965) 'An introduction to modal analysis and control', *Control* (April, May, June).

Friedly, J. C. (1972) *Dynamic Behaviour of Processes*, Prentice-Hall, N.J.

Fuchs, B. A. and Levin, V. I. (1961) *Functions of a Complex Variable*, Pergamon Press, N.Y.

Gantmacher, F. R. (1960) *The Theory of Matrices*, Chelsea Publ. Co., N.Y.

Gibson, J. E. and Tuteur, F. B. (1958) *Control Systems Components*, McGraw-Hill, N.Y.

Gilbert, E. G. (1963) 'Controllability and observability in multivariable control systems', *SIAM J. Control*, **1**, 128.

Gould, L. A. (1969) *Chemical Process Control: Theory and Applications*, Addison-Wesley, Mass.

Gould, L. A. and Murray-Lasso, M. A. (1966) 'On the modal control of distributed systems with distributed feedback'. *IEEE Trans. Auto. Control*, AC-11, No. 4, 729.

Gurel, O. and Lapidus, L. (1969) 'A guide to the generation of Liapunov functions', *Ind. Eng. Chem.* **61** (3) 30.

Harriott, P. (1964) *Process Control*, McGraw-Hill, N.Y.

Hartley, F. T. and Richards, R. J. (1974) 'The hot surface drying of paper – the development of a diffusion model'. *TAPPI* **57** (3) 157.

Hestenes, M. R. (1966) *Calculus of Variations and Optimal Control Theory*, John Wiley, N.Y.

Himmelblau, D. M. and Bischoff, K. B. (1968) *Process Analysis and Simulation: Deterministic Systems*, John Wiley, N.Y.

Hull, D. E. and von Rosenberg, D. U. (1960) 'Radiochemical tracing of fluid catalyst flow', *Ind. Eng. Chem.* **52**, 989.

Hurwitz, A. (1895) 'Ueber die Bedingungen unter welchen eine Gleichung nur Wurzeln mit negativen reelen Thielen besitzt, *Math. Ann.* **46**, 273.

Jackson, R. (1958) 'Calculation of process controllability using error-squared criterion', *Trans. Soc. Inst. Tech.* **10**, 68.

Kalman, R. E. (1960) 'A new approach to linear filtering and prediction problems', *J. Basic Eng.* **82**, 35.

Kalman, R. E. (1963) 'Mathematical description of linear dynamical systems', *SIAM J. Control*, **1**, 152.

Kalman, R. E. and Englar, T. S. (1963) 'Optimal Filters and Control Systems', *RIAS report on ASP.*

Karnaugh, M. (1953) 'The Map method for combinational logic circuits', *Trans. AIEEE, Communications and Electronics,* **72,** Pt. I, 593.

Kochenburger, R. (1950) 'Frequency-response methods for analysis of a relay servomechanism', *Trans. AIEE* **69,** Pt. I, 270.

Koppel, L. B. (1968) *Introduction to Control Theory with Applications to Process Control,* Prentice-Hall, N.J.

Krasovskii, N. N. (1963) *Stability of Motion,* Stanford U.P., Calif.

Kreyszig, E. (1968) *Advanced Engineering Mathemics,* John Wiley, N.Y. (2nd ed.).

Kwakernaak, H. and Sivan, R. (1972) *Linear Optimal Control Systems,* Wiley-Interscience, N.Y.

Laning, J. H. and Battin, R. H. (1956) *Random Processes in Automatic Control,* McGraw-Hill, N.Y.

Lapidus, L. and Luus, R. (1967) *Optimal Control of Engineering Processes,* Blaisdell, Mass.

LaSalle, J. and Lefschetz, S. (1961) *Stability by Lyapunov's Direct Method,* Academic Press, N.Y.

Lee, Y. W. (1960) *Statistical Communication Theory,* John Wiley, N.Y.

Levenspiel, O. (1962) 'Mixed models to represent flow of liquids through vessels', *Can. J. Chem. Eng.* **40,** 135.

Levenspiel, O. and Bischoff, K. B. (1963) 'Patterns of flow in chemical process vessels', *Advan. Chem. Eng.* **4,** 95.

Liapunov, M. A. (1892) 'Dissertation', *Kharkov, USSR.*

Luenberger, D. G. (1971) 'An introduction to observers', *IEEE Trans. Aut. Control,* AC 16 (6), 596.

Lupfer, D. E. and Oglesby, M. W. (1961) 'Applying dead-time compensation for linear predictor process control', *ISA J.* 8 (11), 53.

MacFarlane, A. G. J. (1970) *Dynamical System Models,* Harrap, London.

Mason, S. J. (1953) 'Feedback theory: some properties of signal flow graphs', *Proc. IRE* **41,** 1144.

Mason, S. J. (1956) 'Feedback theory: further properties of signal flow graphs', *Proc. IRE* **44,** 920.

McCluskey, E. J. Jr (1956) 'Minimization of Boolean functions', *Bell System Tech. J.* **35,** (6), 1417.

Meditch, J. S. (1969) *Stochastic Optimal Linear Estimation and Control,* McGraw-Hill, N.Y.

Melsa, J. L. and Schultz, D. G. (1969) *Linear Control Systems,* McGraw-Hill, N.Y.

Mishkin, E. et al. (1961) *Adaptive Control Systems,* McGraw-Hill, N.Y.

Naslin, P. (1965) *The Dynamics of Linear and Nonlinear Systems,* Blackie, London.

Neubert, H. K. P. (1975) *Instrument Transducers, An Introduction to their Performance and Design,* Oxford U.P. (2nd ed.).

Nyquist, H. (1932) 'Regeneration Theory', *Bell System Tech. J.* **11**, 126.

Ogata, K. (1970) *Modern Control Engineering*, Prentice-Hall, N.J.

Oldenburger, R. (ed.) (1956) *Frequency Response*, Macmillan, N.Y.

Papoulis, A. (1965) *Probability, Random Variables and Stochastic Processes*, McGraw-Hill, N.Y.

Parent, J. and Ray, P. H. (1964) *57th Annual Mtg., A.I.Ch.E.*, Boston, Mass.

Poincaré, H. (1882) *Les Methodes Nouvelles de la Mechanique Celeste*,Vol. 1, Gauthier-Villars, Paris.

Pontryagin, L. S. et al. (1962) *The Mathematical Theory of Optimal Processes*, Interscience, N.Y.

Prentice, J. M. (1970) *Dynamics of Mechanical Systems*, Longman, London.

Quine, W. V. (1959) 'On cores and prime implicants of truth functions', *Am. Math. Monthly* **66**, (9), 755.

Rosenbrock, H. H. (1962) 'Distinctive problems of process control', *C.E.P.* **58**, 43.

Routh, E. J. (1877) *A Treatise on the Stability of a Given State of Motion.* Macmillan, London.

Savas, E. S. (1965) *Computer Control of Industrial Processes*, McGraw-Hill, London.

Salukvanze, M. E. (1963) 'On the analytic design of an optimal controller', *Automatic and Remote Control* **24** (4), 409.

Shannon, C. E. (1949) 'Communication in the presence of noise', *Proc. IRE* **37**, 10.

See also 'Guidelines and general information on user requirements concerning D.D.C.', *First Users Workshop on D.D.C.*, Princeton, J. J. 1963.

Silindir, V. (1970) *Modelling and Control Aspects of Heat Exchangers*, Ph.D. thesis, University of Cambridge.

Takahashi, Y., Rabins, M. J. and Auslander, D. M. (1972) *Control and Dynamic Systems*, Addison-Wesley, Mass.

Tou, J. L. (1959) *Digital and Sampled-data Control Systems*, McGraw-Hill, N.Y.

West, J. C. (1962) *Textbook of Servomechanisms*, English U.P., London (4th imp.).

Wills, D. M. (1962) 'Tuning maps for three mode controllers', *Control Eng.* **9** (4), 104.

Young, A. J. (1965) *An Introduction to Process Control System Design*, Longman, London (4th imp.).

Yore, E. E. and Takahashi, Y. (1967) 'Identification of dynamic systems by digital computer modelling in state space', *Trans. ASME* **89**, (2), D, 295.

Ziegler, J. G. and Nichols, N. B. (1942) 'Optimum settings for automatic controllers', *Trans. ASME* **64**, 759.

Index

DATE DUE